AF688856

Electronic Scanned Array Design

Related titles on radar:

Advances in Bistatic Radar Willis & Griffiths
Airborne Early Warning System Concepts, 3rd Edition Long
Bistatic Radar, 2nd Edition Willis
Design of Multi-Frequency CW Radars Jankiraman
Digital Techniques for Wideband Receivers, 2nd Edition Tsui
Electronic Warfare Pocket Guide Adamy
Foliage Penetration Radar: Detection and characterisation of objects under trees Davis
Fundamentals of Ground Radar for ATC Engineers and Technicians Bouwman
Fundamentals of Systems Engineering and Defense Systems Applications Jeffrey
Introduction to Electronic Warfare Modeling and Simulation Adamy
Introduction to Electronic Defense Systems Neri
Introduction to Sensors for Ranging and Imaging Brooker
Microwave Passive Direction Finding Lipsky
Microwave Receivers with Electronic Warfare Applications Tsui
Phased-Array Radar Design: Application of radar fundamentals Jeffrey
Pocket Radar Guide: Key facts, equations, and data Curry
Principles of Modern Radar, Volume 1: Basic principles Richards, Scheer & Holm
Principles of Modern Radar, Volume 2: Advanced techniques Melvin & Scheer
Principles of Modern Radar, Volume 3: Applications Scheer & Melvin
Principles of Waveform Diversity and Design Wicks et al.
Pulse Doppler Radar Alabaster
Radar Cross Section Measurements Knott
Radar Cross Section, 2nd Edition Knott et al.
Radar Design Principles: Signal processing and the environment, 2nd Edition Nathanson et al.
Radar Detection DiFranco & Rubin
Radar Essentials: A concise handbook for radar design and performance Curry
Radar Foundations for Imaging and Advanced Concepts Sullivan
Radar Principles for the Non-Specialist, 3rd Edition Toomay & Hannan
Test and Evaluation of Aircraft Avionics and Weapons Systems McShea
Understanding Radar Systems Kingsley & Quegan
Understanding Synthetic Aperture Radar Images Oliver & Quegan
Radar and Electronic Warfare Principles for the Non-specialist, 4th Edition Hannen
Inverse Synthetic Aperture Radar Imaging: Principles, algorithms and applications Chen & Marotella
Stimson's Introduction to Airborne Radar, 3rd Edition Baker, Griffiths & Adamy
Test and Evaluation of Avionics and Weapon Sytems, 2nd Edition McShea
Angle-of-Arrival Estimation Using Radar Interferometry: Methods and applications Holder
Biologically-Inspired Radar and Sonar: Lessons from Nature Lessons from Nature Balleri, Griffiths & Baker
The Impact of Cognition on Radar Technology Farina, De Maio & Haykin
Novel Radar Techniques and Applications, Volume 1: Real Aperture Array Radar, Imaging Radar, and Passive and Multistatic Radar Klemm, Nickel, Gierull, Lombardo, Griffiths and Koch
Novel Radar Techniques and Applications, Volume 2: Waveform Diversity and Cognitive Radar, and Target Tracking and Data Fusion Klemm, Nickel, Gierull, Lombardo, Griffiths and Koch
Radar and Communication Spectrum Sharing Blunt & Perrins
Systems Engineering for Ethical Autonomous Systems Gillespie
Shadowing Function from Randomly Rough Surfaces: Derivation and applications Bourlier & Li
Photo for Radar Networks and Electronic Warfare Systems Bogoni, Laghezza and Ghelfi
Multidimensional Radar Imaging Martorella
Radar Waveform Design Based on Optimization Theory Cui, De Maio, Farina & Li
Micro-Doppler Radar and its Applications Fioranelli, Griffiths, Ritchie and Balleri

Electronic Scanned Array Design

John S Williams

The Institution of Engineering and Technology

Published by SciTech Publishing, an imprint of The Institution of Engineering and Technology, London, United Kingdom

The Institution of Engineering and Technology is registered as a Charity in England & Wales (no. 211014) and Scotland (no. SC038698).

© John Williams 2020. Reproduced with permission.

First published 2020

This publication is copyright under the Berne Convention and the Universal Copyright Convention. All rights reserved. Apart from any fair dealing for the purposes of research or private study, or criticism or review, as permitted under the Copyright, Designs and Patents Act 1988, this publication may be reproduced, stored or transmitted, in any form or by any means, only with the prior permission in writing of the publishers, or in the case of reprographic reproduction in accordance with the terms of licences issued by the Copyright Licensing Agency. Enquiries concerning reproduction outside those terms should be sent to the publisher at the undermentioned address:

The Institution of Engineering and Technology
Michael Faraday House
Six Hills Way, Stevenage
Herts, SG1 2AY, United Kingdom

www.theiet.org

While the author and publisher believe that the information and guidance given in this work are correct, all parties must rely upon their own skill and judgement when making use of them. Neither the author nor publisher assumes any liability to anyone for any loss or damage caused by any error or omission in the work, whether such an error or omission is the result of negligence or any other cause. Any and all such liability is disclaimed.

The moral rights of the author to be identified as author of this work have been asserted by him in accordance with the Copyright, Designs and Patents Act 1988.

British Library Cataloguing in Publication Data

A catalogue record for this product is available from the British Library

ISBN 978-1-78561-929-8 (hardback)
ISBN 978-1-78561-930-4 (PDF)

Typeset in India by Exeter Premedia Services

Cover image - Photo by Eric Long, Smithsonian National Air and Space Museum (NASM2006-934). Stars courtesy of NASA.

Contents

List of figures

List of tables

Preface

This book presents a personal and applied approach to the subject based upon my experience with radar at leading companies in the United States during a period of rapid change and innovation. The physics and design of radar is not a new concept. It was established and reduced to practice during World War II. Subsequent development was led and funded by government laboratories, at a diminished pace. With the invention of microelectronics and their application to consumer products, investment and innovation surged. Radar design and manufacturing advanced commensurately, today achieving near theoretical levels of performance at affordable prices.

The genesis of this book is an ESA class which I have presented at a number of radar conferences. I thank the many colleagues who contributed to my understanding of this subject and are responsible for the technical insight represented here. Errors and omissions are my responsibility.

The scope of this book is a class of antennas termed electronic scanned arrays (ESAs), which construct a wavefront in their apertures by means of a multiplicity of radiating elements. Element or groups of elements are controlled in amplitude and/or phase. The elements may also incorporate amplifiers in transmit and receive. Advances in hardware have enabled the implementation of this remarkable technology and broadened its applications.

The objective of this book is to provide a basic understanding of ESA design principles and applications. The discussion focuses on antenna hardware, specifically radar antennas, although communications and receive antennas are broadly similar. The antenna itself does not know or care about the signal or modulation.

ESA functionality enables or enhances system performance but that will not be addressed in any detail because requirements development is a different subject, albeit an important one. Specific applications determine detailed specifications including transmit power, noise figure, and bandwidth so this presentation presupposes some knowledge about systems and modes.

Some topics not directly relevant to ESA design have been omitted even though they are enabled or enhanced by the ESA capability. These include adaptive beamforming, STAP, MIMO, and interferometry which involve primarily post-detection signal processing. SAR, arguably a data processing construct, is addressed briefly because it is also a means to construct a very large array and because it is so ubiquitous among radar applications.

This book is organized as follows:

Design principles, approaches, architectures, and partitioning

These chapters describe general design principles of aperture antennas applied to the specific case of ESA design. System applications will be discussed to set the framework for requirements allocation.

1 Introduction
2 Antenna principles
3 Synthetic arrays
4 Antenna figures of merit

ESA features and fixes

Common ESA design issues are described. Numerical examples illustrate performance of design choices. ESAs and reflector antennas with ESA feeds for reflectors are described.

5 Mutual coupling effects
6 Errors and tolerances
7 Grating lobes
8 Thinned arrays
9 Beamwidth and sidelobes
10 Beam shaping and spoiling
11 Reflector applications

Practical design considerations

ESA performance is largely determined by the choice of specific components, notably including radiating elements, T/R modules, monolithic microwave integrated circuits (MMICs), and microwave distribution and packaging. Technology base and cost conclude this section

12 Design practice
13 Radiating elements
14 T/R modules
15 Assembly, packaging, power and thermal management
16 Technology base and cost

Space systems and ESA design example

Recent radar satellite designs illustrate actual performance and design decisions. A conceptual design for an L-band ESA shows the application of the ESA design techniques to a specific system requirement.

17 ESAs in space
18 ESA and reflector comparison

Software tools such as MATLAB® and TICRA GRASP® have eliminated much of the drudgery of the design process and facilitated systematic evaluation of multiple designs. This text relies on a number of MATLAB® scripts which are available from the publisher at https://digital-library.theiet.org/content/books/ra/sbra526e. Appendix B provides a brief overview.

©2018 The MathWorks, Inc. MATLAB and Simulink are registered trademarks of The Math-Works, Inc. See mathworks.com/trademarks for a list of additional trademarks. Other product or brand names may be trademarks or registered trademarks of their respective holders.

About the author

John S. Williams received his bachelor's and master's degrees in physics from Caltech and the University of California at Irvine respectively and is a Senior Member of the IEEE. He worked on a variety of ESAs since 1980 at Hughes Aircraft Corporation, Raytheon and The Aerospace Corporation. He managed T/R module development, ESA demonstration and manufacturing technology programs. He is now retired.

Chapter 1
Introduction

Electronic scanned array (ESA) design is based on fundamental principles known almost since the time of Maxwell. They were only abstract concepts for many years because the component technology did not exist. ESAs offer a variety of benefits to the system designer to justify their additional cost and complexity. In recent times, the reduced cost of advanced electronics together with advanced systems has seen realization of many of these concepts.

The development of ESAs took many years and the efforts of many talented individuals and organizations. It has its roots in military applications, and advances may be credited to World War II during which there was an emphasis on rapid development of critical technologies including a requirement for many and varied radar systems.

ESA development parallels that of modern electronic devices and has relied greatly on those efforts. The invention of the transistor and the integrated circuit initiated a technological transformation replacing vacuum tubes with solid-state components. Both analog and digital domain applications benefited. The latter enabled revolutionary advances in computation, both sophisticated electromagnetic (EM) simulation for ESA design and enormous processing capability required to understand the resulting sensor data.

ESAs entered common use in the latter half of the last century but opinions differ as to the first instance. Milne [1] notes that "The Chain Home stations, installed before the outbreak of war in 1939, had antenna arrays comprising a small number of dipoles, capable of beam steering by phase shifting." Holpp [2] states regarding the 1940 era Mammut 1 "The beam direction could be steered in azimuth by ±50° using helical lines as phase shifters named 'Wellenschieber.'" Following WWII, Schrank [3] in "Array antenna history" dates multifunction phased arrays from the 1960s, citing Texas Instruments' molecular electronics for radar applications (MERA) described by Hyltin [4]. Delaney [5] in "From vision to reality – 50+ years of phased array development" proposes 1962 as the date for first substantial phased arrays with the deployment of the Hughes-developed AN/SPS 32/33 radars.

1.1 Overview

Pioneering work was done by government research laboratories in many countries with a military focus. In the United States, Defense Advanced Research Projects Agency (DARPA), Army Research Laboratory (ARL) in New Jersey, Naval Research Laboratory (NRL) in Washington, DC and outlying locations, Air Force Research

Laboratories (AFRL), primarily in Dayton, OH, and Strategic Defense Initiative Organization (SDIO), now the Missile Defense Agency (MDA), in Huntsville, AL, all played major roles. The National Aeronautics and Space Administration (NASA) sponsored a number of space projects beginning with SeaSAT in 1978 and the shuttle series of radars.

These organizations led and sponsored risk reduction and technology development. DARPA provided a unifying strategy and portfolio management vision as well as funding fundamental technology and high-risk/high-payoff project. The technology received significant assistance from two US government (USG) programs, Very High Speed Integrated Circuits (VHSIC) and Microwave and Millimeter Wave Monolithic Integrated Circuits (MIMIC), funded to about $1 billion. The need for major cost reduction was obvious and motivated several manufacturing technology (ManTech) programs. A number of nonprofit and federally funded organizations were involved, notably MIT Radiation Laboratory, Lincoln Laboratory, Sandia Laboratory, Willow Run Research Center, and Jet Propulsion Laboratory (JPL).

The United Kingdom was instrumental in initial radar developments with its Defence Research Establishments. Development of beamforming was pioneered at Roke Manor, while companies such as GEC (General Electric Company UK not to be confused with American company of same name), e2v (name derived from English Electric Valve), and Marconi provided the industrial base. In Germany, the Max Planck and Fraunhofer societies sponsor related work as do government agencies such as Deutsches Zentrum für Luft- und Raumfahrt (DLR). With the advent of the European Community and the European Space Agency, multinational developments build on the national technologies. The advent of commercial space-borne remote-sensing systems, notably the TerraSAR and COSMO-SkyMed series, are examples of technological leadership. European aerospace companies, subject to the same pressures as US companies, have also undergone substantial consolidation. Kopp [6] provides an interesting overview of the evolution of technology globally.

A number of for-profit contractors developed the required hardware and systems. The initial return on investment was meager, so government investment was the primary source of innovation. The postwar period saw the founding and growth of a variety of corporations, collectively dubbed "Aerospace." Concurrently, an important second tier of suppliers of specialized components and devices grew up.

The early programs established application needs and capabilities and primed the technology pump to the point that it became self-sustaining. In the late 1990s, shrinking defense budgets sparked a trend of aerospace industry consolidation which concentrated the military technology base into a few companies. Meanwhile, commercial applications were proliferating. Ultimately, consumer products, personal computers, mobile telephones, global positioning devices, and wireless networking underwrote the necessary cost reduction with orders of magnitude more investment.

Today, development roles are substantially reversed, with the military reaping the benefits of the commercial R&D budget which dwarfs the relevant military investment. The Internet of Things, 5G mobile telephony, and autonomous automotive developments ensure that the demand will continue to grow. The ever-increasing demand for bandwidth requires frequency reuse such as sophisticated beamforming techniques. Emergence of specialized semiconductor "foundries" and ever larger wafers provided industrial base economies of scale while migration of production to Asia reduced costs to a degree unimaginable even 20 years earlier. For example, GaN microwave components are available in large quantities and low prices, because the commercial market for white LEDs established a supply of low cost substrates. Commercial applications have pioneered manufacturing, packaging, and miniaturization, which enhance the military systems.

1.2 Scope

An antenna is a device to convert incident EM fields into currents and voltages on conductors or vice versa. An ESA is a particular type of antenna which receives or creates the radiated field in an extended, usually planar, source termed the aperture, in contrast to antennas such as wire loops in which the source is approximately a point. The extended source may be approximated by numerous quasi-point sources, termed elements, whose size is of the order of a wavelength in linear dimension, and the resulting radiated field is the linear superposition of all of the element radiation. Much of the work involves the transition between the element and free space.

The unique capability of the ESA is that it may be reconfigured at will, according to the flexibility incorporated into its design. The ESA can generate a large number of different beam patterns either in rapid succession or simultaneously as determined by system commands. This ability greatly enhances the utility of the associated system, justifying the increased cost and complexity of the ESA.

The property of electronic scan requires the ability to adjust the excitation voltages and currents in order to generate a desired antenna beam pattern. This is in contrast to a conventional antenna with an invariant antenna beam pattern established by specific design choices and associated fabrication. Any single-beam pattern generated by the ESA may, in principle, be replicated by a non-ESA antenna. The same analysis applies.

An ESA combines multiple elements with phase or time delays to form a beam in a specified direction in contrast to a mechanically steered antenna, which physically rotates an antenna to point a beam in a specified direction. Gain and phase control not only scan the beam, but also create detailed beam shapes including nulls. Each element requires phase or time delay to scan the beam, and gain control is required for beam shaping. ESAs commonly include amplifiers to overcome distribution and control loss.

The operation of an ESA is based on sampling. Each radiating element can radiate a signal with a nearly arbitrary value of phase and amplitude. In contrast to signal theory with a one-dimensional time-varying signal, an aperture has two spatial variables. When dealing with an array, the time variable is customarily removed and assumed to be sinusoidal or an arbitrary superposition of sinusoids. The signal must be band-limited to avoid aliasing. ESAs are restricted to ±90° which constitutes a band limit. So long as the aliases occur beyond this limit, they have little or no practical effect.

If the signal is undersampled, the aliases occur in real space where they are termed grating lobes (by analogy to optical diffraction gratings). Their effect can range from minor to unacceptable and this is a key design issue. Oversampling provides the opportunity to obtain higher angular resolution but at the expense of efficiency and bandwidth due to excessive currents.

The element-level power amplifiers replace the centralized transmitter power amplifier. ESAs rely on solid-state power amplifiers. For conventional antennas, the solid-state amplifiers require combining to achieve desired power levels. The power combination (in space) is inherent to the ESA. Travelling wave tube amplifiers (TWTAs) used for conventional antennas have power limitations of their own, so ultimately solid state can provide the greater power. Reliability is enhanced because of the parallel amplifiers.

1.2.1 Antenna evolution

Historically, aperture antennas have been based on reflectors. Reflectors provide high gain using the radiation from one or a few radiating elements to illuminate the surface of the reflector, which reradiates the power from its larger surface. A single parabolic reflector with a point source at its focal point generates outgoing parallel rays and it similarly focuses incoming parallel rays to its focal point. More complex systems use two reflectors which provide more design opportunities. Reflector designs are discussed in Chapter 11.

Figure 1.1 compares a radar system based on a reflector antenna and a system based on an ESA with the differences outlined in the dashed box. Figure 1.1a shows a conventional radar. The dashed box represents functionality provided by an ESA. In principle, the backend electronics are unchanged, but in practice they are enhanced to support additional ESA functionality.

Figure 1.1b shows the ESA counterpart to the system shown in Figure 1.1a. The ESA comprises a multitude of radiating elements, each connected to a transmit/receive module (T/R module). With limited field-of-view systems, multiple radiating elements may connect to a single T/R module. Behind the T/R modules are manifolds for signal, power, and logic. Thermal management is also required and cooling, manifolds, and associated structure constitute a significant part of the size and weight of ESAs.

1.2.2 ESA benefits

Table 1.1 summarizes some of the reasons that the difficulties are worthwhile, showing the improvements that the ESA offers relative to non-ESA antennas.

1.2.3 Types of ESAs

The usual type of ESA is an active ESA (AESA) which includes amplifiers in the antenna for transmit, receive, or both. The amplifiers introduce complexity but improve performance by reducing the effects of losses in the antenna. As their availability has improved greatly, AESAs with gain and gain control in transmit and receive are prevalent providing improved beam control and higher combining efficiency of solid-state amplifiers by use of space combining.

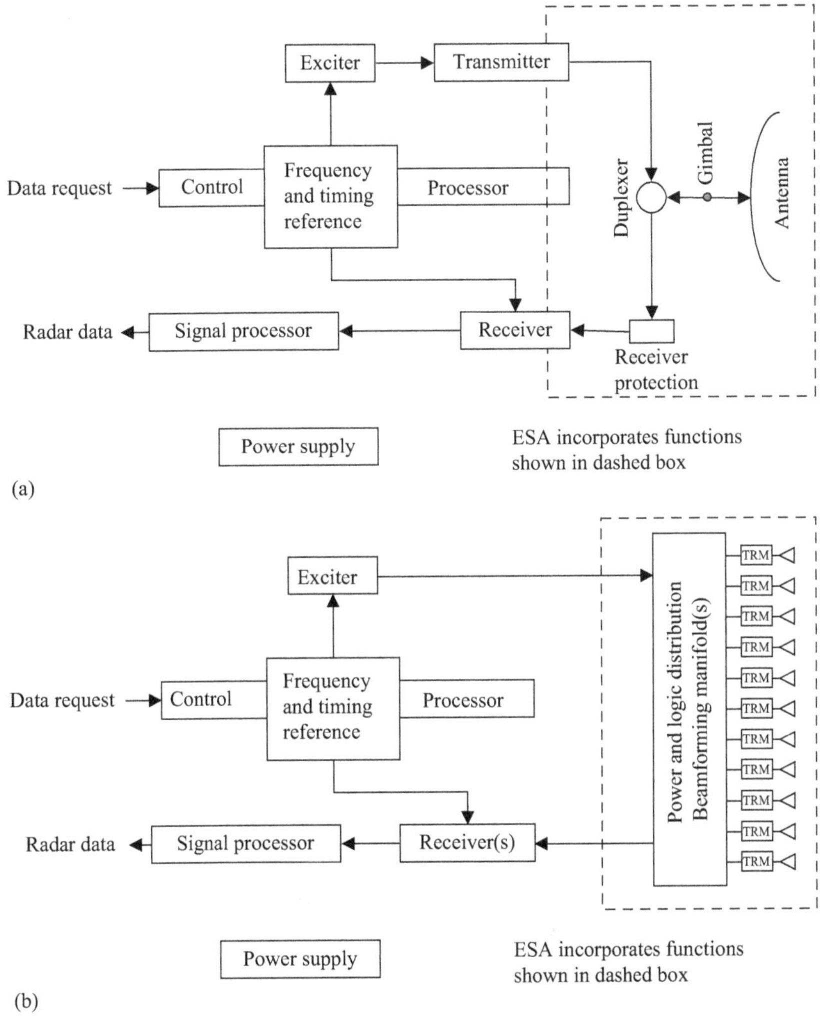

Figure 1.1 Block diagrams of a radar system: (a) reflector and (b) ESA

An ESA without these amplifiers is termed a passive ESA and includes only control provisions such as phase shifters. Passive ESAs provide many of the same benefits and their performance may be analyzed in much the same way as AESAs. They retain a central transmitter and provide phase control at the element or subarray level. They are less common now as T/R modules are readily available.

A planar array which omits both the amplifiers and the control mechanisms is not an ESA. However, with multiple radiating elements, it provides the opportunity to shape a single beam in a similar but unchanging manner. Figure 1.2 shows an early example of a planar array, developed for the Airborne Warning and Control System (AWACS).

Table 1.1 ESA attributes and benefits

Attribute	Benefit
Beam steering (agile beam)	Beam steering is nearly instantaneous, limited by the switching times in the T/R module control elements. Reduces slew and settle time
Mainlobe shaping, sidelobe control, and nulling	Beam shaping and nulling is practically an exclusive benefit of an ESA. Other antennas offer some beamshape control but only at the time of manufacture. Reconfigurability in use is difficult or impossible. Permits clutter and interference mitigation
Antenna deformation	Correct with calibration table
Multiple beams	Enables multiple concurrent radar modes. Toso [7] provides an eloquent description of multiple beams with several examples. Multiple beams are described in Section 2.6.2
Multiple phase centers	Multiple effective antennas in a single ESA enabling moving target indication and for synthetic aperture radar discussed in Chapter 3. *MTI* and multichannel *SAR*/enable angle of arrival measurement
Inherent redundancy	Multiple elements provide graceful degradation
Space combining of solid-state power amplifiers	Lower loss between amplifiers and free space. Enables very high effective radiated power (ERP) and additional modes such as electronic attack
Better match to free space	Higher efficiency, less reflection re-radiation, lower observability

Figure 1.2 AWACS antenna
Courtesy NIST [8]

A reflector antenna may be configured with an ESA feed. This arrangement provides a larger aperture but a reduced scanning capability. Array feeds are being retrofitted into existing reflector antennas or designed into new systems, especially for large radio telescopes where the multiple beams enable higher coverage rates.

Array techniques have been used to design frequency-selective surfaces. They find use in radio astronomy as beam-splitting mirrors, sending different frequency bands to different receivers. Anderson [9] provided the early analysis.

Figure 1.3 Overall view of the VLA
Courtesy NRAO/AUI [10]

Another radio astronomy application is sparse interferometric arrays which provide high angular resolution of a large aperture without the necessity of filling the aperture with receivers. This is feasible because of the small number of objects surveyed and the very low background noise level. Figure 1.3 illustrates such a radio astronomy array. These antennas may be considered examples of thinning, which is discussed in Chapter 8.

1.3 Early ESA development

The development of ESA technology began slowly. The following examples exemplify this process.

1.3.1 F-15 airborne radar

The F-15 entered production in 1972 equipped with a planar array radar. Concurrently, the US Air Force was promoting ESA development with the MERA program cited by Schrank [3] and subsequent design of a solid-state phased array (SSPA) for the aircraft. The SSPA represented the first attempt to apply GaAs to fighter aircraft radar. Weber [11] provides a contemporaneous description of the SSPA background and justification. Its technical goals are shown in Table 1.2.

The SSPA was put out for bid in 1982. Texas Instruments won the contract with a value of about $20 million besting Hughes Aircraft. McQuiddy *et al.* [13] describe the development history in "Transmit/receive module technology for X-band active array radar." Figure 1.4a illustrates the SSPA prototype array delivered by Texas Instruments in May 1988 after about five years of development.

The SSPA program experienced the usual technical and financial difficulties. Nonetheless, according to Marchese [14] in "Breakthrough Technologies Developed by the Air Force Research Laboratory and Its Predecessors":

Table 1.2 SSPA characteristics

Array	**Transmit and receive**	
Operating frequency	X-band	
Aperture diameter	0.81 m	
Number of T/R modules	1,980	
Lattice	Equilateral triangular	
Lattice spacing	1.7 cm	
Array dimensions	0.91 m diameter × 0.33 m deep	
Array volume	0.275 m^3	
Weight	220 kg	
Prime power	9,500 W	
T/R module	**Transmit**	**Receive**
Transmit duty cycle (max)	40%, maximum	
Transmit pulse width (max)	40 μs, maximum	
Efficiency (mean)	14% at maximum duty cycle	
Module peak power (W)	1.55	
Gain (dB)		10.3
Noise figure (dB)		5.4

Source: Kley *et al.* [12]

The SSPA program was successful in demonstrating the superior capabilities of the SSPA over conventional mechanical radars, in particular a mean time between critical failure (MTBCF) in excess of 70,000 hours although the cost of component fabrication, particularly of the T/R modules, was deemed excessive at $12,000 per module.

The stated module cost seems inconsistent with the contract value.

(a) (b)

Figure 1.4 (a) SSPA (1982-1986) and (b) APG-63v(2) (2000)

Source: (a) Courtesy United States Air Force and (b) ©2000 Boeing Corporation

Lingle *et al.* [15] describe the Ultra-Reliable Radar program awarded to Westinghouse in 1985 to fabricate and test a complete ESA radar system using the SSPA. The program led to AESA development and production for the F-22 [16] and F-35. Both systems were developed by a team of Texas Instruments and Westinghouse.

In parallel, Hughes continued AESA development for the F-15 culminating in the APG-63v(2) limited production model in 2000. Figure 1.4b shows one of the 18 F-15C aircraft retrofitted with this ESA. Boeing's press release [17] notes that this upgrade provided the US Air Force the "world's first operational fighter jets with the advanced-technology radar system." Since that time, hundreds of existing F-15 radars have been upgraded with a more advanced AESA discussed in Section 16.4.3.

The two AESAs shown in Figure 1.4 and delivered 14 years apart do not differ greatly in appearance but embody different technology levels.

1.3.2 MESAR and the SAMPSON naval radar

Figure 1.5 illustrates an analogous ESA development in the United Kingdom. The Microwave Electronically Scanned Adaptive Radar (MESAR) radar program began in 1982. It was probably the first demonstration of adaptive beamforming. MESAR 1, shown in Figure 1.5a, had a single-array face and was only partially populated with active elements. MESAR 2 began development in 1995 and had a larger and fully populated array. Table 1.3 summarizes their characteristics.

In 2005, the ARTIST program built a research demonstrator system which was tested in 2009–10 at Wallops Island in the United States in collaboration with the US Navy. Further development led to the SAMPSON radar, developed for the Type 45 destroyer, shown in Figure 1.5b. The first ship, HMS Daring, was launched in 2006 and commissioned in 2009. Stafford [20] relates the development history.

(a) (b)

Figure 1.5 (a) MESAR (1982) and (b) SAMPSON-First of class installation-HMS Daring (2006)

© 2010 R.L. Burr [18]

Table 1.3 MESAR array characteristics

Attribute	MESAR 1	MESAR 2
Frequency	2.7–3.3 GHz	
Wavelength	11.1–9.1 cm	
Array shape	Octagonal	
Type	Thinned	Filled
Number of radiating elements	918	1,264
Number of T/R modules	157	1,264
Module peak power	2 W	10 W
Lattice spacing	5 cm	
Array dimensions	1.8 m × 1.8 m	

Source: Billam [19], Stafford [20], Scott [21], and Bell [22]

1.3.3 Global Protection Against Limited Strike family of radars

In his January 1991 State of the Union address, Ronald Reagan announced the Global Protection Against Limited Strike (GPALS) as an element of his Strategic Defense Initiative. This initiative ultimately resulted in the Terminal High-Altitude Area Defense (THAAD) system (comprising a radar, missile launcher, control, power, and cooling) and the sea-based X-band (SBX).

The demonstration-validation phase started in 1992 with a contract award to Lockheed, with Raytheon as the radar supplier. The engineering and manufacturing development phase was awarded in 2000, also to Lockheed. The first system was deployed in 2008.

1.3.3.1 AN/TPY-2

The AN/TPY-2 built by Raytheon is the THAAD radar and it is illustrated in Figure 1.6a. It incorporates a large ESA with many T/R modules in a relatively tight lattice yielding a large electronic field of regard. It is mounted on a gimbal to increase the coverage volume. Some of its attributes are shown in Table 1.4. Twelve radars have been delivered through 2019 according to the Missile Defense Agency [28].

1.3.3.2 SBX

Figure 1.6b and c shows the SBX system, which began test in 2005 according to MDA [29]. Some of its attributes are listed in Table 1.4. The SBX shares hardware heritage with the THAAD radar, consistent with the initial family of radars concept. Boeing is the system integrator and prime contractor, and Raytheon builds the radar. The radar is designed for exceptionally long range and the aperture is very large, approximately 40× that of THAAD although it has only 2× the number of T/R modules. Being a thinned array, it implements measures against

(a) AN/TPY-2 antenna

(b) SBX platform

(c) SBX array

Figure 1.6 GPALS family of radars
Courtesy Missile Defense Agency [23–25]

Table 1.4 Family of radars

Parameter	AN/TPY-2	SBX
Frequency	X-band	X-band
Array size (m^2)	9.2	384
T/R modules	25,344	45,056
Subarrays (transmit/receive)	72/72	352/352
Scan (azimuth/elevation)	53°/53°	Unspecified
Mechanical elevation	10°–60°	Unspecified

Source: Sarcione *et al.* [26] and Feth [27]

Table 1.5 Personal computer evolution

Attribute	**1982**	**Now**
Price	$1,565	<$1,000
CPU speed	4.77 MHz	2+ GHz
RAM	16 kB/256 kB	>64 GB
Disk drive	160 kB	>1 TB
Video	480 × 640 pixel	4 k resolution

grating lobes including time delay steering and randomizing the element locations to reduce peak sidelobes. One radar was delivered.

1.3.4 Concurrent computer development

Digital computers are both a design tool and an essential component of ESAs. They were also developed in WWII because several projects required unprecedented computing capability. Mainframe computers dominated until the 1970s when personal computers such as Acorn (which spun off ARM) and TRS-80 democratized the technology.

The IBM PC shown in Figure 1.7 was introduced on August 12, 1981, at a price of $1,565. Table 1.5 compares the original PC with current performance metrics. Price reductions have been accompanied by enormous performance improvements. These quantitative improvements dwarf improvements in ESA performance although ESA improvements were more than sufficient to drive large-scale adaptation. Development of personal computers changed the ESA design process both qualitatively and quantitatively.

Figure 1.7 IBM PC 5150
Wikipedia [30]

1.4 ESA applications

This text focuses on ESAs for radars as they exemplify the challenges and benefits of ESAs and because there is a substantial body of published information.

As costs continue to drop, ESAs will find increasing application in communications such as WiFi, mobile telephony, and satellite internet. These applications are far more numerous but they will require still further cost reduction.

1.4.1 Terrestrial ESAs

1.4.1.1 Communications

Mobile telephony has provided much of the investment related to microwave technology, in general, and has in turn used some of the antenna design techniques used for ESAs. Cellular base stations use beamforming technology for increased frequency reuse. The mobile telephone 5G revolution may implement ESAs in handsets.

1.4.1.2 Radar

Some of the first phased arrays were developed for missile defense and operated at UHF frequencies. Built between 1960 and 1980, several are still operating today. Delaney [5] provides an overview.

Tables 1.6 and 1.7 list some of the US military radars which use ESAs. The first systems were developed in the twentieth century and fielded in the twenty-first century. Northrop Grumman airborne systems were largely developed by precursor companies Westinghouse and Norden. The Raytheon airborne systems were largely developed by the precursor company Hughes Aircraft Company.

Space fence (AN/FSY-3) is an interesting application of ESA technology. It was constructed under a $915 million contract [31] awarded to Lockheed Martin

Table 1.6 Airborne ESA systems

Manufacturer	System	Platform
Northrop Grumman/Raytheon	AN/APG-77	F-22
Northrop Grumman	AN/APG-80	F-16E/F Block 60
Northrop Grumman	AN/APG-81	F-35 Joint Strike Fighter
Northrop Grumman	AN/APG-83	F-16
Northrop Grumman	Multirole AESA	Boeing Wedgetail
Northrop Grumman	APY-9	E-2D
Northrop Grumman	SABR-GS	B-1B
Raytheon	AN/APG-63(v)2	F-15C
Raytheon	AN/APG-79	F/A-18E/F
Raytheon	AN/APG-82(v)1	F-15E
Raytheon	AN/APQ-181	B-2
Raytheon	ASARS-2B	ASARS
Raytheon		B-52G

Table 1.7 Ground and sea-based ESA systems

Manufacturer	System	Platform
Raytheon	AN/SPY-3	DD(X), CG(X), and CVN-21
Raytheon	AN/SPY-6	DD(X), CG(X), FFG(X)
Raytheon	National Missile Defense X-Band Radar (XBR)	
MEADS International (MI)	Multifunction Fire Control Radar (MFCR)	
Lockheed Martin/ Raytheon	AN/TPY-2	THAAD system radar
Lockheed Martin	AN/SPY-4	Volume search radar
Lockheed Martin	AN/FSY-3	Space Fence
Lincoln Laboratory	MPAR	MACOM partner

in 2014 and achieved operational status in March 2020. As described by Haimerl *et al.* [32], it operates at S-band using separate transmit and receive ESA arrays with 36,000 and 86,000 elements, respectively. With a peak radiated power of 2.69 MW, each element produces about 75 W peak power. In transmit, groups of eight elements have dedicated RF sources. In receive, digitization occurs at each element. Gallagher et al [33] describe the application of GaN to the system. The use of separate transmit and receive arrays, one square and one diamond in outline, disorient the principal planes and as noted in Figure 7 of Haimerl, "provide low two-way sidelobe levels with reduced weighting losses, minimizing the required power aperture." This design decision was simple but very effective. It is further illustrated in Figure 4.6.

The Multimission Phased-Array Radar (MPAR) began in 2003 with a demonstration based on a single AN/SPY-1 radar in Oklahoma. In 2007, a team of Lincoln Laboratory and MACOM began studies and tests to develop a low-cost replacement for existing air traffic control radars in the United States. Weber *et al.* [34] summarize the requirements and Conway *et al.* [35] and Carlson [36] describe the hardware. Carlson [37] describes various cost reduction methods. It is an 4,864-element, dual polarized, S-band system and the cost target for its T/R module is between $50 and $100 [38]. Its predicted performance is substantially better than the systems it will replace. It is noteworthy for its use of air cooling (see Chapter 15.4 for the use of commercial technology).

1.4.2 Communications and internet satellites

Communications was one of the first satellite services. Early examples were in low orbit but most services today operate at geosynchronous orbit, more than 400 by published reports. They use large deployable reflectors combined with feed systems to illuminate only the desired service area. Rao [39] provides a detailed summary of systems and their designs.

Internet from space has been available for 20 years. Hu and Li [40] provide a tutorial. Recently, several very well-funded companies such as Amazon, Facebook,

Google, and SpaceX have announced plans for constellations of thousands of satellites offering vastly increased aggregate bandwidth at lower cost. The number of satellites and limited spectrum allocation requires careful frequency management and reuse.

1.4.3 Radar satellites

The majority of radar satellites use ESA antennas and have done for a number of years. Figure 1.8 illustrates several earth observation SAR satellites. They will be discussed in more detail in Section 17.2. Croci *et al.* [47] in "Space based radar technology evolution" describe some trends in radar satellites.

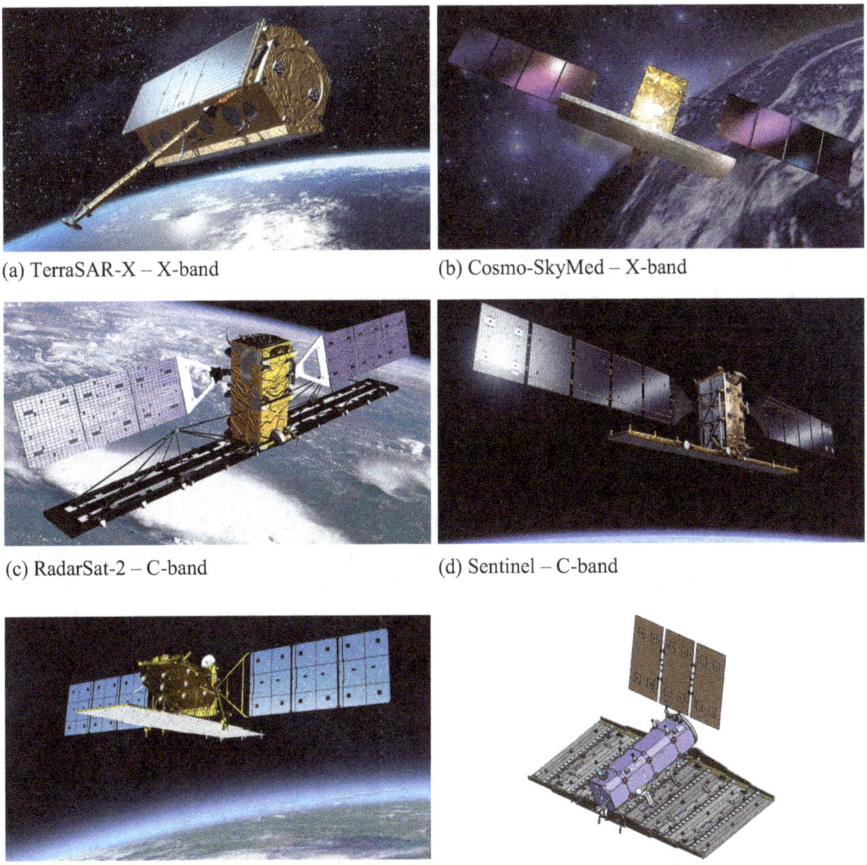

(a) TerraSAR-X – X-band

(b) Cosmo-SkyMed – X-band

(c) RadarSat-2 – C-band

(d) Sentinel – C-band

(e) ALOS-2 – L-band

(f) SAOCOM – L-band

Figure 1.8 Some current radar satellites

Source: (a) © Astrium GmbH [41], (b) © Finmeccanica [42], (c) © Canadian Space Agency [43], (d) © European Space Agency – P. Carril [44], (e) © JAXA [45] and (f) © CONAE [46]

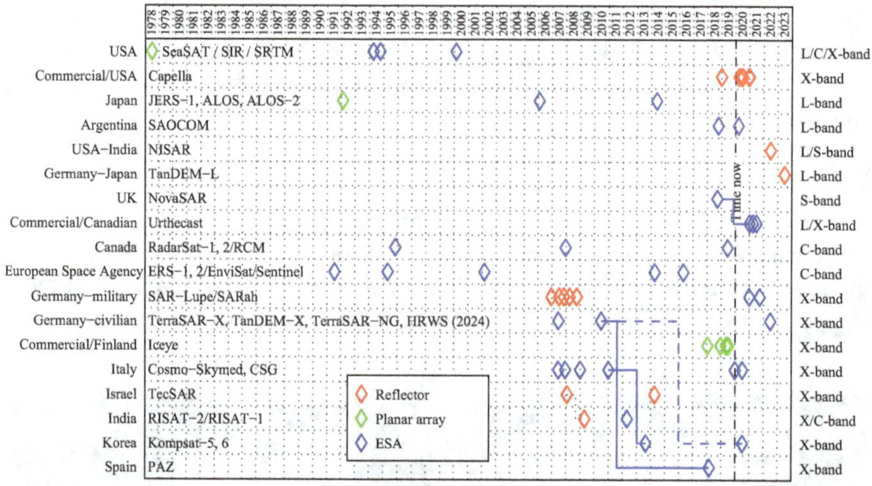

Figure 1.9 On-orbit and planned radar satellites

Figure 1.9 summarizes existing and planned radar satellites using reflector, planar, and ESA antennas. Of the 64 systems shown on the chart, 28% are based on reflector antennas shown as red diamonds, 14% are based on planar arrays shown as green diamonds, and 58% are based on ESAs shown as blue diamonds.

Some systems are derived from others. Suri *et al.* [48] describe how the TerraSAR has been adapted by Thales Spain and becomes the PAZ. Yoon *et al.* [49] describe the Korean Kompsat based on COSMO-SkyMed with the addition of some domestic content. The Indian RISAT-2 is based on TecSAR. Fox *et al.* [50] note that Urthecast utilizes technology developed by Surrey Satellite for NovaSAR.

At X-band, two systems, TerraSAR and COSMO-SkyMed, are based on ESAs and two, TecSAR and SAR-Lupe, are based on reflectors. At L-band, two operational satellites, ALOS-2 and SAOCOM (Satélite Argentino de Observación COn Microondas), are based on ESAs, while two proposed systems, NISAR and TanDEM-L, are based on reflectors. Evidently, both technologies are useful.

The annual launch rates in Figure 1.10 exhibit a distinct positive trend.

1.4.3.1 Commercial radar satellites

Most of the radar satellites to date are government or scientific, some of which recoup some costs with commercial sales. Recently "commercial" systems, e.g., systems which expect to cover all their costs with commercial sales, have emerged. Because TerraSAR and COSMO-SkyMed already provide a high-quality product, these new market entrants compete on the basis of revisit times and lower cost.

The commercial variants are smaller and more numerous than their predecessors to match the needs of their customers. Table 1.8 summarizes some of the existing and proposed satellites. Image quality is largely maintained but power provisions

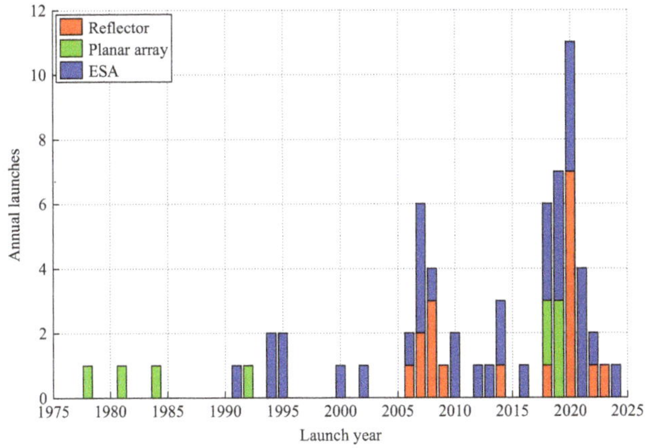

Figure 1.10 Launch schedule

Table 1.8 Commercial radar satellites

	Iceye	Capella	Synspective	Urthecast
Reference	**[51]**	**[52, 53]**	**[54–56]**	**[50, 57, 58]**
Band	X	X	L and X	X
First launch	2017	2018	2020	2021
Constellation	18	36	100	8
Antenna size (m)	0.4 × 3.2		0.7 × 4.9	1 × 1.8
Satellite mass (kg)	<100			
Technology	Planar	Reflector	Planar	ESA
Country	Finland	USA	Japan	Canada

are smaller representing fewer minutes of imaging time per orbit. Very few technical details have been released.

1.4.3.2 Iceye compared to Seasat

Table 1.9 compares the attributes of the 1978 L-band Seasat satellite and the 2019 Iceye X-band satellite. Despite vastly different characteristics, they have comparable performance.

1.4.4 Radio astronomy

Radio telescopes need aperture (area) above all, so they tend to be dominated by reflector systems. Warnick *et al.* [59] discuss the tradeoff in Section 7.1.2 of his book *Phased Arrays for Radio Astronomy, Remote Sensing, and Satellite Communications*. At a certain size, reflectors become too large to move, so the largest reflectors are stationary.

Table 1.9 Seasat and Iceye parametric comparison

Attribute	Seasat	Iceye
Dimensions (m)	1.5D ×21L	0.8 × 0.6×0.5
Mass (kg)	2,290	<100
Array type	Planar	Planar
Array size (m)	10.74 × 2.16	3.2 × 0.4
Frequency (GHz)	1.275	10
Bandwidth (MHz)	19	10–300
Peak power (kW)	1	2–4
Transmit duty	≈5%	tbd

Several radio telescopes have been retrofitted with phased array receivers to increase their coverage. Because the sky has a very low noise temperature, they use low-temperature feeds and arrays to reduce the system noise figure to a very low level. The Green Bank Telescope in West Virginia, USA, is a 100 m diameter offset parabolic reflector and the largest fully steerable antenna. It has active compensation for deformation which extends its useful frequency to 100 GHz. It is being retrofitted with a 19-element cryogenically cooled array feed described by Roshi *et al.* [60]. The Arecibo telescope in Puerto Rico, USA, is a 300 m stationary, zenith pointing spherical reflector. It is being retrofitted with an 80-element cryogenically cooled array feed described by Cortes-Medellin *et al.* [61] and Warnick *et al.* [62].

FAST (Five-hundred-meter Aperture Spherical radio Telescope) [63] in Guizhou province, China, is a stationary, zenith pointing reflector and the largest radio telescope in the world.

The Square Kilometer Array (SKA) will be the largest radio telescope when completed. Dewdney *et al.* [64] state, "The Square Kilometre Array (SKA) will be an ultrasensitive radio telescope, built to further the understanding of the most important phenomena in the Universe, including some pertaining to the birth and eventual death of the Universe itself." The SKA is so large that it defies a single aperture and must be constructed of smaller systems networked or arrayed together. This approach requires many coherently combined antennas which in effect constitute an ESA. Jones [65] describes some of its technical challenges.

Interferometers are arrayed systems, widely separated, and hence extremely sparse. The sidelobes of the sparse array can be tolerated because of the a priori knowledge of the nearly stationary target space. An example is illustrated in Figure 1.3.

References

[1] Milne K. 'Phased arrays in radar'. *IEE Tutorial Meeting on Phased Array Radar*. 1989, pp. 2/1–2/31.

[2] Holpp W. *100 Jahre Radar – 1904-2004*. June 2004. Available from https://www.100-jahre-radar.fraunhofer.de/vortraege/Holpp-TheCenturyofRadar.pdf.

[3] Schrank H.E. 'Some notable firsts in array antenna history'. *2001 Antenna Applications Symposium*. 2001, pp. 231–49.

[4] Hyltin T.M. 'The pioneering of the MERA program'. *2008 IEEE MTT-S International Microwave Symposium Digest*. 2008, pp. 719–22.

[5] Delaney W. 'From vision to reality 50+ years of phased array development'. *2016 IEEE International Symposium on Phased Array Systems and Technology (PAST)*. 2016, pp. 1–8.

[6] Kopp C. 'Evolution of AESA radar technology'. *Microwave Journal*. August 2012.Available from https://www.microwavejournal.com/articles/print/17992-evolution-of-aesa-radar-technology.

[7] Toso G. 'The beauty of multibeam antennas'. *2015 9th European Conference on Antennas and Propagation (EuCAP)*. 2015, pp. 1–5.

[8] NIST. *AWACS Antenna*. Available from https://www.nist.gov/image/awacsjpg.

[9] Anderson I. 'On the theory of self-resonant grids'. *Bell System Technical Journal*. 1975, vol. 54(10), pp. 1725–31.

[10] NRAO/AUI. *VLA*. Available from https://public.nrao.edu/gallery/this-isthe-vla-wye/.

[11] Weber H. 'Solid state phased array technology'. *1983 Allerton*. RADC-TR-84-52. U.S. Government Printing Office; March 1983. pp. 7–11. Available from https://apps.dtic.mil/dtic/tr/fulltext/u2/a142003.pdf.

[12] Kley R.C., Hill W.P., Lamb F.D. 'Solid-State Phased Array (SSPA) performance'. *IEEE International Conference on Radar*. 1990, pp. 298–303.

[13] McQuiddy D.N., Gassner R.L., Hull P., Mason J.S., Bedinger J.M. 'Transmit/receive module technology for X-band active array radar'. *Proceedings of the IEEE*. 1991, vol. 79(3), pp. 308–41.

[14] Marchese J. *"Breakthrough" Technologies Developed by the Air Force Research Laboratory and Its Predecessors*. Tech. Rep. ADA528970. US Air Force Research Laboratory; December 2005. Available from https://apps.dtic.mil/dtic/tr/fulltext/u2/a528970.pdf.

[15] Lingle D.E., Mikszan D.P., Mukai D. 'Advanced technology Ultra Reliable Radar (URR)'. *Proceedings of the IEEE National Radar Conference*. 1989, pp. 1–6.

[16] Malas J.A. 'F-22 radar development'. *Proceedings of the IEEE 1997 National Aerospace and Electronics Conference. NAECON 1997*. 1997, vol. 2, pp. 831–9.

[17] Boeing. *Boeing Delivers Final AESA-Equipped F-15 Aircraft to U*.S. Air Force. 2000. Available from https://boeing.mediaroom.com/2000-12-20-Boeing-Delivers-Final-AESA-Equipped-F-15-Aircraft-to-U.S.-Air-Force.

[18] Burr R.L. *The Decca Legacy, A View from Inside the Radar Company*. Stubbington, Hampshire, UK: E. O. Grove. 2010. Available from https://woottonbridgeiow.org.uk/decca-legacy/index.php.

[19] Billam E.R. 'MESAR-the application of modern technology to phased array radar'. *IEE Tutorial Meeting on Phased Array Radar*. 1989, pp. 5/1–516.
[20] Stafford W.K. 'MESAR, Sampson & Radar Technology for BMD'. *2007 IEEE Radar Conference*. 2007, pp. 437–42.
[21] Scott M. 'SAMPSON MFR active phased array antenna'. *IEEE International Symposium on Phased Array Systems and Technology, 2003*. 2003, pp. 119–23.
[22] Bell C.J. 'Multifunction radar development in the UK'. *2013 IEEE International Symposium on Phased Array Systems and Technology*. 2013, pp. 17–20.
[23] Missile Defense Agency. *THAAD Antenna*. Available from https://www.mda.mil/global/images/system/sensors/fbx-t-2.jpg.
[24] Missile Defense Agency. *SBX Platform*. Available from https://www.mda.mil/global/images/system/sbx/img%5C 2845.jpg.
[25] Missile Defense Agency. *SBX Antenna*. Available from https://www.mda.mil/global/images/system/sbx/sbxin.jpg.
[26] Sarcione M., Mulcahey J., Schmidt D., *et al*. 'The design, development and testing of the THAAD (Theater High Altitude Area Defense) solid state phased array (formerly Ground Based Radar)'. *Proceedings of International Symposium on Phased Array Systems and Technology*. 1996, pp. 260–5.
[27] Feth S. 'Overview of MDA/SN radar systems (5273-11) – UEWR, Cobra Dane, TPY-2 and SBX'. *IDGA 5th Annual Military Radar Summit*. 2012, pp. 1–17.
[28] Missile Defense Agency. *Army Navy / Transportable Radar Surveillance (AN/TPY-2)*. 2016. Available from https://www.mda.mil/global/documents/pdf/antpy2.pdf.
[29] Missile Defense Agency. *A Brief History of the Sea-Based X-Band Radar- 1*. Tech. Rep. 08-MDA-3447. 2008. Available from https://www.mda.mil/global/documents/pdf/sbx booklet.pdf.
[30] de Rijcke R. *IBM Personal Computer*. 2018. Available from https://en.wikipedia.org/wiki/IBM_Personal_Computer.
[31] USAF. *Space Fence Contract Awarded to Lockheed Martin*. 2014. Available from https://www.af.mil/News/Article-Display/Article/485271/space-fence-contract-awarded.
[32] Haimerl J.A., Hudson B., Fonder G.P., Lee D.K. 'Overview of the large digital arrays of the Space Fence radar'. *2016 IEEE International Symposium on Phased Array Systems and Technology (PAST)*. 2016, pp. 1–8.
[33] Gallagher J., Haimerl J.A., Higgins T., Courtney P., Hindle P. 'Space Fence radar leverages power of GaN'. *GaN Power Amplifier Design Solutions*; 2018. pp. 4–9. Available from https://www.microwavejournal.com/articles/31075-gan-poweramplifier-design-solutions.
[34] Weber M.E., Cho J.Y.N., Herd J.S., *et al*. 'The next-generation multimission U.S. surveillance radar network'. *Bulletin of the American Meteorological Society*. 2007, vol. 88(11), pp. 1739–52.

[35] Conway M.D., Russel D.D., Morris A., *et al.* 'Multifunction Phased Array Radar advanced technology demonstrator nearfield test results'. *2018 IEEE Radar Conference (RadarConf18)*. 2018, pp. 1412–5.
[36] Carlson D. *'Tile Arrays Accelerate the Evolution to Next-Generation Radar'*. 2017. Available from http://www.microwavejournal.com/articles/27652-tile-arrays-accelerate-the-evolution-to-next-generation-radar.
[37] Carlson D. 'Breaking through the cost barrier for phased arrays'. *Microwave J.* 2018. Available from https://www.microwavejournal.com/articles/31325-breaking-through-the-cost-barrier-for-phased-arrays.
[38] Weber M. 'Multifunction phased array radar: technical synopsis, cost implication and operational capabilities'. *Proceeding of 23th Conference on International Interactive Information and Processing Systems (IIPS) for Meteorology*. Oceanography and Hydrology; 2007. Available from https://ams.confex.com/ams/pdfpapers/116588.pdf.
[39] Rao S. 'Advanced antenna systems for 21st century satellite communications payloads'. *IEEE APS Distinguished Lecture*. 2015. Available from http://site.ieee.org/scv-aps/files/2015/05/DLRaoSantaClara23Apr2015Distribution.pdf.
[40] Hu Y., Li V.O.K. 'Satellite-based internet: a tutorial'. *IEEE Communications Magazine*. 2001, vol. 39(3), pp. 154–62.
[41] Pitz W., Miller D. 'The TerraSAR-X satellite'. *IEEE Transactions on Geoscience and Remote Sensing*. 2010, vol. 48(2), pp. 615–22.
[42] Ferrando E., Rizzi F., Hazan D., *et al.* 'The COSMO SkyMed solar array'. *2006 IEEE 4th World Conference on Photovoltaic Energy Conference*. 2006, vol. 2, pp. 1990–3.
[43] Maxar. *RADARSAT-2 Data and Products ©MacDONALD, DETTWILER AND ASSOCIATES LTD – All Rights Reserved. RADARSAT is an official mark of the Canadian Space Agency*. 2005. Available from https://asccsa.gc.ca/eng/search/images/watch.asp?id=448&search=RADARSAT-2.
[44] European Space Agency. *Introducing Sentinel-1*. Available from http://www.esa.int/Applications/ObservingtheEarth/Copernicus/Sentinel-1/IntroducingSentinel-1.
[45] Shimada M., Kankaku Y., Watanabe M., Motooka T. 'Current status of the ALOS-2/PALSAR-2 and the CALVAL program'. *CEOS SAR CALVAL Workshop at ASF*. 2011. Available from http://sarcv.ceos.org/sitemedia/media/documents/CurrentStatusofALOS-2M.Shimada.pdf.
[46] Euillades P.A., Euillades L.D., Azcueta M., Sosa G. 'SAOCOM 1A interferometric error model and analysis'. *Advances in the Science and Applications of SAR Interferometry and Sentinel-1 InSAR Workshop*; 2015. Available from http://seom.esa.int/fringe2015/files/presentation201.pdf.
[47] Croci R., Delfino A., Marchetti F. 'Space based radar technology evolution'. *2009 European Radar Conference (EuRAD)*. 2009, pp. 601–4.
[48] Suri S., Peterson L., Kaptein A., *et al.* 'TerraSAR-X/PAZ constellation: CONOPS, highlights and access solution'. *2015 IEEE 5th Asia-Pacific Conference on Synthetic Aperture Radar (APSAR)*. 2015, pp. 178–83.

[49] Yoon J.C., Keum J.H., Shin J.M., *et al.* 'KOMPSAT-5 SAR design and performance'. *2011 3rd International Asia-Pacific Conference on Synthetic Aperture Radar (APSAR)*. 2011, pp. 1–4.

[50] Fox P., Tyc G., Beckett K. 'The UrtheCast SAR-XL multi-band, multi-aperture spaceborne SAR system'. *2017 IEEE Radar Conference (RadarConf)*. 2017, pp. 1761–4.

[51] Antropov O., Praks J., Kauppinen M., *et al.* 'Assessment of operational microsatellite based SAR for earth observation applications'. *2018 2nd URSI Atlantic Radio Science Meeting (AT-RASC)*. 2018, p. 1.

[52] Farquharson G., Woods W., Stringham C., Sankarambadi N., Riggi L. 'The Capella synthetic aperture radar constellation'. *IGARSS 2018 - 2018 IEEE International Geoscience and Remote Sensing Symposium*. 2018, pp. 1873–6.

[53] Farquharson G., Woods W., Stringham C., Sankarambadi N., Riggi L. 'The Capella synthetic aperture radar constellation'. *EUSAR 2018; 12th European Conference on Synthetic Aperture Radar*. 2018, pp. 1–5.

[54] Saito H. 'Development status of compact X-band synthetic aperture radar compatible with a 100 kg-class SAR satellite and its future plan'. *31st Annual AIAA/USU Conference on Small Satellites. SSC17-IX-01*. 2017. Available from https://digitalcommons.usu.edu/cgi/viewcontent.cgi?article=3659&context=smallsat.

[55] Watanabe H. 'Test and development of prototype 1000W X-band microwave solid-state power amplifier for small SAR satellite'. *32nd Annual AIAA/USU Conference on Small Satellites. SSC18-III-08*. 2018. Available from https://digitalcommons.usu.edu/cgi/viewcontent.cgi?article=4079&context=smallsat.

[56] Obata T. 'The development status of the first demonstration satellite of our commercial small synthetic aperture radar satellite constellation'. *33rd Annual AIAA/USU Conference on Small Satellites*. SSC19-V-04. 2019. Available from https://digitalcommons.usu.edu/cgi/viewcontent.cgi?article=4390&context=smallsat.

[57] Gomes L. 'Rich data, cheap satellites'. *31st Annual AIAA/USU Conference on Small Satellites*. SSC17-IX-06. 2017. Available from https://digitalcommons.usu.edu/cgi/viewcontent.cgi?article=3682&context=smallsat.

[58] Gomes L. 'Rich data, cheap satellites'. *31st Annual AIAA/USU Conference on Small Satellites*. SSC17-IX-06. 2017. Available from https://digitalcommons.usu.edu/cgi/viewcontent.cgi?filename=0&article=3682&context=smallsat&type=additional.

[59] Warnick K.F., Maaskant R., Ivashina M.V., Davidson D.B., Jeffs B.D. *Phased Arrays for Radio Astronomy, Remote Sensing, and Satellite Communications. EuMA High Frequency Technologies Series*. Cambridge, UK: Cambridge University Press; 2018.

[60] Roshi D.A., Shillue W., Fisher J.R., *et al.* 'A highly-sensitive cryogenic phased array feed for the Green Bank telescope'. *2017 XXXIInd General Assembly and Scientific Symposium of the International Union of Radio Science (URSI GASS)*. 2017, pp. 1–4.

[61] Cortes-Medellin G., Parshley S., Campbell D.B., *et al.* 'Concept design of an 80-dual polarization element cryogenic phased array camera for the Arecibo radio telescope' in Evans C.J., Simard L., Takami H. (eds.). *Ground-based and Airborne Instrumentation for Astronomy VI.* 9908. International Society for Optics and Photonics. SPIE; 2016. pp. 1622–9.

[62] Warnick K.F., Kunzler J., Cortes-Medellin G. 'Geometry optimization of a phased array feed on the Arecibo telescope for maximum survey efficiency'. *2019 IEEE International Symposium on Antennas and Propagation and USNC-URSI Radio Science Meeting.* 2019, pp. 399–400.

[63] Gibney E. 'Gigantic Chinese telescope opens to astronomers worldwide'. Nature 574 October 2019, pp. 16–17. Available from https://www.nature.com/articles/d41586-019-02790-3.

[64] Dewdney P.E., Hall P.J., Schilizzi R.T., Lazio T.J.L.W. 'The Square Kilometre Array'. *Proceedings of the IEEE.* 2009, vol. 97(8), pp. 1482–96.

[65] Jones D.L. 'Technology challenges for the Square Kilometer Array'. *IEEE Aerospace and Electronic Systems Magazine.* 2013, vol. 28(2), pp. 18–23.

Chapter 2
Antenna principles

2.1 Introduction

Antennas are devised for the purpose of transmission and reception. In transmission, the antenna objective is to create a current/voltage distribution at the aperture surface which will create a specified beam pattern. In reception, the antenna produces an output signal whose magnitude and phase correspond to the incoming signal's magnitude and phase according to the antenna beam shape.

General antenna theory provides concepts and rules for ESA analysis. Antenna analysis assumes linearity and superposition to decompose solutions into frequency and angular components.

2.1.1 Reciprocity

The Lorenz reciprocity theorem states that transmit and receive behaviors are similar so that it is frequently a matter of convenience which mode is analyzed. It generally comports with the physics principle that time usually does not have a preferred direction and receive may be construed as viewing transmit in reverse.

A number of authors discuss this principle including Carson [1, 2], Ballantine [3], and Neiman [4].

2.1.2 Uniqueness

The uniqueness theorem states that for a given farfield antenna pattern, there is a unique aperture distribution of voltage or current. Accordingly, given the same farfield objective function, any synthesis technique should yield the same aperture distribution.

2.1.3 Sampling and Fourier transforms

ESAs provide a sampled representation of almost arbitrary current/voltage distribution notwithstanding which the actual fields must be continuous. The use of large numbers of T/R modules and radiating elements permits almost arbitrary field distributions.

As we will see, the antenna pattern is the Fourier transform of the aperture illumination which provides a very powerful analytic tool. Broadly speaking, the antenna pattern is the Fourier transform of aperture illumination for both discrete (sampled) and continuous cases. In the discrete case, the sample interval is the element spacing and the angular resolution is a function of the number of samples. $\lambda/2$ element spacing assures no grating lobes, a corollary to the Nyquist–Shannon sampling theorem.

2.1.4 *Beam shape and resolution*

The angular distribution of the antenna pattern is a function of the aperture shape. Round apertures tend to have circularly symmetric patterns, while square apertures have discrete pattern ridges, termed principal planes. The angular resolution limit of the antenna is a function of its size, analogous to the Rayleigh criteria in optics.

Antenna gain is normalized to an idealized isotropic radiator radiating into a spherical volume (4π steradians). The ESAs we consider radiate in a hemisphere, so they achieve 3 dB of gain for that reason alone.

2.2 Electromagnetic radiation

The origin of an electromagnetic wave is the motion, specifically the acceleration, of a charged particle. The electric field of a particle with charge q is described by (28.3) and (28.5) of Feynman *et al.* [5], reproduced in (2.1):

$$\boldsymbol{E} = \frac{-q}{4\pi\varepsilon_0}\left[\frac{\boldsymbol{e}_{r'}}{r'^2} + \frac{r'}{c}\frac{d}{dt}\left(\frac{\boldsymbol{e}_{r'}}{r'^2}\right) + \frac{1}{c^2}\frac{d^2\boldsymbol{e}_{r'}}{dt^2}\right]$$

and for large r (2.1)

$$\approx \frac{-q}{4\pi\varepsilon_0}\left[\frac{1}{c^2}\frac{d^2\boldsymbol{e}_{r'}}{dt^2}\right]$$

where $\boldsymbol{E}$ is the resulting field at the measurement point, r' is the distance between the measurement point and the source (at the time the radiation was emitted), and $\boldsymbol{e}_{r'}$ is the unit vector from the measurement point toward the source (at the time the radiation was emitted). While there is no explicit range dependence, the magnitude of the acceleration of $\boldsymbol{e}_{r'}$ has the necessary $1/r$ dependence. Most commonly, the charged particle is an electron although other particles, e.g., atomic nuclei, may radiate through the cyclotron or bremstrahlung mechanism. The formulation in (2.1) is not generally used but provides insight into the underlying radiation mechanism.

The associated magnetic field is given by (2.2), which reproduces (28.4) of Feynman *et al.* [5]:

$$\boldsymbol{B} = -\boldsymbol{e}_{r'} \times \frac{\boldsymbol{E}}{c} \tag{2.2}$$

Electrons have the convenient property, not shared by other charged particles, of traveling in a wire. Accordingly, a wire is the fundamental radiating element from which others are derived. It is possible to devise systems which do not require wires and rely exclusively on waveguides using RF sources such as klystrons, magnetrons, etc., but these are not relevant to our purposes.

Transmission and reception of radiating fields constitutes the transition from free space to a conductor. It results from specific boundary conditions which introduce and/or remove power from free space under the influence of the EM wave on electrons or vice versa. If we knew the trajectories of the electrons, (2.1) might be sufficient. Unfortunately, electron motion is generally a matter of conjecture and the origin of much of the art of microwave design. Each electron creates a field and is acted upon by its own field as well as those of its neighbors. Accordingly, Maxwell's equations constitute a better starting point.

One manifestation of this effect is termed mutual coupling and will be discussed in Chapter 13.

2.3 Maxwell's equations

Maxwell unified the experimental observations and corresponding formulas of electricity and electromagnetic fields in his equations. Feynman *et al.* [5] note in Chapter 28 of volume I of their lectures on physics:

> Perhaps the most dramatic moment in the development of physics during the 19th century occurred to J.C. Maxwell one day in the 1860s, when he combined the laws of electricity and magnetism with the laws of the behavior of light.

Heaviside simplified Maxwell's formulation and created the formulation shown in (2.3) which is now the standard representation:

$$\begin{aligned} \nabla \cdot \boldsymbol{E} &= \frac{\rho}{\varepsilon_0} \\ \nabla \times \boldsymbol{E} &= -\frac{\partial \boldsymbol{B}}{\partial t} \\ \nabla \cdot \boldsymbol{B} &= 0 \\ c^2 \nabla \times \boldsymbol{B} &= \frac{\boldsymbol{j}}{\varepsilon_0} + \frac{\partial \boldsymbol{E}}{\partial t} \end{aligned} \tag{2.3}$$

Solutions include spherical waves whose power density diminishes as the inverse square of distance or evanescent waves which have an exponential relationship with distance. These solutions describe the behavior of electric and magnetic vector fields which are related by Maxwell's equations. In homogeneous regions, the equations for the two quantities may be separated, resulting in a scalar wave equation for each quantity. Linearity and superposition ensure that the ESA beam is the sum of the individual element beams. The free space solution, wave equations in a homogeneous space and without charges and currents, is shown in (2.4) which is the classic wave equation:

$$\nabla^2 \boldsymbol{E} - \frac{1}{c^2}\frac{\partial^2 \boldsymbol{E}}{\partial t^2} = 0$$
$$\nabla^2 \boldsymbol{B} - \frac{1}{c^2}\frac{\partial^2 \boldsymbol{B}}{\partial t^2} = 0 \tag{2.4}$$

These two equations are not independent, and either field, $\boldsymbol{E}$ or $\boldsymbol{B}$, may be calculated from the other. The $\boldsymbol{E}$ and $\boldsymbol{B}$ quantities are vectors. The direction of the $\boldsymbol{E}$ field is termed the polarization. The $\boldsymbol{E}$ and $\boldsymbol{B}$ fields are orthogonal to each other and to the direction of propagation in free space. There is an extensive body of literature dealing with these equations.

2.3.1 Radiating/evanescent solutions

There are several classes of solutions to Maxwell's equations. The static solutions set the time derivatives in (2.3) to zero and correspond to Gauss', Faraday's, and Ampère's laws. It is difficult to find general solutions to these equations. Feynman *et al.* [6] note that:

> Mathematical analysis is not the grand thing it is said to be; it solves only the simplest possible equations. As soon as the equations get a little more complicated, just a shade—they cannot be solved analytically. But the numerical method ... can take care of any equation of physical interest.

This pragmatic advice has been widely adopted because of the development of powerful and inexpensive computers. There is a huge body of open source and proprietary software to solve a wide range of electromagnetic problems.

Closed-form solutions to Maxwell's equations exist for several geometries: planar, cylindrical, and spherical whose $\boldsymbol{E}$-fields follow the respective distance relationships: constant, inverse square root, and inverse. The dynamic solutions may be decomposed into linear combinations of single-frequency oscillatory functions. The fields have a space- and time-dependence characteristic of waves represented by (2.5):

$$e^{j(\omega \cdot t - k \cdot r)} \tag{2.5}$$

where ω is the angular frequency equal to $2\pi f$ and k is the "wavenumber" (spatial frequency) equal to $2\pi/\lambda$.

This dependence is consistently omitted in analysis because the sinusoidal variation does not contribute to pattern calculations and is a constant factor in all calculations. It also simplifies the time derivative which is replaced by multiplication by $j\omega$. Implicit in this formalism is that the solution is valid for a single frequency. Problems involving multiple frequencies such as any pulse shape require solutions that are a function of frequency.

Radiating solutions have a real value of the wavenumber k. These solutions are termed radiating as they may be observed at large distances and their decrease is consistent with constant total power in the wavefront. The radiating solutions to Maxwell's equations are the dominant fields in the farfield where they are all spherical due to the finite size of the antenna. The angular dependence is determined by the aperture design and approximates the Fourier transform of the aperture function.

Additional solutions exist with an imaginary value of the wavenumber k. These solutions increase or decrease exponentially. The increasing exponential solution is discarded as it implies energy creation for which there is no physical basis. The decreasing exponential solution decays more rapidly than inverse distance and is negligible at large distances and is therefore called evanescent. Evanescent solutions are important for analysis of antenna interactions with its local environment. As these decay rapidly with distance, they do not transfer power and represent a reactive load to the antenna.

Equation (2.1) relates the fields to the motion of electrons. The radiating element is where this motion is created or induced. The function of a radiating element is twofold: first to launch the field and second to match the impedance of the element to that of free space Z_0 represented by (2.6):

$$\begin{aligned} Z_0 &= \frac{|E|}{|H|} \\ &= \sqrt{\frac{\mu_0}{\varepsilon_0}} \\ &\approx 377\ \Omega \end{aligned} \tag{2.6}$$

2.3.2 Boundary conditions

The solutions to (2.4) do not explicitly include the source and represent a uniform medium where μ and ε are constant. These assumptions are not accurate in general and the associated waves would not be very useful without a source and a receiver.

The exact form of the radiated solution is determined by the boundary conditions of the source. The simplest case is a uniform source distributed over a closed surface.

The analysis task becomes more complex when there are scatterers such as other sources in the vicinity of the principal source. This is associated with mutual coupling effects described in Chapter 5.

2.3.3 Analysis regions (exact to approximate)

Figure 2.1 illustrates the transformation of the radiated field as it propagates away from the source. These regions correspond to approximations described in Section 2.4, which are used to simplify solutions.

The beam shape develops gradually in the locale of the antenna under the influence of the wave equation. Initially planar and shaped like the aperture, as the wave travels away from the aperture, it transforms into a constant angular width.

Various regions have been defined corresponding to approximations in the governing equations. The near-field region has a nearly constant magnitude and includes both the fields which will propagate and the fields which decay exponentially corresponding respectively to real and imaginary values of u and v. The near field finds use especially in antenna measurement facilities based on near-field

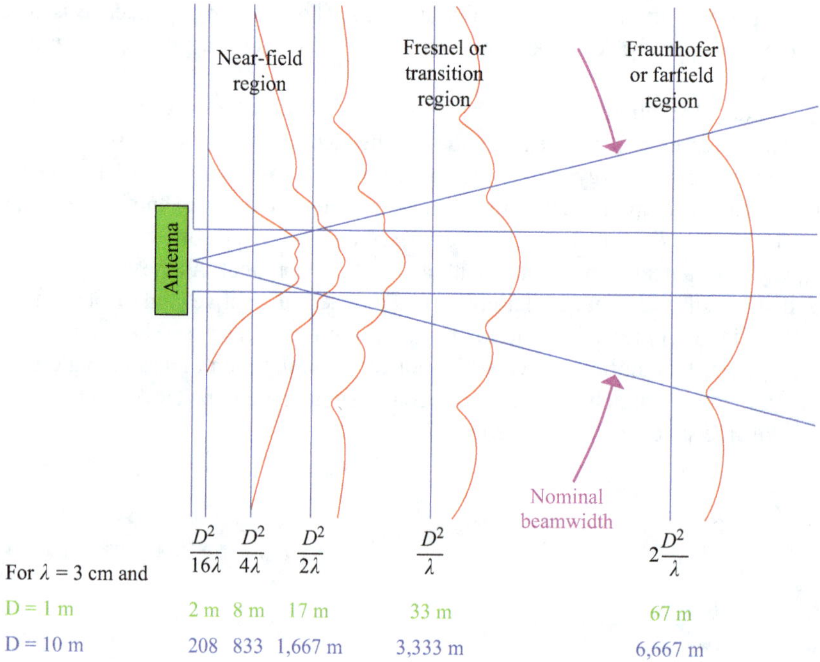

Figure 2.1 Analysis regions – exact to approximate
After Lynch [7], Figure 6.2

scanners which measure the fields and derive the farfield pattern, the one we are usually interested in. All of the beam patterns presented are farfield. The location of transitions between analysis regions is dependent on wavelength and antenna dimension as shown.

Table 2.1 summarizes some of the characteristics of the field regions and also shows the location of the farfield for a laser pointer.

Table 2.1 Radiating regions

	Evanescent	**Near field (Fresnel)**	**Farfield (Fraunhofer)**
Near limit	0	3λ	$2D^2/\lambda$
Far limit	3λ	$2D^2/\lambda$	∞
Power decay	R^{-n}	1	R^{-1}
E and H orthogonal	No	Yes	Yes
$Z_0 = 377\ \Omega$	No	Yes	Yes

Laser pointer
$\lambda = 630$ nm, $D = 1$ mm $\Rightarrow$ farfield at 3 m

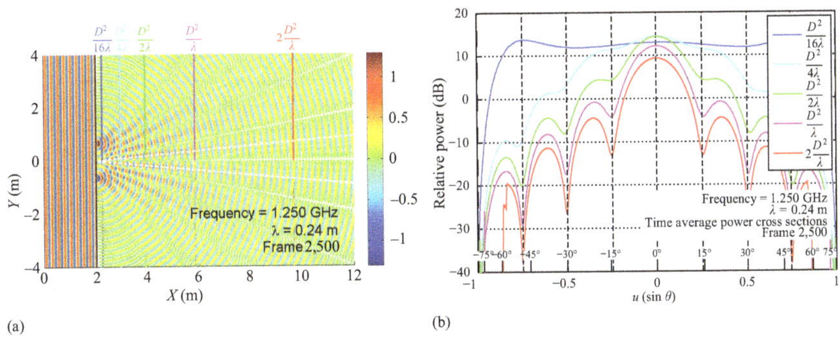

Figure 2.2 *FDTD model of* 4λ *wide slot: (a) field strength and (b) cross sections*
DoubleSlit.m

2.3.3.1 Single four-lambda wide slot

Figure 2.2 shows the field amplitude created by a 4λ wide slot, based upon a finite-difference time-domain method (FDTD) algorithm published by Wieners [8]. Figure 2.2a shows one frame of the simulation showing the phase reversal which accompanies the transition across nulls in the beam pattern. The nulls result from constructive interference across the aperture. Figure 2.2b shows the same data plotted as power (dB) passing through cross sections at various distances.

2.4 Aperture shape

The most common aperture shapes are rectangular and round (elliptical). The shape uniquely determines the overall beam pattern. Figure 2.3a and b illustrates the two cases. These patterns are computed based on a continuous aperture excitation. Sampling introduces additional attributes.

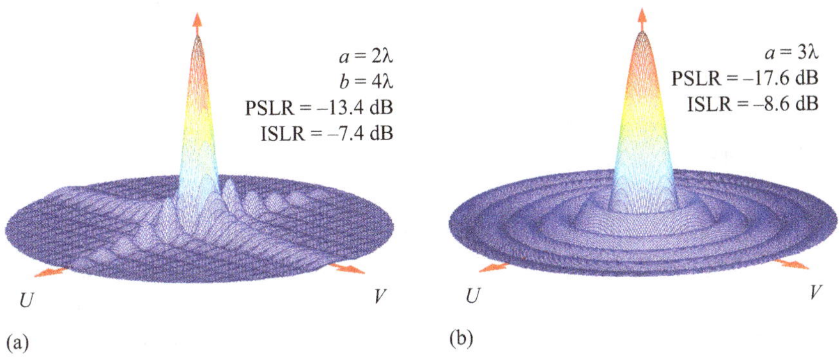

Figure 2.3 *Farfield patterns: (a) rectangular aperture, see (2.16) and (b) circular aperture, see (2.20)*
Apertures.m

Table 2.2 Uniform ***E****-field in aperture*

Attribute	Rectangular	Round
Size	$4 \times 8\lambda$	6λ diameter
First sidelobe	−13.2 dB	−17.8 dB
3 dB beamwidth	$0.886\lambda/D$	$1.03\lambda/D$
First null beamwidth	$\pm\lambda/D$	$\pm 1.22\lambda/D$

See (2.17) and (2.18) for rectangular case derivation

The shape of the beam pattern depends directly on the dimensions of the aperture in units of wavelength. Larger apertures yield narrower patterns and vice versa. With uniform illumination, the size and location of the sidelobes are as shown. The attributes of the first sidelobe are shown on the chart.

The rectangular case has two prominent sidelobe ridges aligned with the axes. These are characteristic for any rectangular aperture. In general, the sidelobe ridges align with straight edges of the aperture and are independent of the element lattice. By contrast, the circular aperture has no straight edges and its sidelobe pattern is rotationally symmetric. It too is independent of the details of the element lattice.

The peak sidelobe ratio (PSLR) is the ratio of the largest sidelobe and the peak of the main beam. The integrated sidelobe ratio (ISLR) is the ratio of the integrated power in the sidelobes and the integrated power in the mainlobe. PSLR and ISLR are discussed in Sections 4.1.2 and 4.1.3, respectively.

Table 2.2 summarizes the properties of two common aperture shapes.

2.4.1 Rectangular aperture

The exact form of the radiated solution is established by the boundary conditions of the source. The simplest case is a uniform source distributed over a rectangular aperture illustrated in Figure 2.4a. This case does not provide the proper boundary condition; specifically, the tangential ***E***-field must be zero at the boundary with a conductor. This introduces an error whose significance depends on the width of the aperture.

The farfield is calculated as the integral of the radiation contribution of the field in the aperture as a continuous integral or a discrete sum over small patches.

2.4.1.1 Uniform distribution on an infinite ground plane

Figure 2.4a represents a rectangular aperture mounted on an infinite ground plane. The rectangular aperture is analyzed in numerous textbooks, notably in Balanis [9] but also in Orfanidis [10], chapter 17.7, and Nikolova [11], Lecture 17, Antenna Engineering, ECE753. To reduce the mathematical complexities, initially the field over the opening will be assumed to be constant and given by (2.7) [Balanis 11-17]:

$$\mathbf{E}_a = \hat{\mathbf{a}}_y E_0, \qquad -a/2 \le x' \le a/2, \qquad -b/2 \le y' \le b/2 \tag{2.7}$$

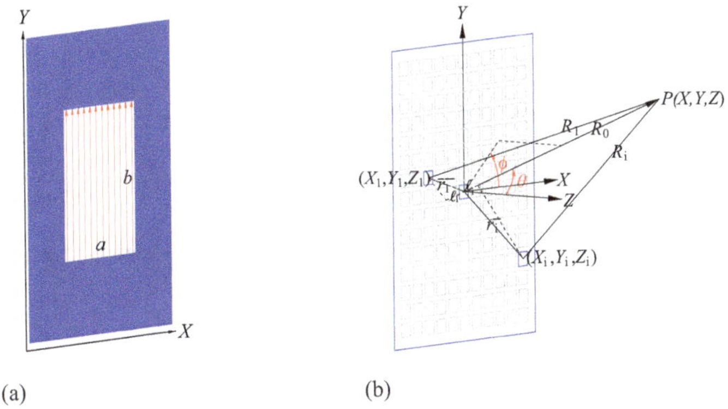

*Figure 2.4 Uniform **E**-field in rectangular aperture: (a) rectangular aperture and (b) discrete patches*

where E_0 is a constant.

The solution is shown in (2.8):

[Balanis 11–23]

$$
\begin{aligned}
E_r &= 0 \\
E_\theta &= j\frac{abkE_0\, e^{-jkr}}{2\pi r}\left[\sin\phi\left(\frac{\sin X}{X}\right)\left(\frac{\sin Y}{Y}\right)\right] \\
E_\phi &= j\frac{abkE_0\, e^{-jkr}}{2\pi r}\left[\cos\theta\cos\phi\left(\frac{\sin X}{X}\right)\left(\frac{\sin Y}{Y}\right)\right] \\
H_r &= 0 \\
H_\theta &= -\frac{E_\phi}{\eta} \\
H_\phi &= \frac{E_\theta}{\eta}
\end{aligned}
\tag{2.8}
$$

with

$$
\begin{aligned}
X &= \frac{ka}{2}\sin\theta\cos\phi \\
Y &= \frac{kb}{2}\sin\theta\sin\phi
\end{aligned}
\tag{2.9}
$$

and

$$
k = \frac{2\pi}{\lambda}
\tag{2.10}
$$

a, b = aperture dimensions

It is sometimes convenient to use the coordinate transformation represented by (C.1). Then (2.9) may be rewritten as (2.11):

$$X = \frac{ka}{2}u$$
$$Y = \frac{kb}{2}v \tag{2.11}$$

For small values of θ and when $\phi = [\pi/2,\ \pi]$, u and v correspond to elevation and azimuth pointing directions.

An interesting special case of (2.11) chooses $a = b = \frac{\lambda}{2}$ which provides Nyquist sampling of the interval. Then (2.9) becomes (2.12):

$$X = \frac{\pi}{2}u$$
$$Y = \frac{\pi}{2}v \tag{2.12}$$

The two senses of polarization are orthogonal, so the magnitude of the ***E***-field may be written as the square root of the sum of the squares as represented by (2.13):

$$E = \frac{abkE_0}{2\pi}\frac{e^{-jkr}}{r}\left[\sqrt{1-\sin^2\theta\cos^2\phi}\left(\frac{\sin X}{X}\right)\left(\frac{\sin Y}{Y}\right)\right] \tag{2.13}$$

The expression under the radical in (2.13) may be expressed as indicated in (2.14) and is graphed in Figure 2.5. It is generally disregarded when describing an antenna pattern as "sin *X*/*X*."

$$\sqrt{1-\sin^2\theta\cos^2\phi} = \sqrt{1-u^2} \tag{2.14}$$

Equation (2.14) reduces to $\cos\theta$ for $\phi = 0, \pi$. This is the well-known foreshortening effect in the *E*-plane. Moreover, the function is a constant for constant values of u, representing the *H*-plane cut.

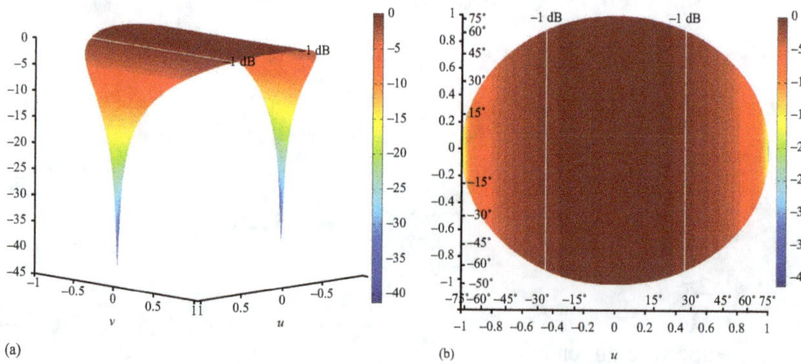

Figure 2.5 Extra factor (plot of $\sqrt{1-u^2}$): (a) orthogonal and (b) normal
Apertures.m

2.4.1.2 Power

It is of interest to consider the total power, both at the aperture and in the farfield. E_0 represents the peak field strength per unit area, so the total power at the aperture is

$$P = \frac{E^2}{2} a\, b \tag{2.15}$$

where the factor of $\frac{1}{2}$ arises from the average of the square of a sinusoid.

Substituting (2.13) into (2.15) and selecting the terms with angular dependence, we obtain (2.16). This expression is the basis for Figure 2.3a:

$$I = \left[\sqrt{1-u^2} \left(\frac{\sin X}{X} \right) \left(\frac{\sin Y}{Y} \right) \right]^2 \tag{2.16}$$

2.4.1.3 Beamwidths

Table 2.2 lists the 3 dB beamwidths derived from (2.13). The 3 dB point occurs when $(\sin X/X)^2 = 1/2$. The corresponding angle is given by (2.17):

$$\begin{aligned}
\left(\frac{\sin X}{X} \right) &= \sqrt{1/2} \\
X &\approx 1.391 \\
&\text{so} \\
\frac{ka}{2} \sin\theta &\approx 1.391 \\
\sin\theta &\approx 1.391 \frac{2}{ka} \\
\sin\theta &\approx 1.391 \frac{\lambda}{\pi\, a} \\
\sin\theta &\approx 0.443 \frac{\lambda}{a} \\
\theta &\approx 0.443 \frac{\lambda}{a}
\end{aligned} \tag{2.17}$$

Since θ represents one-half of the beam width, a factor of 2 is required for the full angle. Additionally, the arcsin is required for large θ. Also, we ignored the contribution of (2.14).

Equation (2.18) shows that the first null occurs for $X = \pi$, with θ representing the half-angle:

$$X = \frac{ka}{2}\sin\theta$$

$$= \pi \qquad (2.18)$$

so

$$\sin\theta = \frac{\lambda}{a}$$

2.4.2 Circular aperture

The circular aperture has a circularly symmetric pattern and corresponds to the airy disk (diffraction limit) of optical design. Balanis' equations, Section 12.6 (12–53), are shown in (2.19):

$$E_\theta = \mathrm{j}ka^2 E_0 \frac{\mathrm{e}^{-\mathrm{j}kr}}{r}\left[\sin\phi\left(\frac{J_1(ka\ \sin\theta)}{ka\ \sin\theta}\right)\right]$$

$$E_\phi = \mathrm{j}ka^2 E_0 \frac{\mathrm{e}^{-\mathrm{j}kr}}{r}\left[\cos\theta\cos\phi\left(\frac{J_1(ka\ \sin\theta)}{ka\ \sin\theta}\right)\right]$$

$$H_\theta = -\frac{E_\phi}{\eta} \qquad (2.19)$$

$$H_\phi = +\frac{E_\theta}{\eta}$$

As previously, substituting (2.19) into (2.15) and selecting the terms with angular dependence, we obtain (2.20). This expression is the basis for Figure 2.3b:

$$I = \left[\sqrt{1-u^2}\left(\frac{J_1(ka\ \sin\theta)}{ka\ \sin\theta}\right)\right]^2$$

$$= \left[\sqrt{1-\left(\frac{2\,X}{ka}\right)^2}\left(\frac{J_1(ka\ \sin\theta)}{ka\ \sin\theta}\right)\right]^2 \qquad (2.20)$$

The magnitude of the field vector is $E = \sqrt{E_\theta^2 + E_\phi^2}$

In this representation, the magnitude of the beam pattern is plotted on the vertical z-axis and the pointing angles θ and φ are plotted in the xy-plane under a translation from angle to (u, v).

Caveat

The representation shown here uses (u, v) space and the circular boundary of the pattern corresponds to radiation which is tangential to the aperture face. Being defined in terms of trigonometric functions, thevalues of u and v cannot exceed unity for real values of angles in the forward hemisphere, hence the circular boundary. However, it is possible to assign various interpretations to values of u and v greater than unity. The region of the uv-plane of radius less than or equal to unity is sometimes termed the real region and the region greater than unity is termed the imaginary region. The imaginary region has a physical manifestation in the reactive (nonradiating) energy in the near-field region.

2.5 Discrete case

Section 2.4 analyzed the farfield pattern of a continuous electric field in the aperture. In the case of an ESA, we consider the electric field as a sample at the lattice position of the radiating element. The field remains continuous, but the variation between lattice points is attributed to the radiating element and we will see that the resulting farfield pattern is the product of the discrete lattice samples and the continuous element field. This is a useful analysis and design convention. This factorization is approximate and may be incorrect in some cases due to the effects of mutual coupling discussed in Chapter 5. Even so, it produces reasonable approximations in most cases and may be adequate in many cases.

Figure 2.6 (adapted from Mailloux) shows a collection of radiating elements located at (x_i, y_i, z_i) and an observer located at (x, y, z). Each radiating element is represented by a square. The radiated field at the observer's location is the sum of the fields of each of the radiating elements as seen at the same location.

2.5.1 Element contribution

Each element i generates the field

$$\boldsymbol{E}_i(r,\theta,\phi) = \boldsymbol{f}_i(\theta,\phi)\frac{\exp(-\mathrm{j}kR_i)}{R_i} \tag{2.21}$$

where $k = 2\pi/\lambda$. Using the identity $R_i = R - \hat{\boldsymbol{r}}\cdot\boldsymbol{r}_i$, we can rewrite the second term as

$$\frac{\exp(-\mathrm{j}kR_i)}{R_i} = \frac{\exp(-\mathrm{j}kR)}{R - \hat{\boldsymbol{r}}\cdot\boldsymbol{r}_i}\exp(+\mathrm{j}k\boldsymbol{r}_i\cdot\hat{\boldsymbol{r}}) \tag{2.22}$$

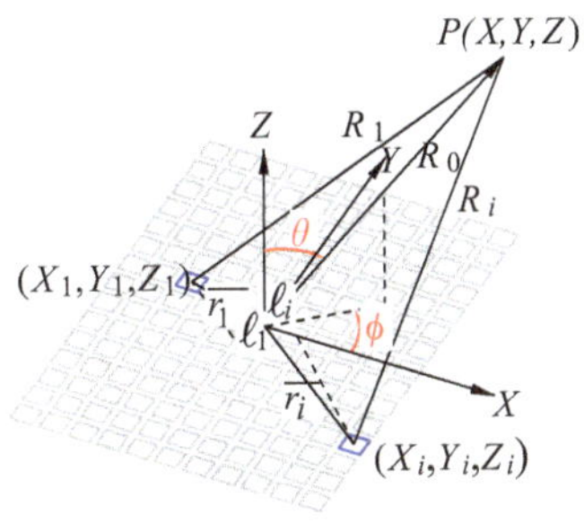

Figure 2.6 Arbitrary field summation
After Mailloux [12]

Each element generates a spherical wavefront, with angular dependence $\boldsymbol{f}_i$. The spherical wave solutions provide a basis set decomposition of the actual radiation pattern; $\boldsymbol{f}_i$ is determined by the antenna design. At large distances, the denominator on the right reduces to R representing the Fraunhofer approximation.

2.5.1.1 Fraunhofer approximation

For distances large compared to the array size, i.e., $R \gg \hat{\boldsymbol{r}} \cdot \boldsymbol{r}_i$,

$$\frac{\exp(-\mathrm{j}kR_i)}{R_i} = \frac{\exp(-\mathrm{j}kR)}{R} \exp(+\mathrm{j}k\boldsymbol{r}_i \cdot \hat{\boldsymbol{r}}) \tag{2.23}$$

so that

$$\boldsymbol{E}_i(r, \theta, \phi) = \frac{\exp(-\mathrm{j}kR)}{R} \boldsymbol{f}_i(\theta, \phi) \exp(+\mathrm{j}k\boldsymbol{r}_i \cdot \hat{\boldsymbol{r}}) \tag{2.24}$$

Adding a complex weight a_i to each element, the resulting antenna pattern is

$$\boldsymbol{E}(\boldsymbol{r}) = \frac{\exp(-\mathrm{j}kR)}{R} \sum_i a_i \boldsymbol{f}_i(\theta, \phi) \exp(\mathrm{j}k\boldsymbol{r}_i \cdot \hat{\boldsymbol{r}}) \tag{2.25}$$

The field at the observer is the sum of each element contribution. We can add a complex weight (amplitude and phase) to each element for control purposes and we have the equation for $\boldsymbol{E}$.

2.5.2 *Array factor*

It is convenient and reasonably accurate to assume that each element has the same pattern $\boldsymbol{f}(\theta, \phi)$ so that the element pattern may be taken out of the sum

$$\boldsymbol{E}(\boldsymbol{r}) = \boldsymbol{f}(\theta, \phi) \frac{\exp(-\mathrm{j}kR)}{R} \sum_i a_i \exp(\mathrm{j}k\boldsymbol{r}_i \cdot \hat{\boldsymbol{r}}) \tag{2.26}$$

This formulation shows that the antenna pattern may be partitioned and is equal to the product of the Element factor, the Space factor, and the Array factor.

2.5.2.1 Pattern separability

We may further simplify the array factor summation if we assume that the radiating elements are arranged in a rectangular grid in the xy-plane such that

$$\boldsymbol{r}_i = \boldsymbol{r}_{mn} = m\Delta x \hat{\boldsymbol{x}} + n\Delta y \hat{\boldsymbol{y}} \tag{2.27}$$

$$m = 0, \pm 1 \pm 2 \pm 3 \ldots, \quad n = 0, \pm 1 \pm 2 \pm 3 \ldots \tag{2.28}$$

$$\hat{\boldsymbol{r}} = \hat{\boldsymbol{x}} u + \hat{\boldsymbol{y}} v + \hat{\boldsymbol{z}} \cos\theta \tag{2.29}$$

$$u = \sin\theta \cos\phi \qquad v = \sin\theta \sin\phi \tag{2.30}$$

Then

$$\boldsymbol{r}_i \cdot \hat{\boldsymbol{r}} = m\Delta x\, u + n\Delta y\, v \tag{2.31}$$

$$E(\boldsymbol{r}) = f(\theta, \phi)\,\frac{\exp(-\mathrm{j}kR)}{R}\sum_i a_{mn}\exp\left(\mathrm{j}k\left(m\Delta x\,u + n\Delta y\,v\right)\right) \tag{2.32}$$

The final step is applicable only to rectangular periodic lattices.

A slightly more complex formulation offsets alternate rows by one-half the lattice spacing to form a triangular lattice. This may be viewed as the superposition of two identical rectangular lattices displaced in both dimensions.

2.5.2.2 Decomposition in *X* and *Y*

Making the restrictive but useful assumption that the complex element weight a_i may be decomposed into x and y components represented by (2.33) , we can separate the sum of element factor into the product of two sums over the two dimensions:

$$a_{mn} = a_m\,a_n \tag{2.33}$$

Then the total array factor is the product of separate functions in x and y shown in (2.34):

$$E(\boldsymbol{r}) = f(\theta, \phi)\,\frac{\exp(-\mathrm{j}kR)}{R}\sum_m a_m\exp\left(\mathrm{j}k\left(m\Delta x\,u\right)\right)\sum_n a_n\exp\left(\mathrm{j}k\left(n\Delta y\,v\right)\right) \tag{2.34}$$

We will use (2.34) in (3.1).

2.5.2.3 Pattern multiplication

The element weights of an array of subarrays represents a convolution of the two arrays. This yields two important consequences. The overall beam pattern is the product of the element pattern and the array pattern of separate arrays in x and y, which may be shown by treating the array factor as a discrete Fourier transform of aperture weights.

2.5.2.4 16 element array = 4 × 4 element array

Figure 2.7 demonstrates construction of a 16 element linear array from the product of a 4 element array by 4 element subarrays.

2.5.3 One-dimensional beamformation (boresight)

We consider a one-dimensional array by examining one of the factors of (2.34) representing M elements equally spaced in a line, each with an element factor a_m:

$$\mathrm{AF} = \sum_{m=0}^{M-1} a_m\,\mathrm{e}^{\mathrm{j}km\Delta x\sin\theta\cos\phi} \tag{2.35}$$

We simplify (2.35) by assuming that a_m are equal and defining $\psi = k\Delta x\sin\theta\cos\phi$. This enables a closed-form representation of the summation shown in (2.36):

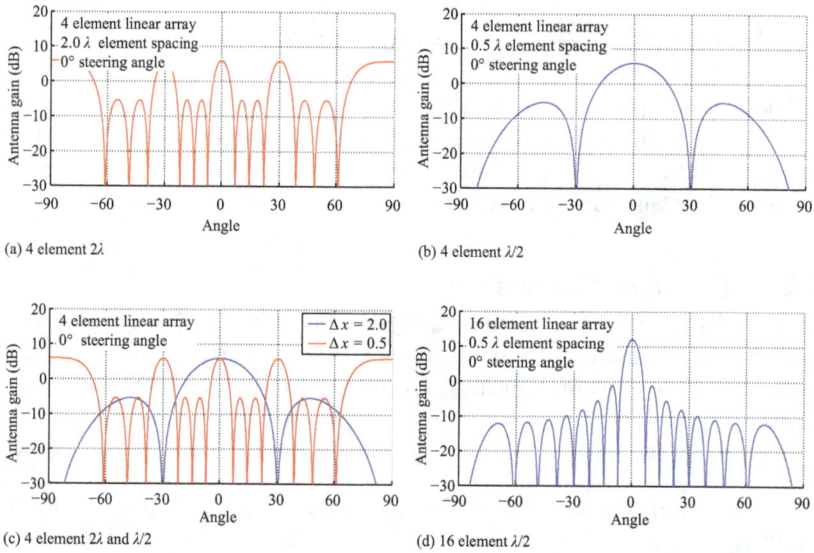

Figure 2.7 16 element array

Array1.mLinearArray.m

$$\mathrm{AF} = a \sum_{m=0}^{M-1} \left(\mathrm{e}^{\mathrm{j}\psi}\right)^m = a \frac{1 - \left(\mathrm{e}^{\mathrm{j}\psi}\right)^M}{1 - \mathrm{e}^{\mathrm{j}\psi}} \tag{2.36}$$

The function is referenced to zeroth element (end of array). If we simplify the equation by moving the phase reference to the center of the array, we obtain (2.37) in which the array factor becomes the product of a real function and a complex phase factor whose magnitude is one.

$$AF = a \left[\frac{\left(\mathrm{e}^{\mathrm{j}\psi}\right)^{M/2}}{\mathrm{e}^{\mathrm{j}\psi/2}} \right] \left[\frac{\left(\mathrm{e}^{\mathrm{j}\psi}\right)^{-M/2} - \left(\mathrm{e}^{\mathrm{j}\psi}\right)^{M/2}}{\mathrm{e}^{-\mathrm{j}\psi/2} - \mathrm{e}^{\mathrm{j}\psi/2}} \right]$$

$$= a\,\mathrm{e}^{\mathrm{j}\psi(M-1)/2} \frac{\sin(M\psi/2)}{\sin(\psi/2)} \tag{2.37}$$

$$\text{so } |AF| = \left| \frac{\sin[M\psi/2]}{\sin(\psi/2)} \right|$$

2.5.3.1 Maximum gain

From the closed-form solution, we see that the maximum gain is equal to the number of elements M in the limit that $\psi/2 \Rightarrow n\pi$ or $\psi = 2n\pi,\ n = 0, \pm 1, \ldots$.

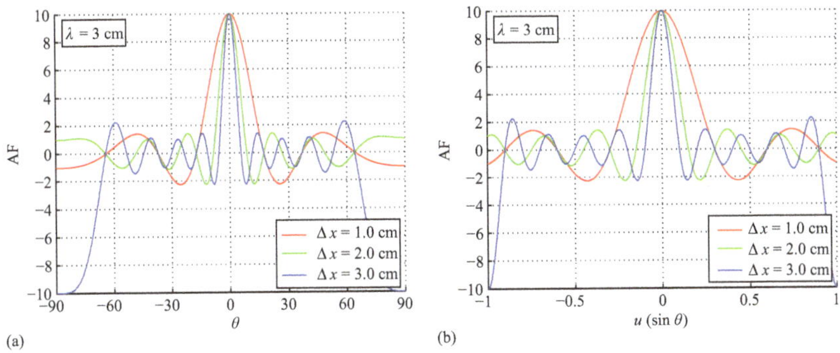

Figure 2.8 Effect of element spacing – boresight: (a) AF(θ) and (b) AF(u)
MaxGain.m

This result offers the possibility of more than one maximum which makes it difficult to determine the angle of arrival of the signal. These maxima, apart from the desired maximum, are termed grating lobes and they are discussed in Chapter 7.

2.5.3.2 Selected boresight case ($M = 10$)

Figure 2.8 illustrates the boresight case ($M = 10$).

Figure 2.8 shows the antenna pattern for a linear array for combinations of specific wavelength (3 cm or X-band) and specific element spacing of 1, 2, and 3 cm. The largest element spacing is one wavelength, so we see the expected grating lobe at 90° scan (blue curve). The figure shows amplitude (not magnitude or power) in linear units as a function of scan angle. Maxima occur at $\theta = \arcsin\left(2n\pi/k\Delta x\right) = \arcsin\left(n\lambda/\Delta x\right)$.

Figure 2.8a shows the beam pattern as a function of angle in degrees. Figure 2.8b uses a coordinate transformation $u = \sin\theta$ which is convenient for analytic purposes. For instance, the nulls are equally spaced as a function of u. This will be useful for beam synthesis. Appendix C.1 describes the two-dimensional counterpart coordinates (u, v).

2.6 Beam steering

The previous sections evaluated the boresight beam pattern. The maximum value of field intensity is in the direction $\theta = 0$ for any value of ϕ in accordance with (2.8) and (2.19).

A defining attribute of ESAs is the ability to direct the beam in arbitrary directions (θ, ϕ). Beam steering is accomplished by progressive signal delays across the aperture to correspond to a plane wave propagating in the desired direction. This variation is termed phase slope and it establishes a linear relationship between position in a planar aperture and desired phase value. The resulting beam direction is normal to the plane of constant phase.

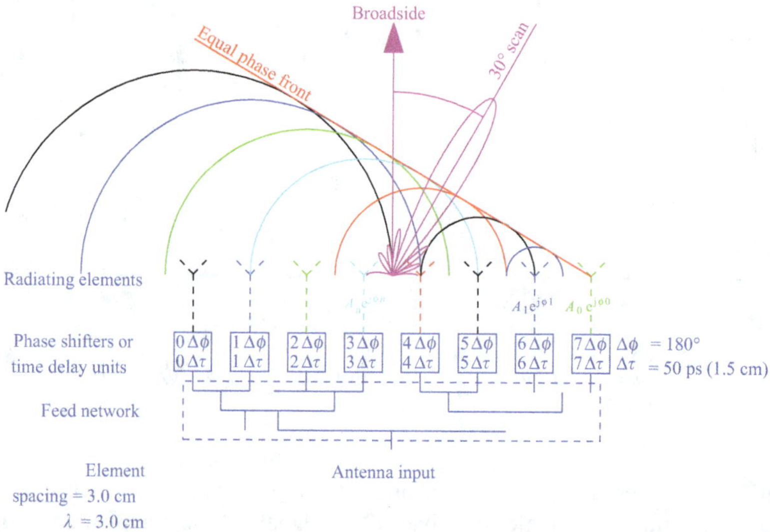

Figure 2.9 Field front representation

Figure 2.9 illustrates the beam steering for a one-dimensional case. Circles represent radiation from individual elements, which start at different times (or with different phases). The illustration shows eight elements equally spaced along a line with constant phase (or time) differences between the individual elements. Element 1 (left-most) radiates a signal beginning at time zero. Element 2 radiates at time τ, element 3 at time 2τ, and so on. The result is that the individual wavefronts from each element merge into a plane wave in direction θ. Element spacing and wavelength are equal, so a grating lobe at 30° would be expected in addition to the main lobe at −30°.

This illustration corresponds to Huygens' principle. In 1678, Huygens proposed that every point which a luminous disturbance reaches becomes a source of a spherical wave; the sum of these secondary waves determines the form of the wave at any subsequent time. Baker and Copson [13], in *The Mathematical Theory of Huygens' Principle*, provide a detailed discussion.

Figure 2.10 illustrates the calculation of required phase shift or time delay. The element spacing is constant with spacing Δx. The beam is steered off boresight by an angle θ and the distance ℓ represents the incremental delay required between adjacent elements in order to align the phase fronts of the elements according to (2.38).

The delay ℓ may be implemented using time delay or phase delay. The amount of time delay τ is shown in (2.38) which is simply distance divided by velocity (c or speed of light). The equivalent phase shift $\Delta\phi$ is shown in (2.39). Section 14.5 describes circuits to implement phase shift or time delay. Phase shift is easier to implement than time delay as it can be accomplished with small LC networks, while

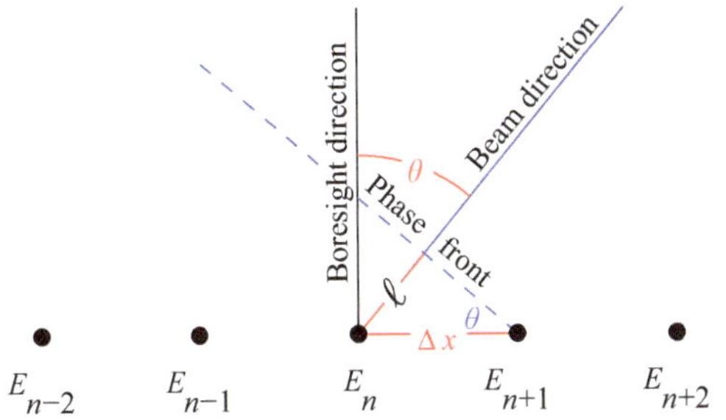

$$\begin{aligned}\theta &= \arcsin(\ell/\Delta x)\\ \ell &= \Delta x \sin\theta\\ \tau &= \frac{\ell}{c}\\ &= \frac{\Delta x \sin\theta}{c}\end{aligned} \tag{2.38}$$

Figure 2.10 Beam steering

time delay requires line length to provide the delay. Phase shift is satisfactory for modest bandwidths. For octave bandwidths, the band may be subdivided to enable use of phase shifters:

$$\begin{aligned}\Delta\phi &= 2\pi \cdot \frac{\tau}{\text{Period}}\\ &= 2\pi\tau \cdot \text{frequency}\\ &= 2\pi\left(\frac{\Delta x}{\lambda}\right)\sin\theta\end{aligned} \tag{2.39}$$

so

$$\sin\theta = \frac{\lambda\Delta\phi}{2\pi\Delta x}$$

This alignment may be accomplished using phase shift or time delay at each element or group of elements.

The choice between time delay or phase shifters to implement the necessary delay is a fundamental design decision. In the analog domain, phase shifters are prevalent because they are smaller, lighter, and cheaper than time delay units (TDUs). After the signal has been digitized, time delay is easily implemented in software for two or more digitized signals. Phase shifter and TDU implementation is discussed in Section 14.5.

2.6.1 Beam-steering geometric construction

Figures 2.11–2.13 represent a snapshot of the RF radiation from each element in a specified direction (30°). The elements are spaced at one wavelength, so grating lobes are expected; however, only the construction for the specified direction is shown.

Figure 2.11 represents a linear array steered with time delay. The proper time delay (50 ps) between adjacent elements generates beam in desired direction (30°). Figure 2.12 represents the same 30° beam direction for an unsteered linear array. The peaks and valleys of waves from adjacent elements coincide which creates a pattern minimum (or null) in this direction.

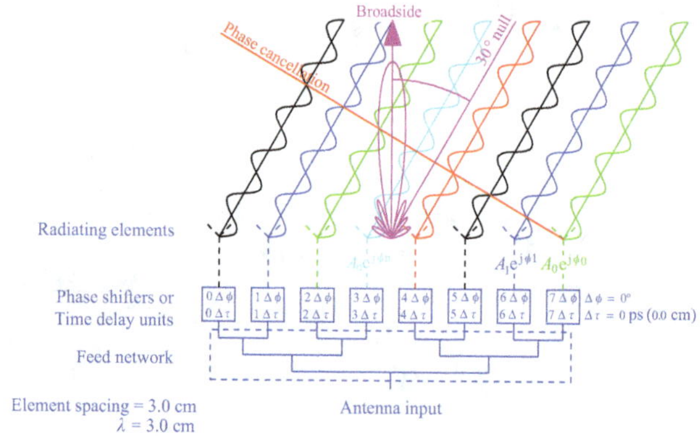

Figure 2.11 Linear phase array – unsteered, null 30° off boresight

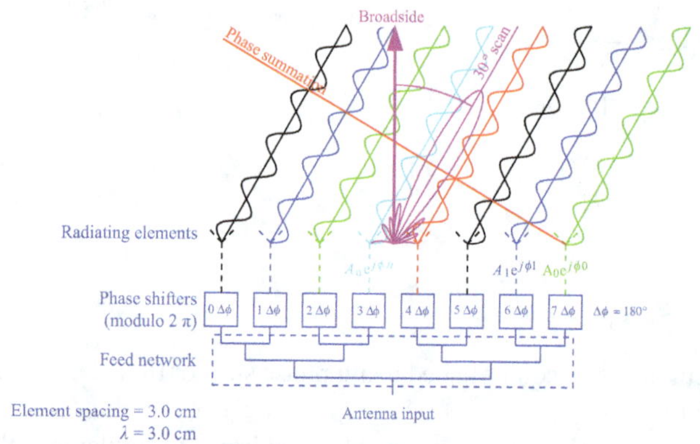

Figure 2.12 Linear phase array with phase shifter – steered 30° off boresight

Figure 2.13 introduces a $\lambda/2$ time delay between adjacent elements for a linear array which realigns the peaks and nulls from adjacent elements; the null at 30° is replaced by a mainlobe.

2.6.2 One-dimensional beamformation (steered)

To steer the beam, we apply a linear phase (only) slope in the element weights. To steer the beam of a linear array, it is sufficient to apply a linear phase slope to the element weights shown in (2.35):

$$a_m = e^{-jmk\Delta x \sin\theta_s \cos\phi_s} = e^{-jm\psi_s} \tag{2.40}$$

$$\begin{aligned} \psi &= k\Delta x \sin\theta \cos\phi \\ \psi_s &= k\Delta x \sin\theta_s \cos\phi_s \end{aligned} \tag{2.41}$$

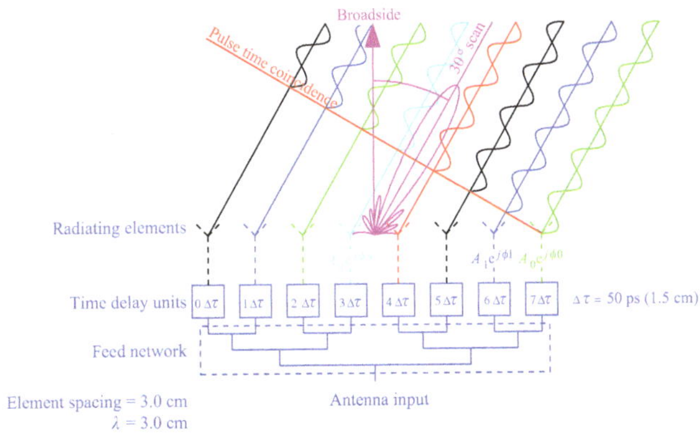

Figure 2.13   Linear phase array with time delay – steered 30° off boresight

2.6.2.1 Scanned array factor

Making the substitutions of variables from (2.41) into (2.40) yields (2.42):

$$\begin{aligned} AF &= \sum_{m=0}^{M-1} e^{jkm\Delta x(\sin\theta\cos\phi - \sin\theta_s\cos\phi_s)} \\ AF &= \sum_{m=0}^{M-1} e^{jm(\psi-\psi_s)} \end{aligned} \tag{2.42}$$

which reduces to (2.43–2.45):

$$AF = \sum_{m=0}^{M-1} e^{jm(\psi-\psi_s)} \tag{2.43}$$

$$AF = \frac{\sin[M(\psi-\psi_s)/2]}{\sin[(\psi-\psi_s)/2]} e^{jM(\psi-\psi_s)/2} \tag{2.44}$$

$$|AF| = \left|\frac{\sin[M(\psi-\psi_s)/2]}{\sin(\psi-\psi_s)/2}\right| \tag{2.45}$$

The beam shape is the same as the boresight case except that it has been shifted by an angle ψ_0.

2.6.2.2 Selected steered (30°) case (*M* = 10)

Figure 2.14 illustrates the steered (30°) case ($M = 10$). Maxima occur at $\theta = \arcsin\left(\frac{2n\pi}{k\Delta x}\right) = \arcsin\left(\frac{n\lambda}{\Delta x}\right)$ and grating lobes for $\Delta x = 3$ cm.

Figure 2.14a is similar to Figure 2.8a and the graph of amplitude versus u in Figure 2.14 is identical to Figure 2.8a except for a shift in pointing direction. For element spacing of 3 cm, we have the main lobe at +30° and the grating lobe at −30° . Element spacing of 2 cm shows a grating lobe at −90°.

In this situation, it is impossible to distinguish a target return in the main lobe from a target return in the grating lobe. Accordingly, this array must be restricted to scan less than 30° to avoid the ambiguity.

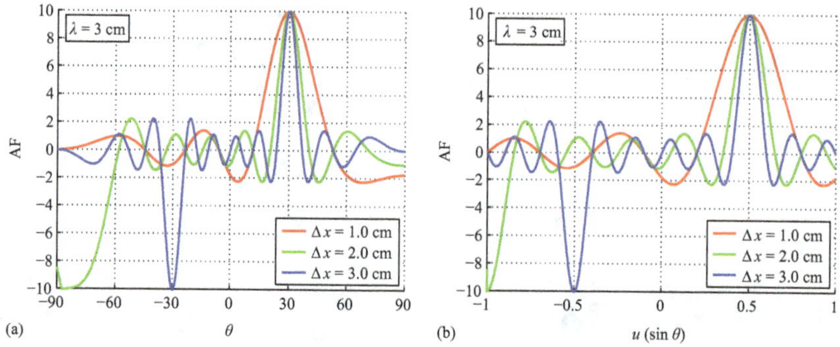

Figure 2.14 *Effect of element spacing – 30° scan: (a) AF(θ) and (b) AF(u)* *MaxGain.m*

2.6.3 Beam squint

Beam squint describes the movement of beam pointing direction as a function of frequency. This is usually an undesired effect although in the early days, some radars relied on this effect to achieve electronic beam positioning.

ESA performance is commonly described at a single frequency, the design center frequency. This is usually sufficient for bandwidths of a few percent. For larger bandwidths, it is necessary to evaluate performance at multiple frequencies across the band. Beam performance typically degrades as the frequency varies from the design frequency. The degradation is also proportional to the beam steering angle and so there is no degradation at boresight at nominal frequency. The bandwidth requirement is discussed in Section 4.3.

Beam steering requires a specific time delay at each element to orient the beam. Time delay has no frequency dependence, but we are interested in a phase shift $\Delta\phi$ providing the equivalent time delay. The substitution of phase shift for time delay introduces a frequency dependence and provides the correct time delay at only a single frequency. As frequency increases, wavelength decreases and a unit of phase represents a smaller amount of time and conversely as frequency decreases. Haupt [14] discusses these tradeoffs. Garakoui *et al*. [15] and Romanofsky [16] provide detailed analyses.

The phase delay equates this time delay to a fraction of the signal period as shown in (2.39). Accordingly, the required phase shift is proportional to frequency. In turn, this direct frequency dependence limits the usable bandwidth of the system since the phase shift is correct only for a single frequency. .

The entire beam pattern also changes (expands or shrinks) in response to frequency since the beam pattern depends on the size of the antenna in wavelengths. As the frequency increases, the mainbeam becomes narrower and the sidelobes move closer to boresight. Since the nuls are narrower than the peaks, they are more sensitive to frequency. Nulling techniques for ESAs are less effective for wideband signals.

Sum and difference patterns are used for target tracking. The difference pattern is created by combining two halves of the antenna out of phase creating a null at antenna boresight. The null experiences little degradation with frequency since it is unsteered.

An early application of array methodology was sidelobe nulling with a primary antenna with a fixed antenna pattern (not an ESA). An auxiliary antenna is aligned with the primary antenna and single unwanted signal may be removed by taking the difference of the two antenna patterns weighted appropriately. The technique requires that the second antenna have lower gain in the mainlobe of the primary antenna and higher gain in the sidelobes of the primary antenna. The nulling is only perfect at a single frequency because the two antennas have different frequency behavior.

2.6.3.1 Phase shift and bandwidth

While time delay provides the exact delay required, phase control may be substituted in many circumstances at considerable savings in cost and complexity. The required phase is simply $\Delta\phi = \frac{\Delta\tau}{\lambda c}$ and will usually exceed 360°. The phase relationship repeats every 360°, so it is convenient to use the phase shift modulo 360° and restrict the selected phase shift to a range of 0°–360°. This phase wrapping introduces artifacts discussed in Chapter 7.

With the use of phase shifters, the beam pointing direction moves toward boresight as frequency increases, because the fixed phase shift corresponds to a decreasing time delay. The amount of deviation is proportional to the desired beam direction. Since beam width is inversely proportional to array size, this effect is more noticeable and more problematic for larger arrays. It represents a fundamental tradeoff between array size and bandwidth. It does not occur with a boresight beam and the error may be controlled by restricting the operational scan limit.

It may be mitigated by altering the phase shift on the fly as a function of the transmitted frequency which is possible with linear frequency-modulated waveforms in transmit. This technique is less applicable in receive because the time of arrival of a given frequency depends on the range to the target. The principal effect of beam squint is that ESA and system bandwidth are constrained beyond the usual bandwidth limitations of components themselves.

The required phase shift at each element, $n \cdot 2 \cdot \pi \cdot \Delta x/\lambda$, is dependent on frequency. To assess the change in beam direction resulting from a change in wavelength $\Delta\lambda$, we use (2.39) to form the derivative of beam direction with respect to wavelength as shown in (2.46).

$$\cos\theta\ \Delta\theta = \frac{\Delta\lambda}{2\pi}\frac{\Delta\phi}{\Delta x} \tag{2.46a}$$

Eliminating $\Delta\phi$ using (2.39), we obtain

$$\Delta\theta = \frac{\Delta\lambda}{\lambda}\frac{\sin\theta}{\cos\theta} \tag{2.46b}$$

The antenna beamwidth is approximately $K\lambda/D$, where D is the dimension of the array and K is a factor approximately equal to 1 relating to the power level at which we define the beamwidth. If we equate $\Delta\theta$ to the beamwidth, we have (2.47) :

$$\frac{K\lambda}{D} = \frac{1}{\lambda}\frac{\sin\theta}{\cos\theta}\Delta\lambda$$

$$\text{or} \tag{2.47}$$

$$\frac{\Delta\lambda}{\lambda} = \frac{\lambda}{D}\frac{K}{\tan\theta}$$

Equivalently

$$f = \frac{c}{\lambda}$$

so

$$\Delta f = -\frac{c\Delta\lambda}{\lambda^2}$$

$$\Delta f = -f\frac{\Delta\lambda}{\lambda} \tag{2.48}$$

$$\frac{\Delta f}{f} = -\frac{K\cdot\lambda}{D\cdot\tan\theta}$$

For the specific case of a 1.0 m antenna at X-band, we consider a bandwidth of 600 MHz which is the ITU allocation (Table 17.1) for radar. For $L = 0.5$ m, $\lambda = 3$ cm, and $\theta = 30°$, (2.47) determines a fractional bandwidth of 12%.

Figure 2.15 illustrates the boresight case at nominal frequency and 10% off nominal. There is no frequency dependence of the beam pointing direction. The beamwidth decreases with increasing frequency but that is not an issue for most applications.

Figure 2.16 illustrates the case for a phase-steered beam pointed 30° off boresight. Although the beam remains well formed, it does not stay on the target,

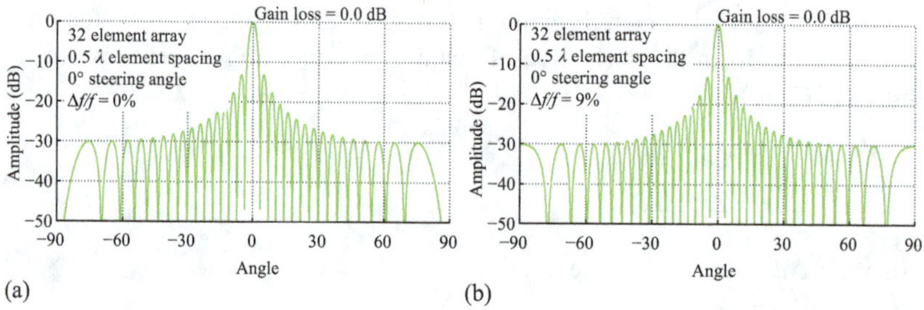

Figure 2.15 No beam squint on boresight: (a) $\Delta f/f = 0$ and (b) $\Delta f/f = 10\%$ LinearArray.m

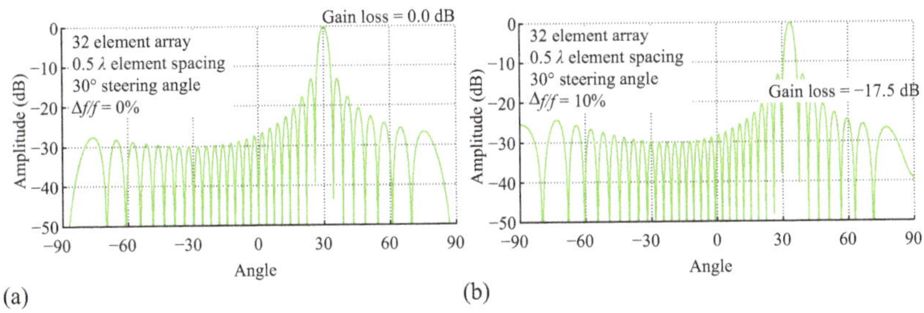

Figure 2.16 Beam squint limits bandwidth: (a) Δf/f = 0 and (b) Δf/f = 10%
LinearArray.m

constituting an amplitude weighting dependent on frequency and degrading the range resolution. The beamwidth also decreases with increasing frequency.

2.6.3.2 Time delay

The general solution provides a time delay for each element aligning the phase front of all elements in the desired direction. Beam steering using time delay units (TDUs) is invariant with frequency and so does not constrain bandwidth. TDUs are larger and more expensive than the equivalent phase shifter and the bandwidth limitation of phase shifters is frequently deemed preferable to the disadvantages of the time delay units.

The maximum time delay is proportional to the sine of the pointing direction and the size of the array in the plane of the pointing angle. The time delay may be provided by appropriate transmission line length in the signal path to and from the element. Alternatively, the time delay may be inserted digitally to the bit stream of the ADC or DAC.

2.6.3.3 Time delay steering eliminates beam squint

If the beam squint is not acceptable for whatever reason, scan volume or bandwidth, time delay steering may be used. This may be done at the element or the subarray level although time delay at the element level tends to be prohibitive for size and weight constraints.

Figure 2.17 illustrates the case for a time delay-steered beam pointed 30° off boresight in an array that is 16λ in length. The beamwidth also decreases with increasing frequency.

The TDU required for each element must have a maximum time delay equal to the antenna length times sine of the scan angle and a minimum time delay set by the number of beams desired.

One feasible configuration is to use the arrays shown in Figures 2.15–2.17 and combine them using time delay. Then the number of TDUs is greatly reduced and

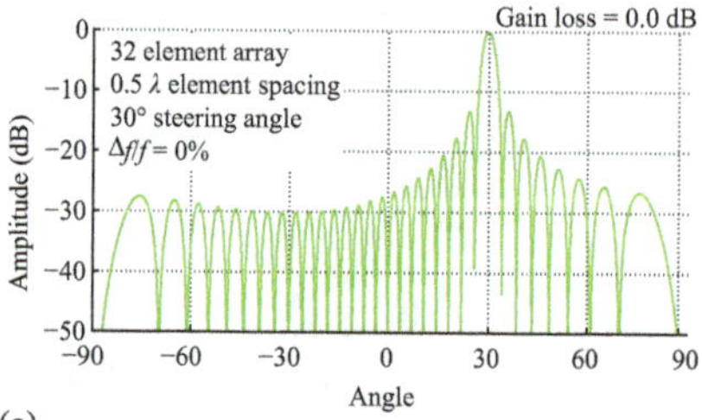

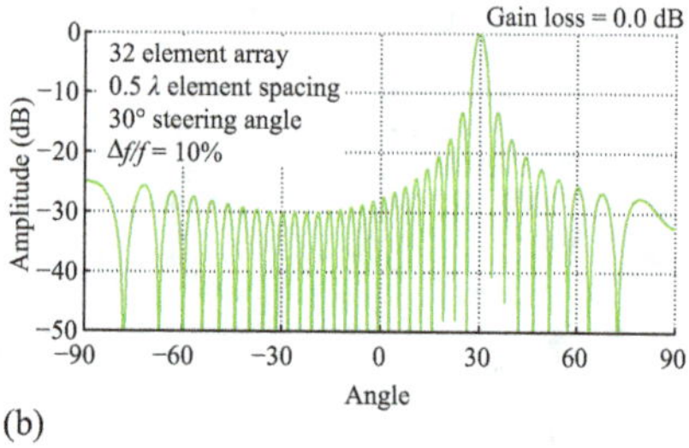

Figure 2.17 Time delay eliminates beam squint: (a) Δf/f = 0 and (b) Δf/f = 10%
Array1.mLinearArray.m

larger arrays are feasible. The COSMO-SkyMed radar system utilizes this technique, including TDUs at the subarray level to support wider electronic scan.

Another approach is to use multiple receivers instead of TDUs. This is more practical at the subarray level although as the cost and performance of ADCs improve, it becomes more feasible to use receivers at the element level. Time delay in memory is straightforward but requires considerable memory because the digitized signal must be stored in order that the properly delayed signals may be recombined to form a beam.

The tradeoff is size, weight, and power (SWAP) and the cost of a receiver versus the cost of a TDU. The cost of receivers has dropped dramatically but they are still too large and expensive to implement on an element level and power consumption is an issue. With the subarray approach, however, the bandwidth/scan volume trade is determined by subarray size instead of array size.

2.7 Feed and beamforming networks

Beamformers are required to distribute RF power from the signal generator to each radiating element and to collect received RF power from each element and pass it back to the receiver. Ideally, the beamformer does not add noise of distortion to the process.

Usually, there are fewer receivers and exciters than there are T/R modules. Some means is required to distribute analog signals between the receiver/exciter(s) and the T/R modules. This is termed a feed network or a beamforming network with some difference in meaning. Generally, a feed network does not introduce any change to the signals so the corresponding antenna beam will be a boresight, equal-illumination pattern absent any gain or phase control at the element. A beamforming network introduces an additional amplitude, phase, or time difference at the element or subarray level which changes the antenna beam shape or direction and may provide multiple, simultaneous beams.

For a rectangular lattice, we can separate the two-dimensional array factor into the product of two one-dimensional array factors. This factorization extends to the beamformers which may be configured in rows and then columns. The Iridium antenna, discussed in Section 17.1, uses this technique to advantage with its Butler beamforming matrices.

In general, however, each beam requires its own network. To create multiple beams, multiple beamformers are necessary. Analog beamformers tend to be bulky, so there is a fairly low limit on the number of beams feasible. Digital beamformers can be implemented in software but require digitization at the element or subarray level. The Space Fence described in Section 1.4.1.2 implements receive digitization at the element level, so it is capable of a very large number of receive beams. A similar approach has been proposed for the Square Kilometer Array (SKA) described in Section 1.4.4.

2.7.1 Feed networks

Usually feed networks are made up of waveguides, planar transmission lines, or coaxial cable. A few systems have used space feeds, in which the signal is distributed over the air. The AN/MPQ-53 radar of the Patriot system uses such arrangement. Space feeds are very light but subject to interference. Constrained feeds are heavier but better controlled and mostly immune to interference.

Feed networks may be series or parallel. Series-fed arrays are associated with the slotted waveguide array which is fed from one end. The progressive delay of the signal produces an inherent frequency dispersion which limits its bandwidth but has excellent narrowband performance and is not expensive. One complication is that the power ratios to feed each element are unequal. Because, the path length to different elements is different, its use is typically restricted to small arrays or subarrays. The example shown has a load termination to the traveling wave. This load reduces efficiency. Other designs substitute a reflection which produces a resonant system with higher efficiency but narrower bandwidth.

Two such arrays may be combined back-to-back so that the beam squint occurs in opposite directions, maintaining the beam direction. A null develops in the center of the beam if the frequency deviation is too high. Elliott and Kurtz [17] describe design techniques for slotted waveguide array. Figure 2.18 illustrates the end-fed and center-fed series feeds.

The parallel (corporate) beamformer ensures equal lengths between the transmit/receive electronics and each radiating element. It is called corporate because it resembles a corporation organization chart (where each manager has two subordinates). Because of the multiple power dividing circuits, it is lossy (up to 0.5 dB per divider). It has no inherent frequency dispersion aside from the stripline or waveguide from which it is constructed. Figure 2.19 shows this configuration.

2.7.2 Beamforming networks

Several approaches exist to form sets of fixed beams selectable by switching between dedicated antenna ports. Toso [18] describes their utility. Hall and Vetterlein [19] provide a good overview of techniques for multiple fixed beams. They describe two types: lens based or circuit based. Some of the lens-based

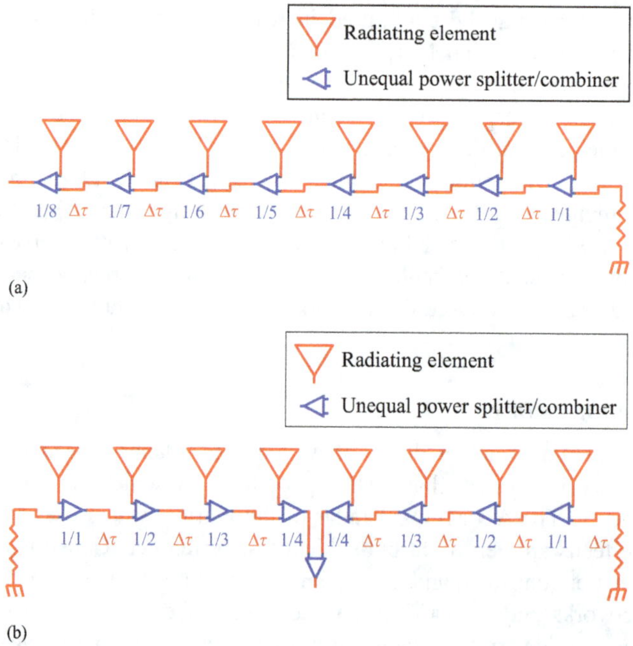

Figure 2.18 Series-fed feeds: (a) end fed and (b) center fed

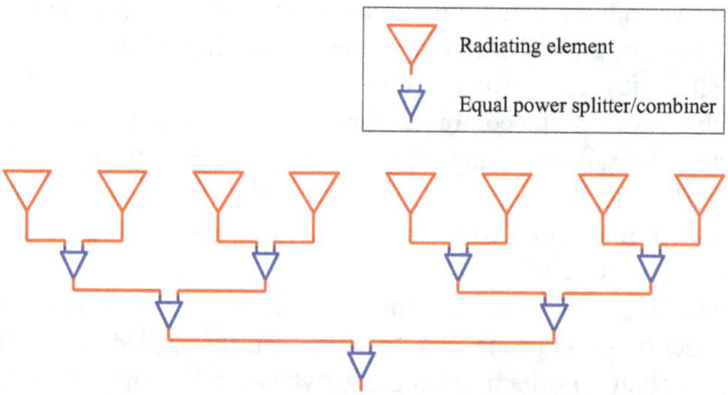

Figure 2.19 Corporate feed

techniques will be considered in Chapter 11. This section considers the circuit-based approaches. Fixed beam systems use signal splitters and combiners in combination with phase or time delay. They may be implemented in the analog section of the beamformer or with the digital equivalent after the receiver. Allen [20] contrasts some aspects of the circuit-based approaches.

Fixed beam systems are useful when the desired beams are known in advance and are invariant, for instance, to survey a fixed sector or a fixed volume. This is the case for the SKA (Khlebnikov *et al.* [21], "All-digital wideband space-frequency beamforming for the SKA aperture array"), the recently inaugurated "Space Fence system" (Haimerl *et al.* [22], "Overview of the large digital arrays of the space fence radar"), communications satellites including Iridium, and scanning SAR (SCORE [SCan On Receive] or SweepSAR). Nonfixed (agile) beams are more commonly associated with ESAs and are appropriate when an area of the full scan volume must be selected dynamically.

Some SAR system designers have devised modes that utilize multiple fixed beams to good benefit. Multiple azimuth beams improve SAR resolution and multiple elevation beams improve stripmap area rate. SCORE is a specific implementation of multiple elevation beams particularly suited to reflector systems. Currie and Brown [23] provide a good discussion of multiple azimuth and multi-elevation beams application to SAR.

Figure 2.20 illustrates some beam combinations available with ESA beamforming techniques. These provide the effect of multiplying the sensor coverage. For receive purposes, the performance of each beam is comparable to the original beam. For transmit purposes, however, it is necessary to increase the transmitted power proportionate to the number of beams to retain the performance of the single beam system for each beam. The power capability of solid-state devices has been significantly improved, by the availability of SiC and GaN devices which facilitates multiple beam applications. Available prime power and cooling are often the limiting factor of beam proliferation.

2.7.2.1 Butler matrix beamformer

The Butler matrix is an analog representation of a fast Fourier transform to provide multiple fixed beam ports to the transmit/receive electronics. It has no inherent loss mechanism but it requires multiple couplers and phase shifters which will incur some loss. It has N inputs (beam ports) and N outputs (antenna ports) which are combined and recombined with appropriate phase shifts to provide multiple simultaneous orthogonal beams. It is a reciprocal device, providing multiple beams in both transmit and receive, useful for many applications.

Butler [24] invented this beamformer which is an analog implementation of the fast Fourier transform (FFT). It was invented independent of the contemporaneous development of the FFT by Cooley and Tukey [25]. A large body of literature exists, see, for instance, [26–30].

Figure 2.21 illustrates a Butler matrix with 16 elements and 16 beams requiring 32 directional couplers and 24 fixed (nonzero) phase shifters. Sections may be cascaded as far as desired to realize any power-of-two number of beams/radiating elements although the complexity of the circuit eventually overtakes its utility. Delaney [31] provides (2.49) relating the number of components to the number of beams.

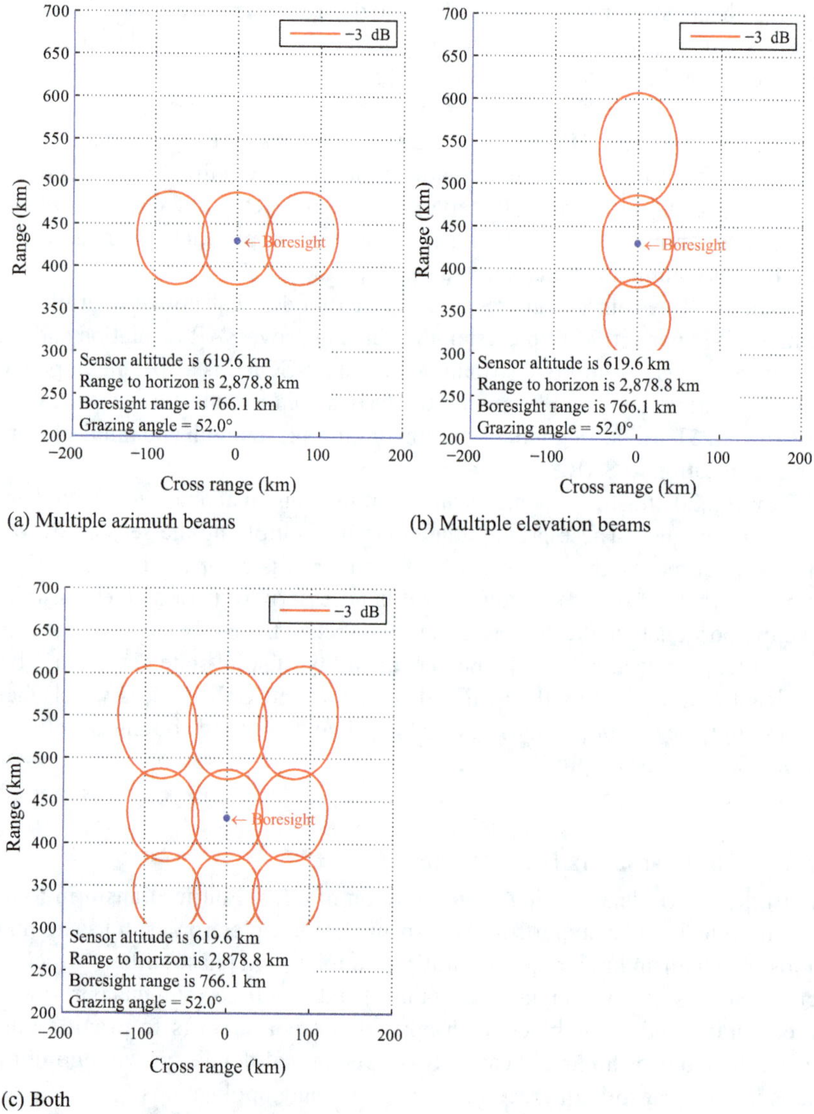

(a) Multiple azimuth beams

(b) Multiple elevation beams

(c) Both

Figure 2.20 Multiple beams increase area coverage
MultipleBeam.m

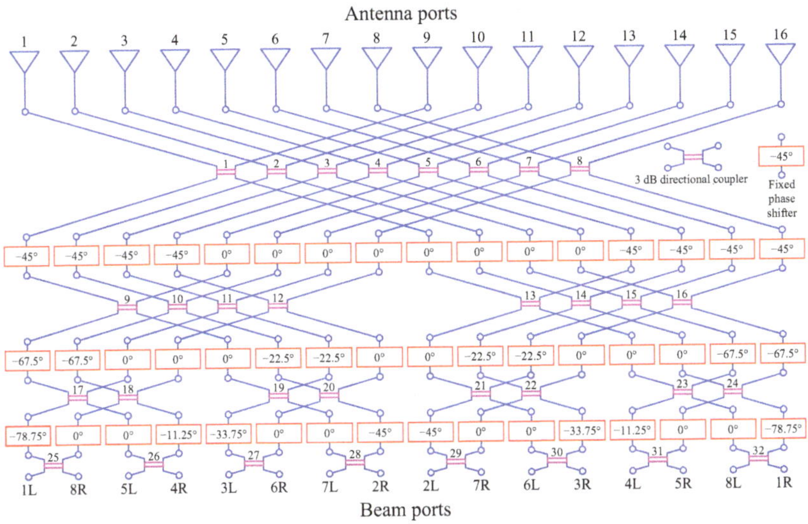

Figure 2.21 *Sixteen-beam Butler matrix*
After Delaney [31]

$$\begin{aligned} \text{Number of couplers} &= \frac{N\log_2 N}{2} \\ \text{Number of phase shifters} &= \frac{N(\log_2 N - 1)}{2} \end{aligned} \tag{2.49}$$

The Iridium satellite (Section 17.1) uses a 16-beam Butler matrix beamformer in elevation and an 8-beam Butler matrix beamformer in azimuth.

2.7.2.2 Blass matrix beamformer

The Blass matrix is a more complex technique to provide arbitrary multiple simultaneous beams. It has N inputs (beam ports) and M outputs (antenna ports) and $N \times M$ couplers whose coupling value varies. It also incorporates loads to absorb unused signal power. The inputs are combined and recombined in different ratios with different path lengths to provide multiple simultaneous beams.

The Blass [32] matrix beamformer offers more flexibility than the Butler matrix, with no restriction to a power-of-two for the beams or the elements. However, it requires complex four-way power splitter/combiners and entails some loss in efficiency.

Figure 2.22 illustrates a Blass matrix. In transmit mode, the signals originate at the beam ports, and some fraction is coupled to the columns representing radiating elements. The direction and shape of the beam is determined by the time delay/phase shift in the total path length and the amplitude weighting provided by the coupling coefficient. Being series fed, it will be subject to fairly narrow

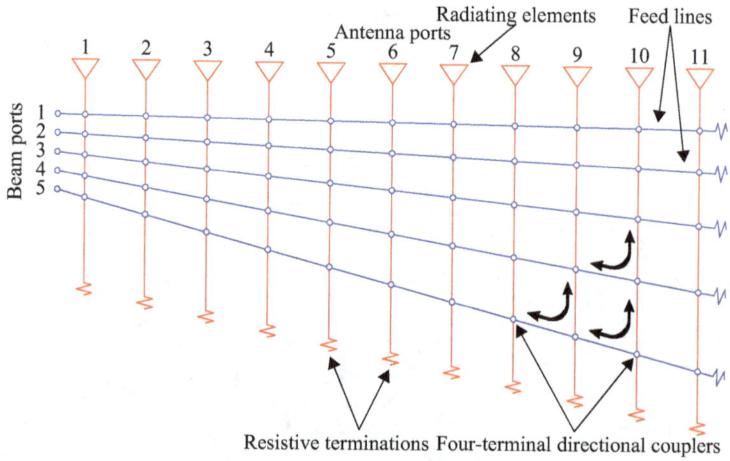

Figure 2.22 Five-beam, 11 element Blass matrix
After Blass [32]

bandwidth capability. One of its design challenges is that the various ports are not isolated. The loads absorb unused power to avoid reflections which would alter the beam shape and direction.

Chen [33] provides a detailed guide to coupler modeling and design. Casini *et al.* [34] provide a general procedure for computing the required coupling values which involves solving the $n \times m$ matrix as well as a high-performance coupler, implemented with waveguide using cross couplers.

2.8 Subarray partitioning and recombination

In order to enhance and facilitate antenna performance as well as integration and test, it is often convenient to construct arrays from building blocks of smaller arrays (subarrays). These subarrays are frequently identical, simplifying manufacture and reducing cost.

The separability of the element pattern and the array factor described in Section 2.5.2 can be extended to constructing arrays of arrays where subarrays take the role of elements in the pattern calculation.

Each subarray requires a feed network. Time delay units or individual receivers may be associated with each subarray. The subarrays may be phase-steered as usual but combined using time delay to retain the wider instantaneous bandwidth associated with the smaller subarray but realize the performance of a much larger array. It is also possible to use a separate receiver for each subarray and do beamforming digitally.

For rectangular lattices and subarrays, the performance may be assessed by using the subarray pattern as the element pattern in the analysis. The array will have a lattice spacing $\gg \lambda/2$ which would ordinarily create excessive sidelobes. The concept of pattern multiplication applies and the nulls in the element pattern tend to coincide with the grating lobes of the array, at least on boresight.

2.8.1 Array partitioning

Figure 2.23 illustrates partitioning of a linear array at various levels of the feed network, described in Section 2.7.1. The example applies phase control to each element, then combining groups of elements by means of a corporate feed, and then digitized by a receiver. The receivers may be placed at different levels of the summation,

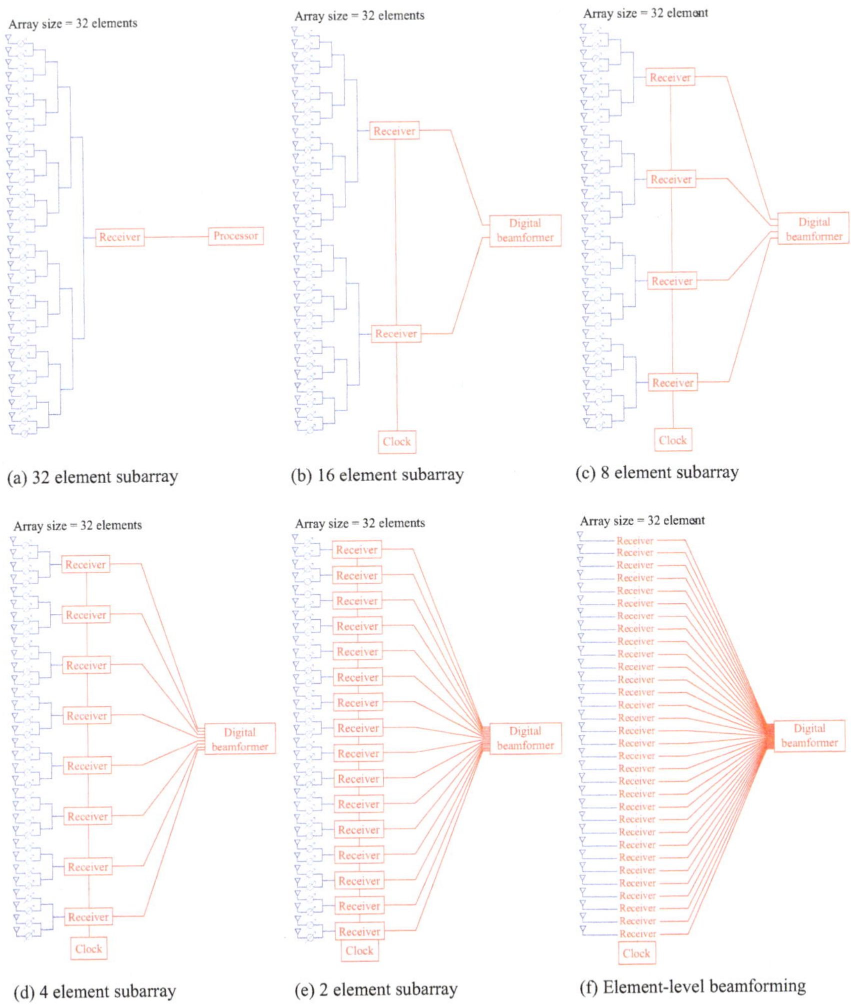

Figure 2.23 Partitioning of 32 element array

providing a tradeoff between cost/complexity and bandwidth. The receiver may provide digital time delays or could be replaced by a time delay unit. The receivers must be synchronized in time to a fraction of a cycle of the signal to provide usable time delay accuracy.

We next consider three cases of subarray partitioning. Each case is shown at boresight and steered to 10°, 20°, and 30° and for nominal frequency as well as 10% deviation from nominal frequency making eight combinations. These eight cases are plotted over a 20° span of scan angle. A ninth case, boresight and nominal frequency, is plotted over ±90° scan.

The first case shows a phase-steered 0.5 m array. The second case is a phase-steered 4.0 m array (effectively eight instances of the 0.5 m array). The third case is the same as the second case but uses time delay to steer the eight subarrays instead of phase delay. This provides wider scan capability and/or wider bandwidth.

2.8.1.1 Beam squint revisited phase-steered 0.5 m array

Figure 2.24 shows the 0.5 m phase-steered case. On boresight the beam is ideal at nominal frequency as well as with a 10% deviation in frequency. It retains its shape with phase steering. It shows considerable gain loss with a 10% deviation in frequency when scanned ranging from −1.4 dB to −17.5 dB for angles between 10° and 30°.

2.8.1.2 Beam squint revisited phase-steered 4 m array

Figure 2.25 shows the 0.5 m phase-steered case. On boresight, the beam is ideal at nominal frequency as well as with a 10% deviation in frequency. It retains its shape with phase steering. It shows massive gain loss (−18.1 dB) with a 10% deviation in frequency when scanned to 10°. Accordingly, this configuration is only usable with narrow bandwidth or narrow scan limits.

2.8.1.3 Eight time-delayed 0.5 m subarrays

Figure 2.26 shows the 0.5 m phase-steered subarrays, combined with time-delayed steering at the array level. On boresight, the beam is ideal at nominal frequency as well as with a 10% deviation in frequency. It retains its shape with phase steering. It shows similar gain loss behavior with scanning and frequency deviation as the 0.5 m phase-steered subarray.

A 4.0 m array formed from eight time-delayed subarrays provides the higher gain of the larger array with the beam squint of the subarray, so its effective scan range is greater. However, the grating lobes of the array factor restrict the useable scan range because they introduce ambiguities into the target direction. In consequence, the gain reduction of the beam squint is not the limiting factor in beam scan; it is grating lobes that restrict beam scan. Ideally, the gain reduction and the grating lobes reduction would be better matched.

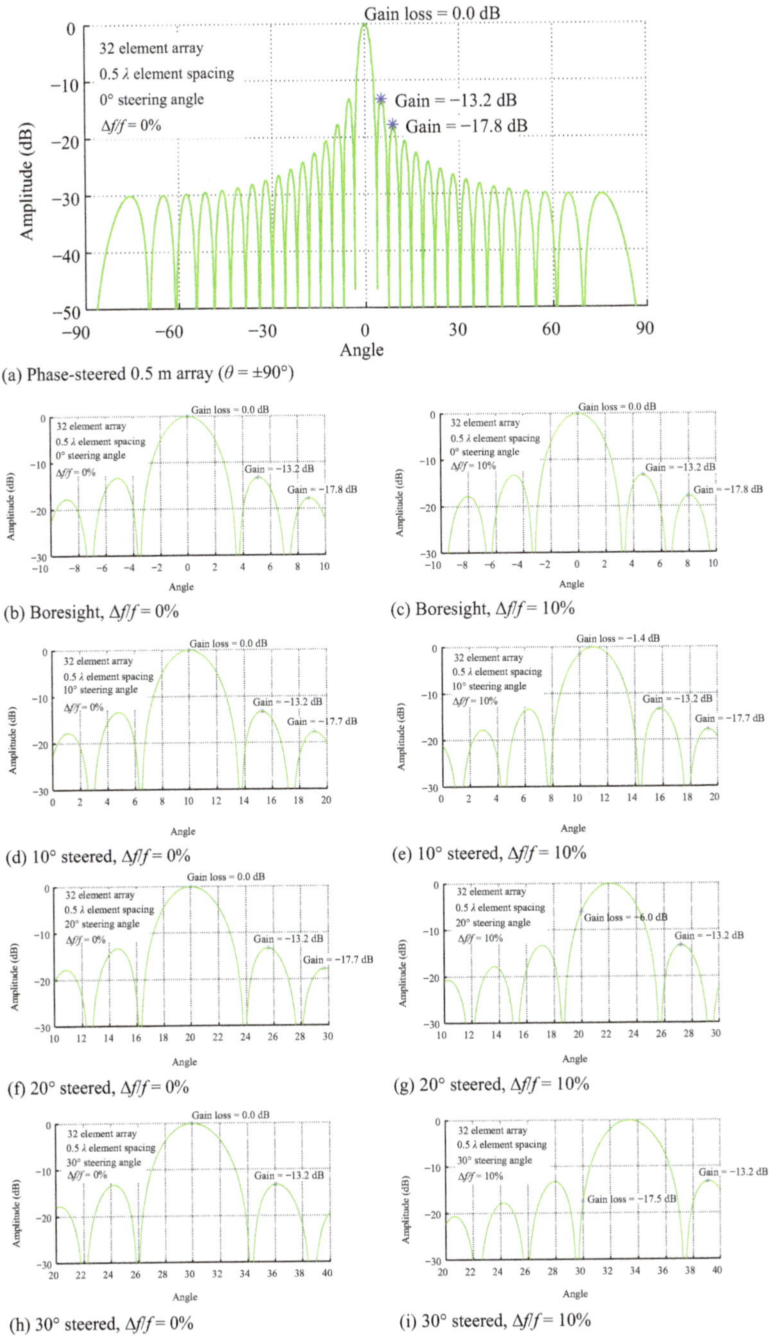

(a) Phase-steered 0.5 m array ($\theta = \pm 90°$)

(b) Boresight, $\Delta f/f = 0\%$

(c) Boresight, $\Delta f/f = 10\%$

(d) 10° steered, $\Delta f/f = 0\%$

(e) 10° steered, $\Delta f/f = 10\%$

(f) 20° steered, $\Delta f/f = 0\%$

(g) 20° steered, $\Delta f/f = 10\%$

(h) 30° steered, $\Delta f/f = 0\%$

(i) 30° steered, $\Delta f/f = 10\%$

Figure 2.24 Phase-steered 0.5 m array (detail)

LinearArray.m

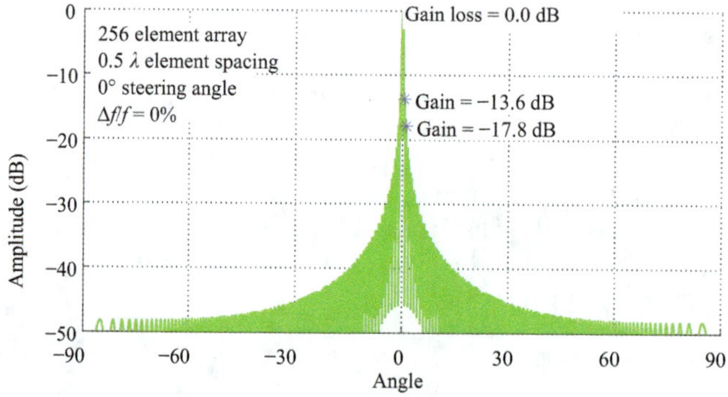

(a) Phase-steered 0.54 meter array ($\theta = \pm 90°$)

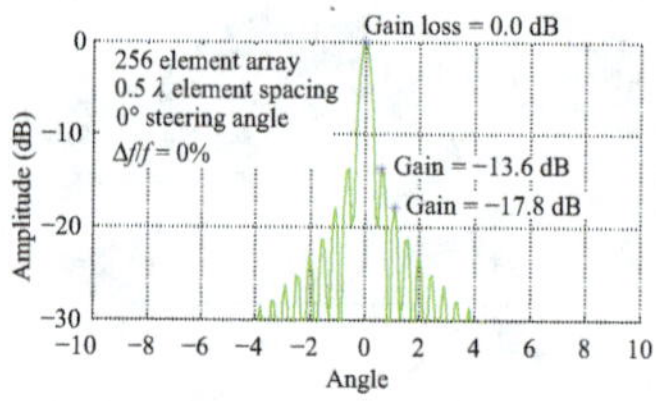

(b) Boresight, $\Delta f/f = 0\%$

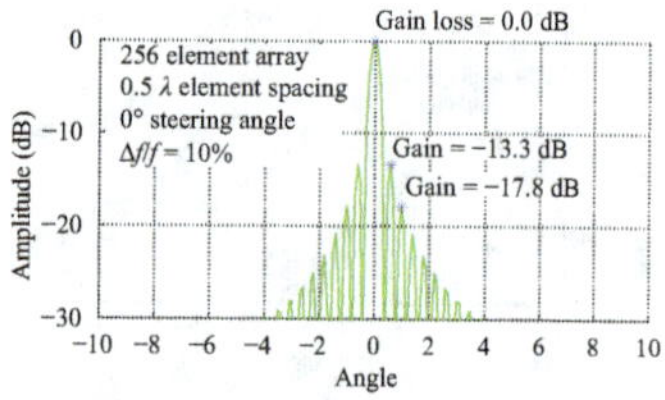

(c) Boresight, $\Delta f/f = 10\%$

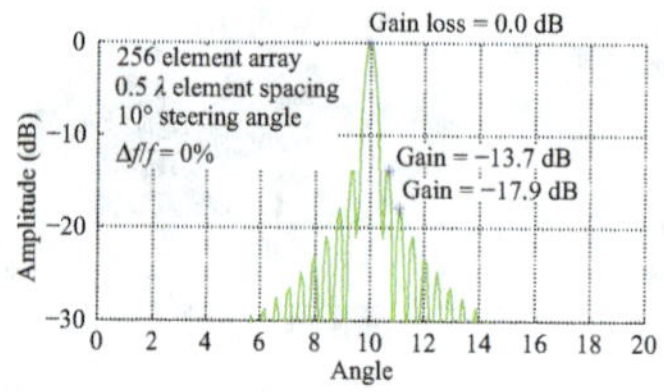

(d) 10° steered, $\Delta f/f = 0\%$

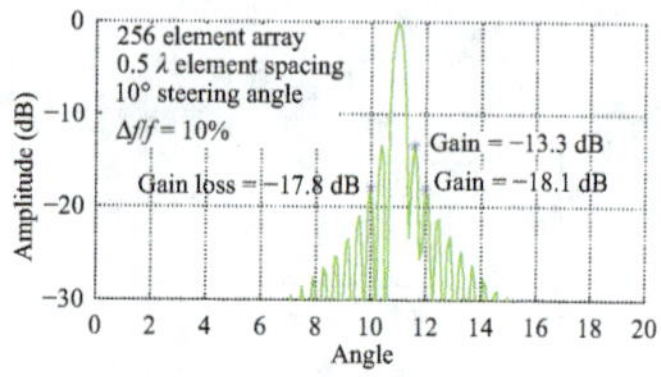

(e) 10° steered, $\Delta f/f = 10\%$

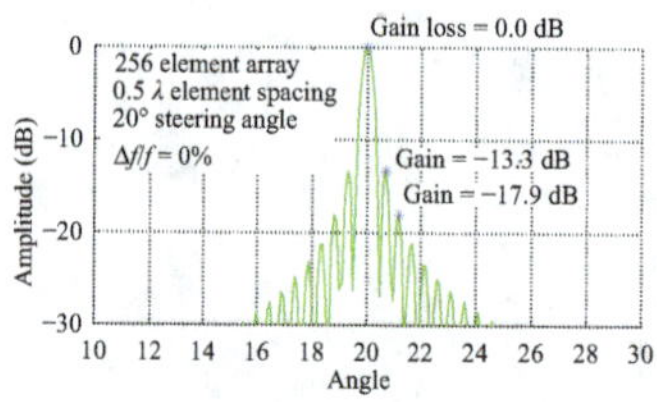

(f) 20° steered, $\Delta f/f = 0\%$

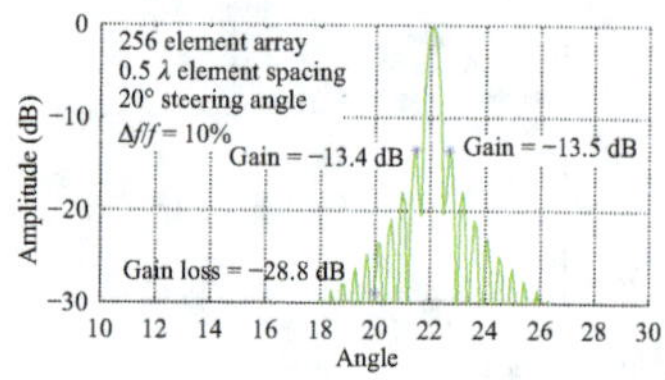

(g) 20° steered, $\Delta f/f = 10\%$

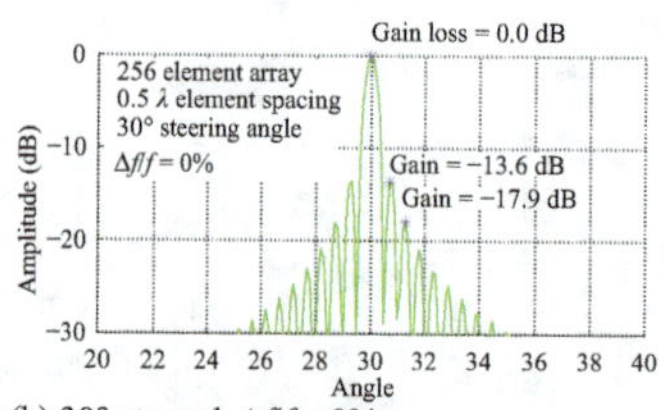

(h) 30° steered, $\Delta f/f = 0\%$

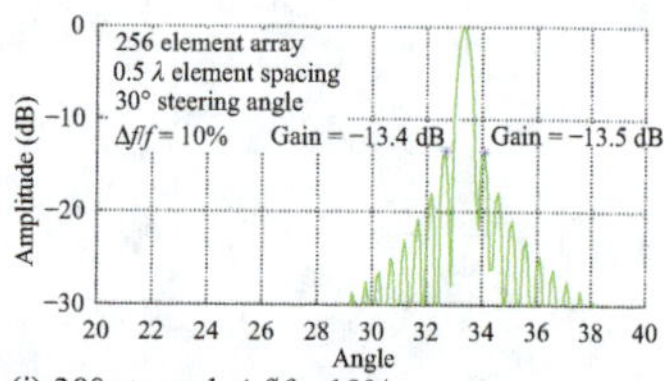

(i) 30° steered, $\Delta f/f = 10\%$

Figure 2.25 Phase-steered 4.0 m array (detail)

LinearArray.m

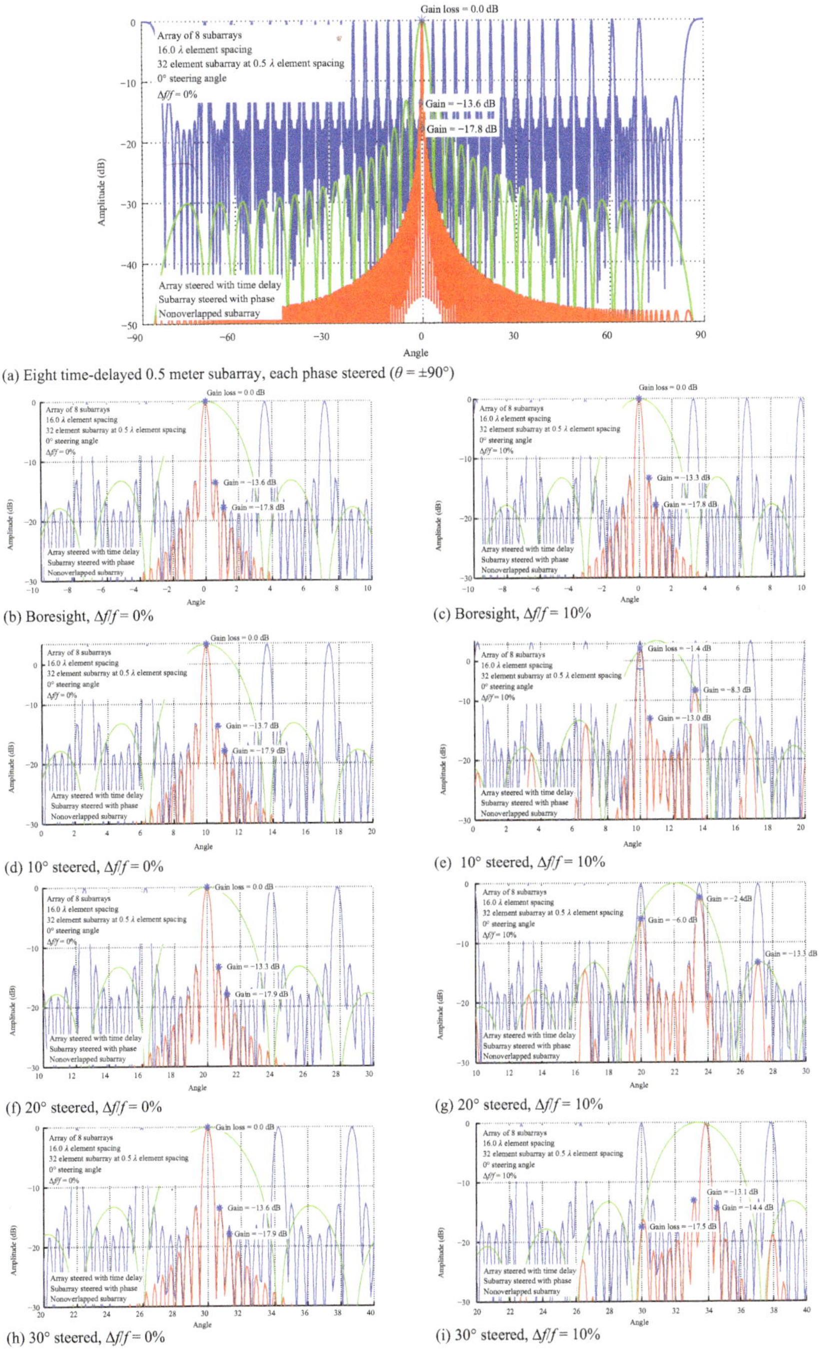

(a) Eight time-delayed 0.5 meter subarray, each phase steered ($\theta = \pm 90°$)

(b) Boresight, $\Delta f/f = 0\%$

(c) Boresight, $\Delta f/f = 10\%$

(d) 10° steered, $\Delta f/f = 0\%$

(e) 10° steered, $\Delta f/f = 10\%$

(f) 20° steered, $\Delta f/f = 0\%$

(g) 20° steered, $\Delta f/f = 10\%$

(h) 30° steered, $\Delta f/f = 0\%$

(i) 30° steered, $\Delta f/f = 10\%$

Figure 2.26 Eight-time delayed 0.5 m subarray, each phase steered (detail)
LinearArray.m

2.8.2 Overlapped subarray

The prior examples show that the subarray pattern is not effective in suppressing the grating lobes of the array factor. A larger subarray would have a narrower beam pattern which would be more effective in suppressing the grating lobes.

At the expense of some complexity, the subarray electrical size can exceed its physical size by borrowing elements from adjacent subarrays. This provides a larger subarray by summing adjacent elements to form a larger effective element. In order to improve the beam-steering capability of a tight lattice, each element is utilized multiple times creating overlapping subarrays.

A suitable approach uses a triangular taper which combines both the array of interest and portions of its neighbors. Triangular weighting is described in Section 9.3. A triangle is the autocorrelation of a rectangle, so the element pattern is square pattern of uniform illumination, i.e., –26 dB first sidelobe. For this analysis, we will simply assume the result and show its derivation later.

This approach provides improved sidelobe control of scanned beam with a reasonably straightforward passive network. Notably, it has no effect on the boresight beam. The example shows three-element subarray but the complexity increases considerably with larger subarrays. Gain blocks may be required to compensate for splitting/combining loss.

The summation must occur before application of beam steering. A linear array example is illustrated in Figure 2.27 with each element contributing to three subarrays by means of power dividers and combiners. Since these are lossy, it might be necessary to add gain to the system to preserve the noise figure. This approach may also be applied to a two-dimensional array but it is considerably more complex.

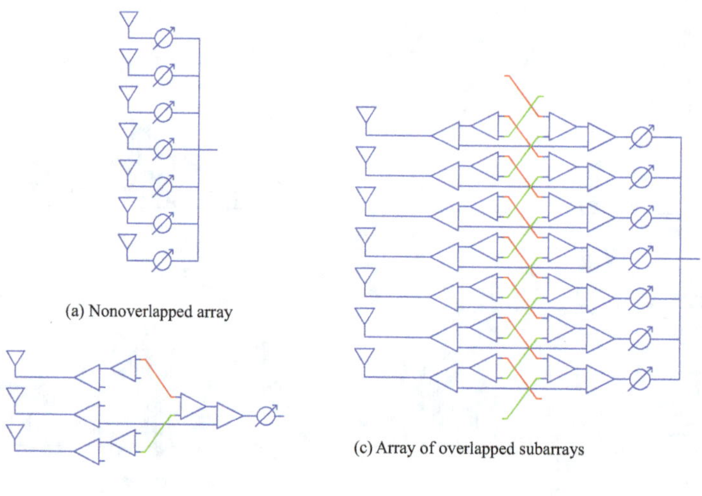

(a) Nonoverlapped array

(b) Single overlapped array

(c) Array of overlapped subarrays

Figure 2.27 Overlapped subarray beamformer

Overlapping subarrays produce a narrower mainbeam which helps suppress grating lobe but also reduces gain as beam is scanned. This is an acceptable tradeoff when grating lobe scan limit is more stringent than the element gain limit. Section 11.5 describes a practical example of a reflector feed where an overlapped subarray may be used to good advantage. An ideal subarray pattern would be rectangular in shape. We will see in Chapter 10 how difficult that is to achieve.

Petrolati *et al.* [35] describe lossless overlapped subarrays. Avser *et al.* [36] describe a commercial application.

Figure 2.28 illustrates the illumination pattern for an array of eight subarrays. Figure 2.28a illustrates uniform illumination across each subarray, while Figure 2.28b applies a triangular weight to each element in the subarray and borrows from neighboring subarrays to construct the triangular weight. The end subarrays need to borrow from nonexistent neighbors introducing an error whose significance depends on the number of subarrays.

If we recompute the patterns in Figure 2.26 using triangular weighted subarrays, we obtain the patterns in Figure 2.29.

2.8.3 Scan constraint summary

We have considered a few examples showing the tradeoff between bandwidth and scan volume. We found that the beam squint limit was –3 dB gain, while the grating lobe limit was –10 dB. Overlapping subarrays recover scan volume limited by grating lobes because with a narrower subarray pattern the grating lobe is suppressed to

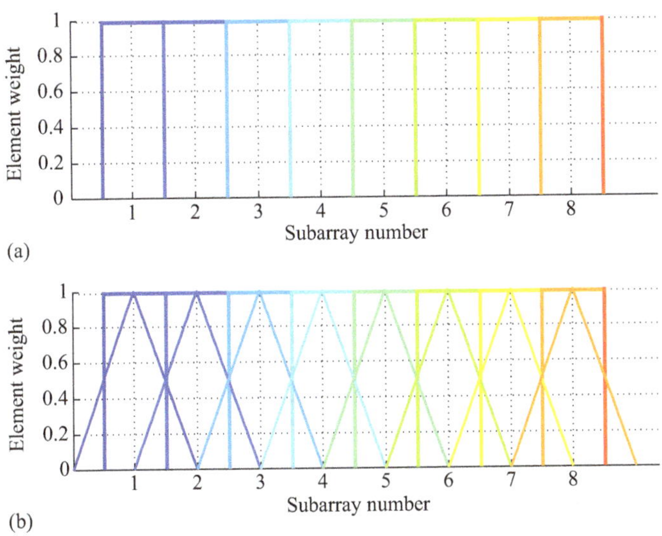

Figure 2.28 Eight subarrays – conventional and overlapping: (a) equal weight subarray and (b) triangular weight subarray

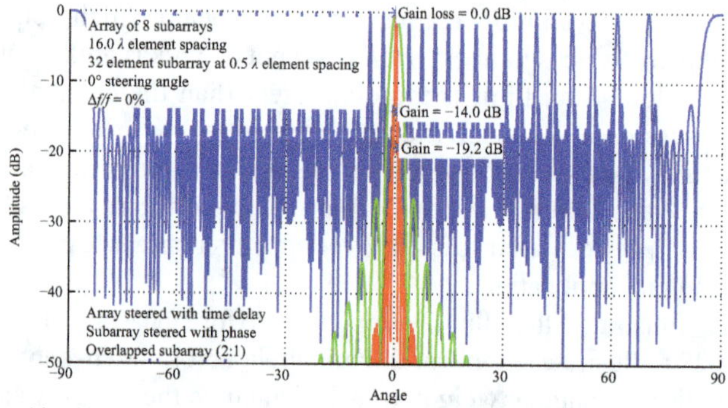

(a) Eight time-delayed 0.5 meter overlapped subarray, each phase steered ($\theta = \pm 90°$)

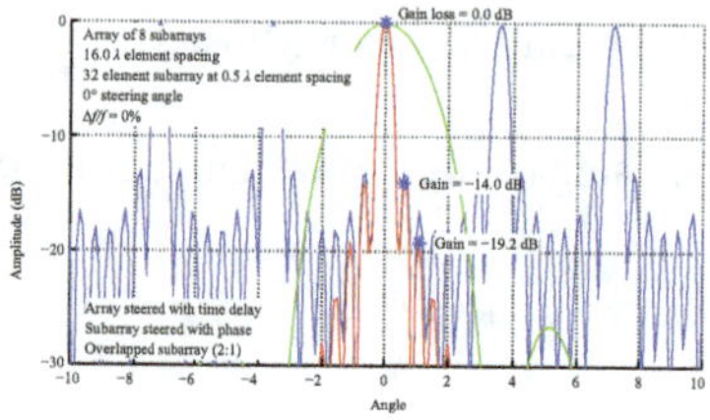

(b) Boresight, $\Delta f/f = 0\%$

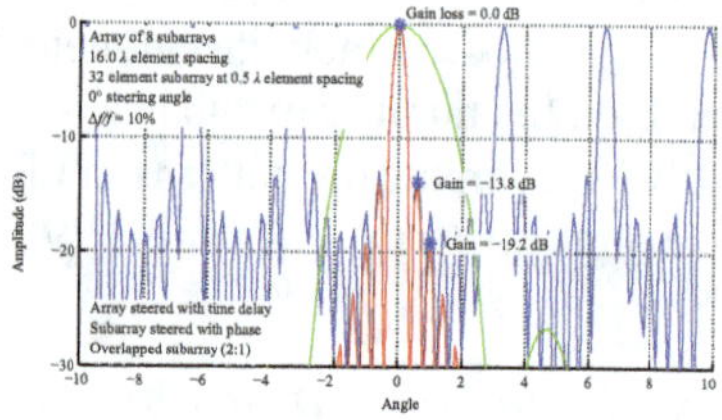

(c) Boresight, $\Delta f/f = 10\%$

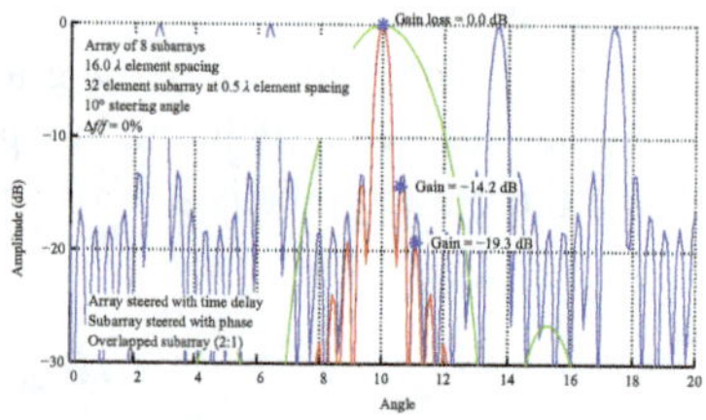

(d) 10° steered, $\Delta f/f = 0\%$

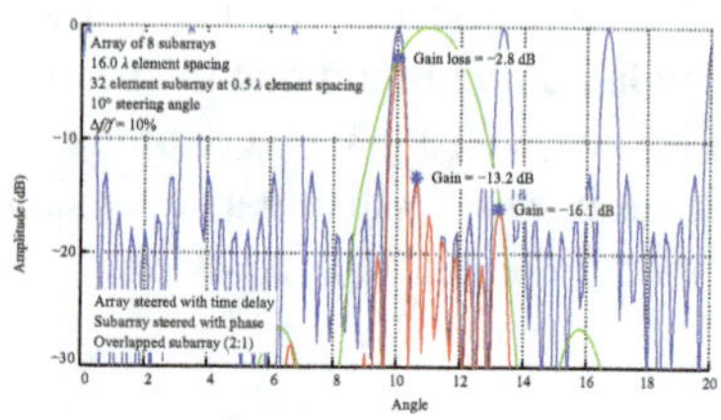

(e) 10° steered, $\Delta f/f = 10\%$

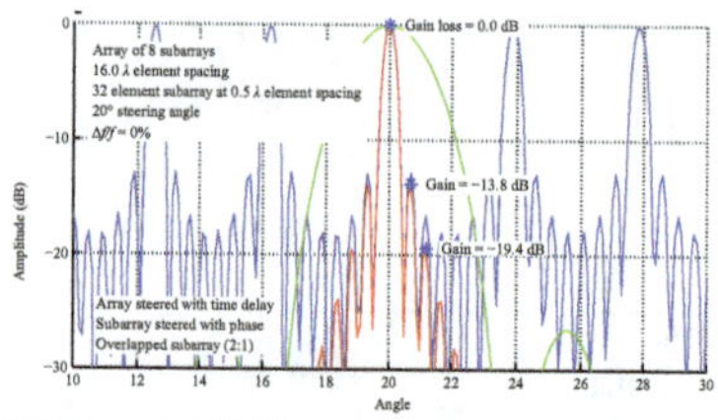

(f) 20° steered, $\Delta f/f = 0\%$

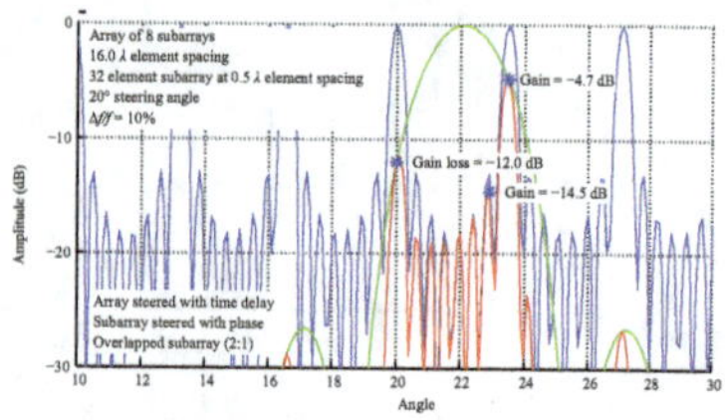

(g) 20° steered, $\Delta f/f = 10\%$

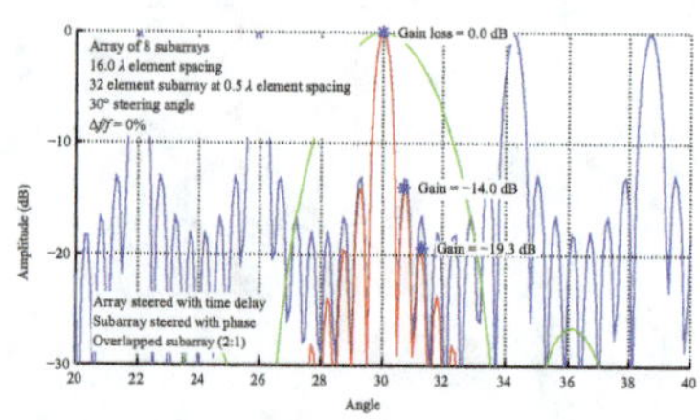

(h) 30° steered, $\Delta f/f = 0\%$

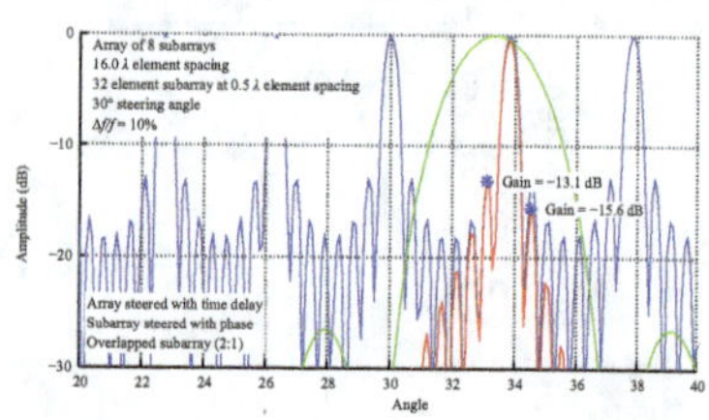

(i) 30° steered, $\Delta f/f = 10\%$

Figure 2.29 Eight-time delayed 0.5 m overlapped subarray, each phase steered LinearArray.m

Table 2.3 Bandwidth constraint summary

Scan angle	**10°**	**20°**	**30°**
0.5 m array phase scan			
Squint	>10%	≈8%	≈6%
Grating lobe	N/A	N/A	N/A
4.0 m array phase scan			
Squint	≈2%	≈1%	<1%
Grating lobe	N/A	N/A	N/A
4 m array			
0.5 m subarray nonoverlapped			
Squint	>10%	≈8%	≈6%
Grating lobe	≈7%	≈4%	≈3%
4.0 m array			
0.5 m subarray 2:1 overlapped			
Squint	>10%	≈6%	≈4%
Grating lobe	>10%	≈6%	≈4%

a wider scan angle. Table 2.3 summarizes the performance of the configurations analyzed in Sections 2.6.2 and 2.8.2. It shows the achievable scan volumes for different approaches. We see that the overlapped approach has up to 50% greater bandwidth than its nonoverlapped counterpart.

References

[1] Carson J.R., Carson J.R. 'A generalization of the reciprocal theorem'. *Bell System Technical Journal*. 1924, vol. 3(3), pp. 393–9.

[2] Carson J.R. 'Reciprocal theorems in radio communication'. *Proceedings of the Institute of Radio Engineers*. 1929, vol. 17(6), pp. 952–6.

[3] Ballantine S. 'The Lorentz reciprocity theorem for electricwaves'. *Proceedings of the Institute of Radio Engineers*. 1928, vol. 16(4), pp. 513–8.

[4] Neiman M.S. 'The principle of reciprocity in antenna theory'. *Proceedings of the Institute of Radio Engineers*. 1943, vol. 31(12), pp. 666–71.

[5] Feynman R.P., Leighton R.B., Sands M. 'Chapter 28: Electromagnetic radiation'. *The Feynman Lectures on Physics*. Vol. I. Addison–Wesley; 1964. Available from http://www.feynmanlectures.caltech.edu/I28.html.

[6] Feynman R.P., Leighton R.B., Sands M. 'Chapter 25, Linear systems and review'. *The Feynman Lectures on Physics*. Vol. I. Addison–Wesley; 1964. Available from http://www.feynmanlectures.caltech.edu/I25.html.

[7] Lynch Jr D. *Introduction to RF Stealth*. Edison, NJ, USA: SciTech Publishing, an imprint of the IET; 2004.

[8] Wieners C. *A Short Introduction to Numerical Methods for Maxwell's Equations*. 2018. Available from http://www.mathematik.uni-karlsruhe.de/user/Xwieners/MaxwellCourse.pdf.

[9] Balanis C.A. *Antenna Theory: Analysis and Design*. 4th edn. New York, NY, USA: Wiley-Interscience; 2016.

[10] Orfanidis S.J. *Electromagnetic Waves and Antennas*. Rutgers University. August 2016. Available from http://eceweb1.rutgers.edu/Xorfanidi/ewa/ ewa-1up.pdf.

[11] Nikolova N.K. *Notes on Antenna Engineering 2018*. McMaster University, Canada. 2018. Available from http://www.ece.mcmaster.ca/faculty/nikolova/antenna dload/currentlectures/LectureNotesAntennasNikolova.pdf.

[12] Mailloux R.J. *Phased Array Antenna Handbook*. 3rd edn. Boston, MA, USA: Artech House; 2018.

[13] Baker B.B., Copson E.T. *The Mathematical Theory of Huygens' Principle. Radar and Antennas*. Oxford University Press; 1939.

[14] Haupt R.L. 'Factors that define the bandwidth of a phased array antenna'. *2019 IEEE International Symposium on Phased Array Systems and Technology (PAST)*. October 2019.

[15] Garakoui S.K., Klumperink E.A.M., Nauta B., *et al.* 'Phased-array antenna beam SQUINTING related to frequency dependency, of delay circuits'. *2011 8th European Radar Conference*. 2011, pp. 416–9.

[16] Romanofsky R.F. Array Phase Shifters: Theory and Technology. Tech. Rep. TM–2007-214906. Cleveland, OH, USA: NASA Glenn Research Center; 2007. Available from https://ntrs.nasa.gov/archive/nasa/casi.ntrs.nasa.gov/20080001449.pdf.

[17] Elliott R., Kurtz L. 'The design of small slot arrays'. *IEEE Transactions on Antennas and Propagation*. 1978, vol. 26(2), pp. 214–9.

[18] Toso G. 'The beauty of multibeam antennas'. *2015 9th European Conference on Antennas and Propagation (EuCAP)*. 2015, pp. 1–5.

[19] Hall P.S., Vetterlein S.J. 'Review of radio frequency beamforming techniques for scanned and multiple beam antennas'. *IEE Proceedings H - Microwaves, Antennas and Propagation*. 1990, vol. 137(5), pp. 293–303.

[20] Allen J. 'A theoretical limitation on the formation of lossless multiple beams in linear arrays'. *IRE Transactions on Antennas and Propagation*. 1961, vol. 9(4), pp. 350–2.

[21] Khlebnikov V.A., Zarb-Adami K., Armstrong R.P. 'All-digital wideband space-frequency beamforming for the SKA aperture array'. *2010 IEEE International Symposium on Phased Array Systems and Technology*. 2010, pp. 911–6.

[22] Haimerl J.A., Hudson B., Fonder G.P., *et al.* 'Overview of the large digital arrays of the space fence radar'. *2016 IEEE International Symposium on Phased Array Systems and Technology (PAST)*. 2016, pp. 1–8.

[23] Currie A., Brown M.A. 'Wide-swath SAR'. *IEE Proceedings F Radar and Signal Processing*. 1992, vol. 139(2), pp. 122–35.

[24] Butler J., Lowe R. 'Beam forming matrix simplifies design of electronically scanned antennas'. *Electronic Design*. 1961, pp. 170–3.

[25] Cooley J.W., Tukey J.W. 'An algorithm for the machine calculation of complex Fourier series'. *Mathematics of Computation.* 1965, vol. 19(90), pp. 297–301.

[26] Nester W. 'The fast Fourier transform and the Butler matrix'. *IEEE Transactions on Antennas and Propagation.* 1968, vol. 16(3), pp. 360–360.

[27] Shelton J.P. 'Fast Fourier transforms and Butler matrices'. *Proceedings of the IEEE.* 1968, vol. 56(3), pp. 350–350.

[28] Tudosie G. 'Analysis and design of compact passive distribution networks for microwave applications'. Eng. PhD thesis. 2009. Available from https://www.research-collection.ethz.ch/handle/20.500.11850/151510.

[29] Mehron M.E. 'Octave bandwidth multiple beam array'. *Proceedings of the Antenna Applications Symposium 1952–2008.* 1963.

[30] Brigham E.O. *The Fast Fourier Transform.* Prentice-Hall; 1974.

[31] Delaney W.P. 'An RF multiple beam-forming technique'. *IRE Transactions on Military Electronics.* 1962, vol. MIL-6(2), pp. 179–86.

[32] Blass J. 'Multidirectional antenna - A new approach to stacked beams'. *1958 IRE International Convention Record.* 1960, vol. 8, pp. 48–50.

[33] Chen J. 'Cross-guide couplermodeling and design (NASA)'. *The Telecommunications and Data Acquisition Report 42-95.* 1988, pp. 82–8.

[34] Casini F., Gatti R.V., Marcaccioli L., *et al.* 'A novel design method for Blass matrix beam-forming networks'. *2007 European Radar Conference.* 2007, pp. 232–5.

[35] Petrolati D., Angeletti P., Toso G. 'A lossless beam-forming network for linear arrays based on overlapped sub-arrays'. *IEEE Transactions on Antennas and Propagation.* 2014, vol. 62(4), pp. 1769–78.

[36] Avser B., Frazita R.F., Rebeiz G.M. 'Interwoven feeding networks with aperture sinc-distribution for limited-scan phased arrays and reduced number of phase shifters'. *IEEE Transactions on Antennas and Propagation.* 2018, vol. 66(5), pp. 2401–13.

Chapter 3

Synthetic arrays

To this point, the discussion has dealt with physical antennas. Real beamforming uses samples collected at one point in time. Resolution and beam shape are limited by the size of the antenna and the number of elements. Real antenna angular resolutions are on the order of 1°, so resolution at 50 km is almost 1 km which is too large for many purposes.

The array concept may be extended to arrays that are synthetic. Figure 3.1 illustrates such an arrangement. Synthetic arrays create the effect of a very long antenna with very small azimuth resolution by coherently combining data from a relatively short antenna which occupies multiple positions sequentially. Usually, but not necessarily, it operates in continuous motion. Target motion during data collection will degrade the image. Targets in uniform motion may be imaged, and many SAR papers are devoted to treating with moving objects in a SAR data set.

The synthetic array concept is to utilize the motion of the antenna platform to collect data samples over a larger region than accessible by the physical size of the antenna. This permits near-photographic resolution images of the ground but requires (1) very accurate knowledge of the antenna position and orientation over time, (2) a stationary target, and (3) considerable additional signal processing. Guarnieri *et al.* [1] provide a nice overview in *Electromagnetic Imaging*.

While a typical antenna has a ratio of height to width on the order of 1 (between 0.1 and 10.0), a synthetic array width may be thousands of times longer than its height. The azimuth beamwidth is decreased by a similar ratio but the elevation beamwidth is unaffected with the result that the synthesized beam is a fan beam. The range resolution may be improved with wide bandwidth waveforms so that the azimuth (along-track) and range (cross-track) resolutions are comparable, producing a "square pixel." The resulting image is a good replica of the scene if the scene contains no significant elevation differences but a nonplanar scene creates various artifacts.

3.1 Synthetic aperture radar

A synthetic aperture resembles a conventional ESA sampled over time but there is a significant difference. Each element of the array in time receives a signal transmitted only from itself, not from the ensemble of elements. It may be viewed as an extreme

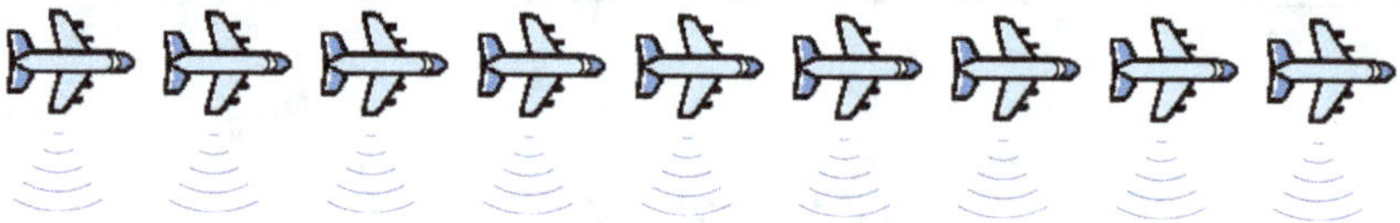

Figure 3.1 An array in time

case of a multiple-input, multiple-output system where the off-diagonal elements are zero. Its performance is only meaningful for two-way performance of the array in contrast to real arrays whose response may be evaluated with respect to an incoming plane wave or evaluation of a transmitted field.

The Doppler shift experienced by virtually all SAR systems is is also experienced by a real array on a moving platform. A real array functions equally well whether stationary or moving. SAR has been demonstrated on rail-mounted systems for which the antenna is stationary for each measurement and moves between pulses. This will be discussed further in Section 3.1.1. Tsynkov [2] discusses the role of the Doppler shift.

Figure 3.2 illustrates the signal paths. Figure 3.2a shows that in a conventional ESA, every element participates in the transmit–receive process and notably receives power transmitted by all elements. Figure 3.2b shows that each element transmits and receives in isolation so that the full aperture is never active simultaneously. For the SAR case, the real array serves the function of the element.

Precise antenna position is required to combine data collected at different times. The antenna position must be determined to a fraction of a wavelength and timing controlled to a fraction of a cycle although it is possible to perform minor corrections with optimization algorithms. The platform inertial measurement unit (IMU)

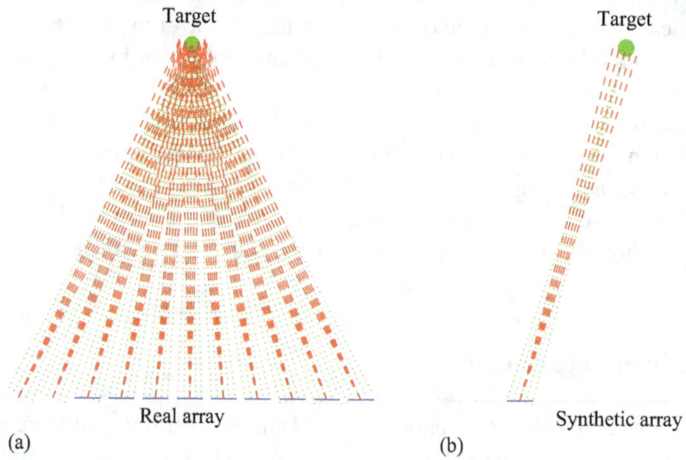

Figure 3.2 Synthetic array collects less information. (a) Conventional ESA and (b) synthetic aperture array–one of M positions of the physical array

or global positioning systems (GPS) is useful for this purpose and, for best accuracy, the IMU or GPS may be mounted on the ESA structure.

We may analyze the synthetic beamwidth using (2.34), choosing a one-dimensional case because, in general, the synthetic array is constructed in only one dimension. This is a one-way pattern so, for our purposes, we consider the product of a transmit pattern, with a receive beam pattern. The product of the transmit and receive beams may be written as

$$\mathrm{AF}_{\text{two-way}} = \left[\sum_{m=0}^{M-1} a_m e^{jkm\Delta x \sin\theta \cos\phi}\right]\left[\sum_{n=0}^{N-1} a_n e^{jkn\Delta x \sin\theta \cos\phi}\right] \tag{3.1}$$

or equivalently (analogous to (2.35))

$$\mathrm{AF}_{\text{two-way}} = \sum_{m=0}^{M-1}\sum_{n=0}^{N-1} a_m a_n e^{jk(n+m)\Delta x \sin\theta \cos\phi} \tag{3.2}$$

If we make the substitution $\psi = k\Delta x \sin\theta \cos\phi$ as before, we obtain (3.3):

$$\mathrm{AF}_{\text{two-way}} = \sum_{m=0}^{M-1}\sum_{n=0}^{N-1} a_m a_n e^{j(n+m)\psi} \tag{3.3}$$

This is the general representation valid for real arrays. To evaluate the SAR case, we will assume that the diagonal terms are equal, i.e., $a_m = a_n \neq 0$ and the cross-terms are zero, i.e., $a_m \times a_n = 0$ for $m \neq n$. We also assume that a_m is a constant a, to obtain (3.4) (analogous to (2.36)):

$$\begin{aligned}\mathrm{AF}_{\text{two-way}} &= a^2 \sum_{m=0}^{M-1} \left(e^{j\psi}\right)^{2m} \\ &= a^2 \left(\frac{1 - e^{j(2M\psi)}}{1 - e^{j2\psi}}\right) \\ &= a^2 \left(e^{j(M-1)\psi}\right)\left(\frac{\sin(M\psi)}{\sin(\psi)}\right)\end{aligned} \tag{3.4}$$

A factor of 2 appeared when we equated m and n. Comparing (3.4) with (2.37), we see that the two-way antenna pattern of a SAR is the same as the one-way antenna pattern of an ESA whose element spacing is twice that of the SAR sample interval. This includes peak sidelobes of the two-way pattern of −13.2 dB. The similarity to a 2× element spacing of a normal ESA extends to the presence of grating lobes at ±90°.

It is not necessary to increase the sample rate, as Freeman *et al.* [3] confirm in a different context. However, with the undersampling, the along-track resolution will be halved, upwards of one-half the transmit power will be wasted, being out of band, and it will be necessary to implement a very good low-pass filter in front of the receiver to avoid aliasing the high Doppler returns into the low Doppler band. Accordingly, it is customary to collect two samples per antenna length.

We may make two substitutions in (3.4) to correspond to a sample rate corresponding to one-half the length; we replace M by $2M$ and ψ by $\psi/2$ to obtain (3.5):

$$\begin{aligned}\mathrm{AF}_{\text{two-way}} &= a^2\left(e^{j(2M-1)\psi/2}\right)\left(\frac{\sin(2M\psi/2)}{\sin(\psi/2)}\right)\\ &= a^2\left(e^{j(2M-1)\psi/2}\right)\left(\frac{\sin(M\psi)}{\sin(\psi/2)}\right)\end{aligned} \tag{3.5}$$

We compare the real and synthetic two-way patterns in Figure 3.3 based on a SAR using 11 samples. This case is unrealistic but produces patterns sufficiently broad that the sidelobes are visible. The equivalent real array pattern is shown in blue with a 3 dB beamwidth of 6.7° and a first sidelobe at −26 dB. The SAR pattern based on one sample per array length (3.4) has a 3 dB beamwidth of 4.6° and a first sidelobe at −13 dB as well as grating lobes at ±90°. When we double the sample rate in accord with (3.5), the pattern near boresight is virtually unchanged but the grating lobes disappear. The two main beamwidths are the same because the length of the synthetic array is unchanged.

3.1.1 SAR and Doppler

The role of the Doppler shift in SAR is a subject of considerable discussion. It is of particular relevance when comparing SAR with Doppler beam sharpening (DBS). With respect to SAR, Sullivan [4] in Section 17.4 states:

> There are at least two mathematically equivalent ways of considering SAR. As we have developed the concept so far, the crossrange resolution may be considered to result from the Doppler shifts resulting from the different apparent line-of-sight velocities of different parts of the scene. However, the crossrange resolution may also be considered to result from the large synthetic aperture, much as fine crossrange resolution would also result from a large real aperture.

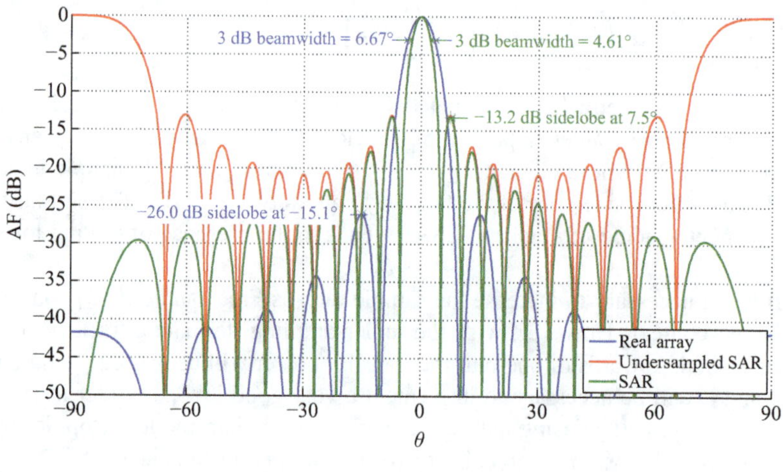

Figure 3.3 Real array and synthetic array two-way patterns
Pattern.m

DBS was invented by Wiley [5, 6] in 1951. SAR processing development was almost concurrent and somewhat independent. A conventional pulsed Doppler system provides the range and Doppler shift of all objects within its beam. DBS requires, like SAR, that the target is stationary and that the entire Doppler shift is attributed to the motion of the radar. With this assumption, the range-Doppler map has a unique correspondence with ground coordinates, permitting a transform from range-Doppler to ground coordinates, creating an image within the ground footprint of the beam from a single pulse or group of pulses. The resolution limits are determined by bandwidth (range resolution) and the number of Doppler bins (dependent on the signal-to-noise ratio). The lines of constant range and constant Doppler are not a Cartesian grid, so resolution varies across the scene. The resolution is better than the normal Rayleigh limit and may be considered a "super-resolution" technique. Because the signal magnitude decreases in the same proportion as resolution increases, additional transmit power or integration of multiple pulses is required to preserve the signal-to-noise ratio. Since the radar must be moving to effect the Doppler shift, additional pulses originate from equally spaced points along a path, thus creating the synthetic aperture.

There are two conceptual differences between the techniques: (1) SAR requires multiple samples (pulses), while DBS could, in principle, function with a single pulse and (2) DBS requires motion, while SAR could, in principle, function even if the antenna were stationary at each measurement point. The two techniques converge because DBS ordinarily uses multiple pulses and SAR data collection normally involves motion.

We might give the last word to Jakowatz *et al.* [7] whose treatise on SAR processing states somewhat acerbically:

> It's worth noting here that nowhere in our development of SAR image reconstruction using beamforming principles did the concept of "Doppler" ever appear. That is, of course, because neither beamforming nor SAR imaging has anything to do with the Doppler effect.

3.1.2 SAR and holography

SAR is sometimes compared to holography and there are many similarities. Hungarian-British scientist Gábor [8] invented holography in 1947 for which he received the Nobel prize in 1971. Both techniques use coherent illumination and reception of electromagnetic waves reflected by the area of interest to reconstruct the image scene with high fidelity. In radar parlance, the hologram is a continuous wave system, while SAR is usually a pulsed system. Both systems image objects which exist in three dimensions although SAR renders only a two-dimensional image.

The fundamental difference between the two techniques is that the hologram records a three-dimensional volume with all three dimensions encoded by phase. The SAR image is a two-dimensional slice of a three-dimensional volume with the cross-range dimension derived from phase and the range dimension derived from the round-trip time of a pulse. The slice is constructed in the slant plane and projected onto the surface (nominally planar) of interest which requires a priori knowledge of local topography to minimize errors. The projection is usually mapped into geographic coordinates (latitude,longitude, elevation), for instance. Gábor himself was very familiar with SAR developments by Leith and others at the University of Michigan and stated in his Nobel lecture that "This [SAR] was in fact two-dimensional holography with electro-magnetic waves. ... Their results were brilliant."

The near-field scanner (Section 3.2) is an example ot synthetic aperture techniques in two dimensions, so it is closely analogous to holography. The analogy is almost complete with the spherical near-field ranges which collect data across a hemisphere (see, for instance, Hansen [9]).

In the early years of SAR, images were produced using optical processing of data recorded on photographic film; Cutrona et al. [10] provide a good overview. This technique was necessary as the digital data recording and data processing techniques then available were not adequate. The optical technique constituted a holographic reconstruction from the photographic interference fringes.

3.2 Near-field scanning

A near-field scanner samples the radiation pattern with a single element which is translated across the surface of a plane, cylinder, or sphere. In fact, the surface can be arbitrary although the mathematics becomes more complex. It is similar to a hologram because it samples in two spatial dimensions.

A near-field scanner operates in the near field of the device under test (DUT) to measure the amplitude and phase of the electromagnetic field. The fields are sampled on a grid spacing of $\lambda/2$ or less over a region which encompasses the entire field, typically quite a bit larger than the DUT. These measurements may be either continuous wave or pulsed. The near-field measurement is extrapolated into the farfield of the DUT by numerical techniques. The farfield may be inverse Fourier transformed to yield an approximation of the field in the plane of the DUT for diagnostic purposes (failed elements, other gain/phase errors, etc.). The representation of the aperture plane RF distribution is often referred to as a hologram, although only loosely related to those described in Section 3.1.2. An example is shown in Figure 13.5. The resolution of the hologram is limited by the wavelength of measurement, so the unit cell is blurred. Rose [11] discusses accuracy and Johnson [12] discusses the near-field to farfield transformation.

Figure 3.4 illustrates a large near-field scanner manufactured by Near Field Systems. Figure 3.4a shows the mechanism responsible for positioning the probe accurately. Figure 3.4b shows the mechanism installed in a large anechoic chamber with the DUT, an AN/TPY-2 antenna. Table 3.1 summarizes some of the specifications for this system.

3.3 Interferometry

Interferometric SAR uses one or more radars, separated in space or time to provide rudimentary information about the third dimension, with a slightly reduced requirement for

(a) (b)

Figure 3.4 *Near-field range example. (a) Scanner mechanism and (b) Scanner and antenna in anechoic chamber*
©2009 NSI-MI [13], used with permission

a relatively flat scene. Cross-track interferometric SAR produces two slices from which altitude may be inferred. A cross-track interferometry might be termed 2.5-dimensional imaging because the height dimension is sampled very coarsely and has numerous aliases. The elevation ambiguities inherent in this approach constitute grating lobes in the elevation plane. Accordingly, it requires either a smoothly varying target surface or a priori knowledge of the surface topography. It may be considered a technique to form a very sparse (two-element) array in the elevation plane and it may require a priori knowledge of the surface profile to resolve altitude ambiguities.

Along-track interferometry images a target multiple times with a short time interval to detect moving targets in the scene. Displaced phase center antennas (DPCAs) were an early interferometric technique which used different portions of a single antenna to sample the target at a very short time interval to assess whether the target is moving.

Repeat pass interferometry images a target at different times from the same location to infer changes in the object topography. Repeat pass uses a single sensor following the same trajectory at multiple times, such as a satellite in a repeating orbit. They are useful to detect surface motion, land, ice, etc.

Table 3.1 *Near-field scanner capability*

Parameter	**Specification**
XY scanner size	7 m × 3.7 m
XY scanner weight	11,340 kg
Scan speed	0.25 m/s
Scanner planarity (rms)	0.1 mm
Scanner planarity – corrected rms	0.025 mm

Source: NSI-MI [13]

Radio astronomy interferometers provide enhanced angular resolution and tolerate grating lobes because the target field is very sparse.

References

[1] Guarnieri A.M., Hansen J.E., Jensen F. *Electromagnetic Imaging*. Politecnico di Milano, Dipartimentodi Elettronica, Informazione e Bioingegneria; 2015. Available from http://home.deib.polimi.it/monti/ftp/srl/SARI.pdf.

[2] Tsynkov S. 'The Doppler effect for SAR'. *2016 AFOSR Electromagnetics Contractors Meeting016 AFOSR Electromagnetics Contractors Meeting*. January 2016. Available from https://community.apan.org/wg/afosr/m/kathy/153015/download.

[3] Freeman A., Johnson W.T.K., Huneycutt B., *et al*. 'The "myth" of the minimum SAR antenna area constraint'. *IEEE Transactions on Geoscience and Remote Sensing*. 2000, vol. 38(1), pp. 320–4.

[4] Sullivan R. 'Synthetic aperture radar' in Skolnik M. (ed.). *Radar Handbook*. New York, NY, USA: McGraw-Hill Education; 2008. pp. 171–3.

[5] Wiley C.A. *Pulsed Doppler Radar Methods and Apparatus*. US 3,196,436; July 1965. Available from https://patents.google.com/patent/US3196436.

[6] Wiley C.A. 'Synthetic aperture radars'. *IEEE Transactions on Aerospace and Electronic System*. 1985, vol. AES-21(3), pp. 440–3.

[7] Jakowatz Jr. C.V., Wahl D.E., Yocky D.A. 'Beamforming as a foundation for spotlight-mode SAR image formation by backprojection' in Zelnio E.G., Garber F.D. (eds.). *Algorithms for Synthetic Aperture Radar Imagery XV*. 6970. International Society for Optics and Photonics. SPIE; 2008. pp. 192–206.

[8] Gabor D. 'Holography, 1948–1971'. *Science*. 1972, vol. 177(4046), pp. 299–313.

[9] Hansen J.E., Jensen F. 'Spherical near-field scanning at the technical University of Denmark'. *IEEE Transactions on Antennas and Propagation*. 1988, vol. 36(6), pp. 734–9.

[10] Cutrona L.J., Leith E.N., Porcello L.J., Vivian W.E. 'On the application of coherent optical processing techniques to synthetic-aperture radar'. *Proceedings of the IEEE*. 1966, vol. 54(8), pp. 1026–32.

[11] Rose C.A. 'Accuracy estimation of microwave holography from planar near-field measurements'. *Antenna Measurement Techniques Association*. 2000. Available from https://www.nsi-mi.com/images/TechnicalPapers/2000/Accuracy%20Estimation%20of%20Microwave%20Holography%20from%20Planar%20Near-Field%20Measurements.pdf.

[12] Johnson R.C., Ecker H.A., Hollis J.S. 'Determination of far-field antenna patterns from near-field measurements'. *Proceedings of the IEEE*. 1973, vol. 61(12), pp. 1668–94.

[13] NSI-MICustomer Installation. Available from http://www.nearfield.com/Sales/customerinstallations/Raytheon.htm.

Chapter 4
Antenna figures of merit

4.1 Beam shape, sidelobes, and nulls

An ESA design is a collection of sampled gain and phase states of its elements. There are numerous design techniques to achieve a desired result. In principle, any specific beam pattern has only one parametric realization. Accordingly, various design techniques striving for the same performance should generate similar designs.

This section will discuss optimization of the mainbeam and sidelobes. Improving one unfortunately tends to degrade the other. Mainbeam shaping is usually a matter of obtaining the desired beamwidth. Sidelobes are rarely beneficial as they are responsible for mainbeam gain reduction, small target masking, false target insertion, and other signal artifacts.

Most techniques for sidelobe reduction are applicable to both transmit and receive antenna patterns but often penalize the transmit efficiency. Since the system performance is a function of the product of the transmit and the receive antenna patterns, it is desirable to apply as much of the sidelobe mitigation as possible to the receive function.

4.1.1 Amplitude weighting

We may adjust gain and phase at each element to optimize performance, improving the mainbeam and reducing sidelobes according to particular criteria. This procedure is called weighting, taper, or apodization, a term used in optics to describe the same process with the same purpose. In transmit and receive, sidelobes may be reduced by reducing the power near the edge of the array. However, this reduces the effective size of the aperture and broadens the beam. Nonuniform weighting in transmit is problematic as element amplifiers operate near saturation and operation at reduced power degrades efficiency. In both modes, the reduction in effective area is termed aperture efficiency or aperture taper efficiency.

Weights are applied to individual elements in arrays or to entire apertures for continuous sources. Weights may be real numbers in which case they affect only the amplitude or complex in which case they affect both amplitude and phase. Weighting of arrays is straightforward; variable weights may be implemented using gain and phase control which also allows beam steering. Weighting techniques are discussed in more detail in Chapters 9 and 10.

The mainlobe peak gain is greatest with a uniform weighting. This condition is frequently used for transmit as it makes best use of available power. This leaves the receive mode with the task of reducing sidelobes as the two-way pattern (product of transmit and receive patterns) determines the system performance.

If we are dealing with a two-way pattern as, for instance, radar, we may choose a receive weighting different from the transmit weighting with the objective of aligning the minima of one with the maxima of the other to reduce two-way peak sidelobe levels.

Sidelobes arise from discontinuities (edges). Reflector antennas are usually under illuminated to reduce or eliminate edge effects. The same approach is readily achievable in ESAs. This technique has the effect of reducing the effective area of the antenna, which reduces gain and broadens the mainbeam. This flexibility comes at the cost of inefficient utilization of each radiating element, for instance, by radiating less than the full radiating power capability.

The gain loss may be quantified by aperture efficiency which compares the gain of the array to the sum of the gain of the individual elements (the uniform illumination case). A similar formula for reflector antennas evaluates the illumination of the reflector compared to uniform illumination.

Doerry [1] provides a comprehensive list and analysis of many weighting designs. Some of these will be discussed in Chapter 9. Doerry's report describes window functions applied to time series signals, which are one-dimensional (time). Our interest is two-dimensional, which adds some complexity although the general principles are the same. Lynch [2, p. 315] provides some examples of antenna weightings.

4.1.2 Peak sidelobe ratio

Sidelobes are ubiquitous in antenna patterns and almost always a nuisance. The problem relates to ambiguity in target detection and image processing. For target detection, large targets in sidelobes resemble average targets in the mainbeam and may mask small targets. For image processing, sidelobes may replicate the scene image offset from the actual scene at low image, producing confusing image artifacts. Additionally, sidelobes provide a path for noise to enter the system without commensurate signal. Much effort is devoted to suppressing them.

The computation of the PSLR is straightforward. It is the gain of the largest sidelobe compared to the gain of the mainlobe. Its value is –13.2 dB for the uniformly illuminated rectangular aperture and –17.6 dB for the uniformly illuminated round aperture. Typically smaller numbers are desired, in the range of –30 to –50 dB.

For antenna patterns without grating lobes, the peak sidelobe is commonly adjacent to the mainbeam. With grating lobes, however, the maximum sidelobe may occur at a considerable distance from the mainlobe.

4.1.3 Integrated sidelobe ratio

The ISLR is a more complex metric. It is defined as the total sidelobe power divided by the total mainlobe power. A similar metric might be the percent of total radiated power which is contained in the mainlobe which could be written as $\frac{1}{1+\text{ISLR}}$. For instance, an ISLR of −10 dB would equate to 91% in the mainlobe.

ISLR distributes the interfering or noise signal across the antenna field of view and may be more or less detrimental than PSLR.

Antenna patterns are two-dimensional, so the calculation is performed over solid angle, and 2π steradians in the case of a hemispherical pattern. The angular pattern may be sampled at equal intervals of θ in which case, to compute associated power in each sample, it is necessary to use a factor of $\sin\theta$ in the integral. If the angular pattern is sampled at equal intervals of $\sin\theta$, it is necessary to use a factor of $\frac{1}{\cos\theta}$ in the integral.

ISLR is an important metric for SAR performance as it introduces background noise in the image. ISLR used for SAR image quality usually assesses the two-way antenna pattern. SAR image quality is discussed at some length in Chapter 6 of Melvin and Scheer [3].

ISLR requires a boundary between the mainlobe and the sidelobe for its calculation. Martinez and Marchand [4] enumerate some definitions. In this book, I define the mainlobe as the smallest contour around the region of maximum gain which retains a compact shape and remains generally convex. For uniform illumination, the boundary is equivalent to the locus of the first nulls surrounding the mainlobe. This definition is the most generous to the mainlobe, so the ISLR values reported will be slightly better (lower) than values computed using other definitions. The computation uses sampled patterns, so the computed value will vary based on the sample interval. The calculation is implemented in a MATLAB® script *ISLR.m*.

Figure 4.1 shows an idealized array pattern with contours at 3 dB intervals to illustrate the ISLR definition. Figure 4.1a illustrates the candidate contours for a uniformly illuminated 16×16 element array with $\lambda/2$ element spacing. In this

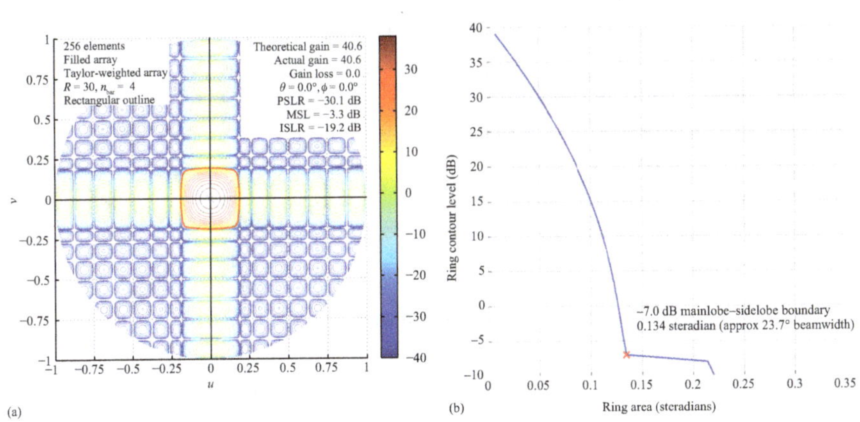

Figure 4.1 Mainbeam definition: (a) 16×16 element array contours mainbeam represented by thick red line and (b) central contour ring area as a function of contour level

ThinnedSquare.m

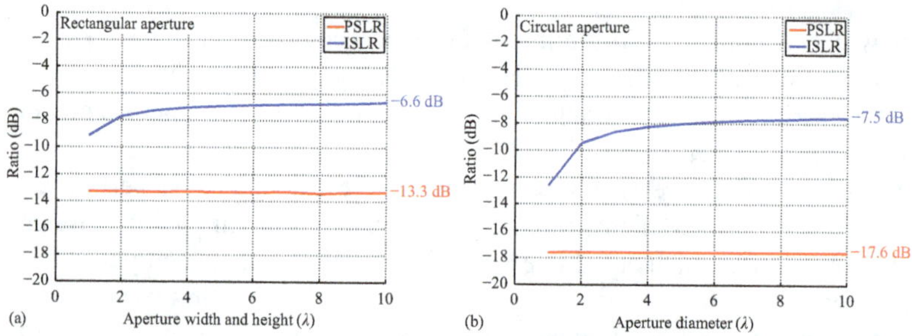

Figure 4.2 *One-way PSLR and ISLR as a function of aperture size: (a) rectangular aperture and (b) circular aperture*
Apertures.m

illustration, the contour is shown by a thick red line. It is taken at the –48 dB level and is roughly square with rounded corners. This choice is slightly smaller than the exact null position, which would yield a perfectly square contour. The error is negligible.

Figure 4.1b shows the relationship between the solid angle of the mainbeam corresponding to the contour level selected. The curve is essentially vertical at the selected level, demonstrating that lower contour levels would not materially affect the solid angle of the mainbeam.

An equivalent beamwidth is calculated based on the solid angle of a cone of the same solid angle. This beamwidth is roughly equivalent to the null-to-null beamwidth which is greater than the customary 3 or 10 dB beamwidth.

Small apertures will have broad mainbeams and relatively few sidelobes. Accordingly, the numerical ISLR will be lower than for a large aperture with most of its sidelobes in real space. Figure 4.2 shows the dependence of PSLR and ISLR on aperture size. PSLR is virtually constant, and the numerical values of ISLR approach an asymptote for large aperture size. The improved values for small apertures result from a smaller fraction of sidelobe power appearing in real space. When small elements are arrayed into larger structures, the distant sidelobes are suppressed.

Table 4.1 summarizes the one-way parameter for three instances of a 64×64 square array with elements spaced at λ/2. The additional instance utilizes Taylor thinning in transmit; this technique will be described in Section 8.3.

Table 4.1 *One-way array performance*

Weighting	Number	BW* (°)	PSLR (dB)	ISLR (dB)
Uniform filled	4,096	1.6/2.6	−13.3	–6.2
Taylor filled	4,096	2.0/3.5	–30.3	−19.7
Taylor thinned	960	2.4/4.4	–32.7	2.0

*Beamwidth shown for −3 and −10 dB levels.

4.1.4 Two-way patterns

Sidelobe ratios are commonly computed based on one-way (transmit or receive) antenna patterns. Figure 4.3 shows the gain as a function of direction cosines (u and v) for 64×64 element arrays with $\lambda/2$ spacing in the x and y directions. Figure 4.3a is based on uniform amplitude weighting and Figure 4.3b is based on Taylor weighting with R = 30 dB and n-bar = 4. The sidelobe ridges are located in the principal planes.

From a system point of view, notably radar, it is frequently the two-way pattern which is of interest. This would be the square of the one-way pattern when the transmit and receive patterns are the same or the product of the transmit and receive patterns where they are different. Antenna gain and peak sidelobe values are also the product of their one-way values but the ISLR, being the integral of the product over all angles, is not.

Figure 4.4 shows the two transmit–receive cases formed as the product of the one-way patterns in Figure 4.3. The gain and peak sidelobes are twice the one-way values. The mean sidelobe level (MSL) and ISLR do not have this simple relationship.

Table 4.1 summarizes array performance in either transmit or receive (one-way). The PSLR for thinned and unthinned apertures is about the same but the ISLR increases by 20 dB in the thinned case.

The two-way transmit and receive performance (PSLR and ISLR) is shown in Tables 4.2 and 4.3 for various pairings of the configurations in Table 4.1. The configuration used for transmit is shown in the row labels and the configuration used for receive is shown in the column labels. The difference between the filled and thinned ISLR and PSLR is between 5 and 10 dB.

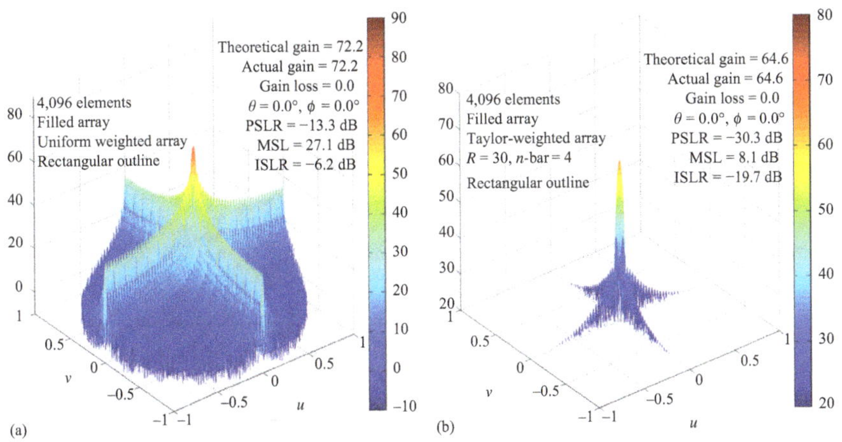

Figure 4.3 One-way 64 × 64 element array factors: (a) uniform weighted array and (b) Taylor-weighted array

ThinnedSquare.m

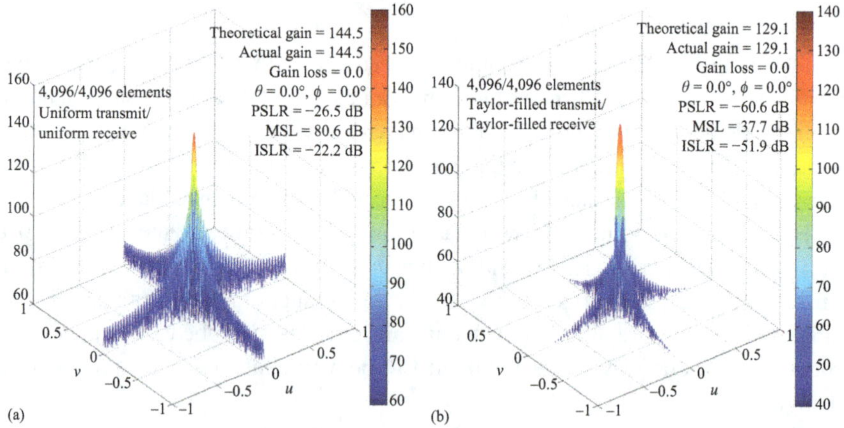

Figure 4.4 *Two-way 64 × 64 element array factors: (a) uniform weighted transmit and receive array and (b) Taylor-weighted transmit and receive array*

ThinnedSquare.m

Table 4.2 Two-way array PSLR (dB)

	Receive technique	
Weighting	**Uniform filled**	**Taylor filled**
Transmit technique		
Uniform filled	–26.5	–33.8
Taylor filled	–	–60.6
Taylor thinned	–	–50.9

Table 4.3 Two-way array ISLR (dB)

	Receive technique	
Weighting	**Uniform filled**	**Taylor filled**
Transmit technique		
Uniform filled	–22.2	–30.4
Taylor filled	–	–51.9
Taylor thinned	–	–46.7

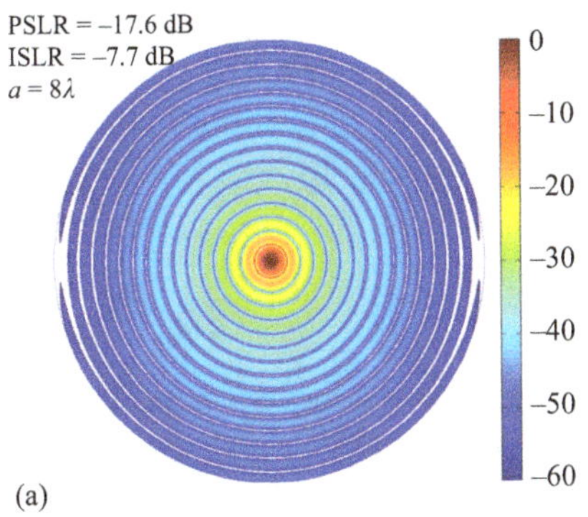

Figure 4.5 *Circular array sidelobes: (a) one-way performance and (b) two-way performance*
Apertures.m

Circular arrays

Section 4.1.3 described rectangular apertures. We may also consider circular apertures as shown in Figure 4.5. The one-way PSLR of the circular aperture is about 4 dB better than a rectangular array, while the ISLR is only about 1.5 dB better. The two-way PSLR of the circular aperture is about 9 dB better, while the ISLR is 4 dB better.

Bistatic case

Finally, it is of interest to consider a bistatic case, used by the Space Fence, described in Section 1.4.1.2. This provides an additional degree of freedom for system optimization. Figure 4.6 illustrates the concept. The example is unweighted for both

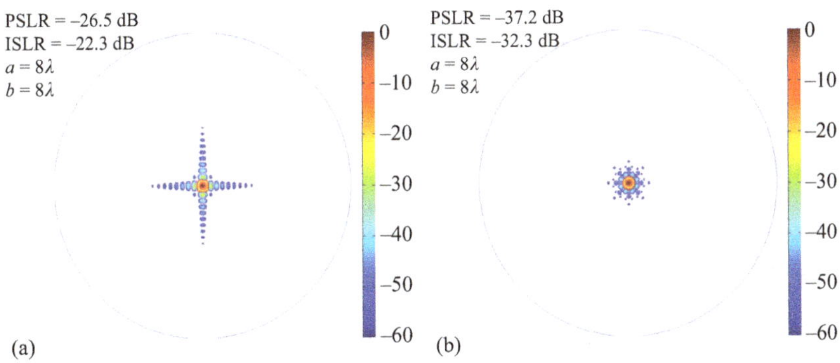

Figure 4.6 *Square array sidelobes without and with rotation: (a) receive array aligned with transmit and (b) receive array rotated 45° to transmit array*
Apertures.m

transmit and receive. Since sidelobes are highest in the principal plane, for a two-way performance improvement, the two arrays may have a relative rotation, in this case 45°. This provides a 10 dB improvement in both PSLR and ISLR compared to the nonrotated case as well as better than the circular aperture case.

4.2 Efficiency and scattering

An ideal antenna converts 100% between free space and the antenna port but real antennas reflect and scatter some energy, so not all the power supplied to an antenna element is utilized productively. Efficiency denotes the ratio of radiated power to input power (transmit) or conversely the percentage of power incident on the array that is collected and provided to the receiver port. The element may be regarded as a two-port device: one port being the connector and the other port being free space. The return losses (S_{11}/S_{22}) represent the voltage rejected and reflected back to the source, while the transmission terms, S_{21} or S_{12}, represent the voltage passed to the other port. As we are generally interested in power, the relevant parameter is the square of the scattering parameter, e.g., $|S_{12}|^2$. This parameter is frequently reported in decibels.

In transmit mode, some power is reflected back to the source, and in receive mode, some power is reradiated. This is termed element efficiency and is characterized by reflection coefficients including the voltage standing wave ratio (VSWR), Γ, and S parameters. Signal losses are caused by impedance mismatch, characterized by "return loss" or loss caused by return (reflection of energy).

Transmit power is either radiated or reflected back to the amplifier. This may be measured at the antenna port with a directional coupler. Receive power is either captured by the antenna or reflected back to free space. The scattered energy is more difficult to measure as it is distributed in space. This is a case where reciprocity is often invoked.

In both transmit and receive, stray power may be coupled into adjacent elements and reradiated or passed to the antenna port. This process is termed "mutual coupling" and discussed further in Chapter 5. The radiator scattering parameters S_{11} and S_{22} quantify this reflection.

Efficiency and scattering are complementary. Efficiency is the fraction of incident power that is absorbed by the antenna; the remainder is scattered. Poor efficiency degrades signal-to-noise ratio (SNR) and radiated power. Additionally, the lost energy may interfere with active components and can interfere destructively with desired signal. Besides degrading system performance, it increases the radar cross section of the system.

Hannan [5] demonstrates that a phased-array antenna can be impedance matched for all scan angles using matching circuits to improve the match of each element to its power source.

The optical analog of 100% efficiency is a black body which is perfectly absorptive (and emissive) at all frequencies/wavelengths. Pozar [6] provides an example of an array with 100% efficiency. Kwon and Pozar [7] provide an analysis without invoking reciprocity. Best and Kaanta [8] provide a comprehensive overview in "A tutorial on the receiving and scattering properties of antennas."

4.2.1 Aperture (utilization) efficiency

Aperture efficiency is another efficiency parameter, conceptually distinct from element efficiency, as it represents an artificial loss resulting from nonuniform aperture illumination. The nonuniformity may be inadvertent, as is the case with reflector antennas or waveguides, or it may be deliberate as is the case with aperture weighting or thinning. It is nominally unity or 100% with each element equally weighted. A number of beamforming techniques utilize aperture weighting (reduce the power transmitted or received by certain elements), in order to improve sidelobes or satisfy other criteria.

Aperture efficiency was originally associated with reflector antennas because the feed pattern on the reflector does not illuminate reflector evenly and may be smaller or larger than the reflector extent. The illumination is usually tapered (reduced) so that at the edge of the reflector, the illumination is 10% or less than at the center. The feed taper reduces edge illumination which in turn reduces sidelobes.

Arrays have no intrinsic aperture efficiency loss. They are capable of capturing all of the power but amplitude weighting will incur some efficiency loss. Aperture efficiency is readily calculated for array antennas according to (4.1). Hopkins *et al.* [9] discuss the aperture efficiency of amplitude weighting for arrays:

$$\mathrm{ATE} = \frac{\left(\sum_{n} |A_n|\right)^2}{N\sum_{n} |A_n|^2}$$

or regular two-dimensional lattices (4.1)

$$\mathrm{ATE} = \frac{\left(\sum_{m,n} |A_{mn}|\right)^2}{M\sum_{m,n} |A_{mn}|^2}$$

If the aperture weights are separable, then the aperture efficiency is also separable so that

$$\mathrm{ATE} = \frac{\left(\sum_{m} |a_m|\right)^2}{M\sum_{m} |a_m|^2}\frac{\left(\sum_{n} |b_n|\right)^2}{N\sum_{n} |b_n|^2} \tag{4.2}$$

where we make the substitution

$$A_{mn} = a_m b_n$$

4.3 Resolution and bandwidth

Bandwidth determines range resolution in radars and information capacity in communication systems.

There are several factors affecting bandwidth. The analog front ends must be capable of passing the bandwidth of interest without excess noise or distortion and the digital back ends must deal with the large data rates produced by the analog-to-digital converters. The constituent components have the same requirement but as the system response is the cascade of the component responses, each component has to be considerably better than the system requirement. For the transmit path, this is a particular challenge as the high-power components are more efficient in the nonlinear (saturated) signal range but this produces spurious signals which may be in-band of the system. Bandwidth is usually established by the frequency response of its components which may include bandpass filters.

Besides the usual bandwidth limits of electronic circuits, ESAs deal with beam squint caused by phase shift steering described in Section 2.5.2 and phase shifter circuit bandwidth described in Section 14.5.

Range resolution ΔR is related to bandwidth Δf according to the relation $\Delta R = 2c/\Delta f$. In the case of SAR, there is an additional factor of $1/\sin\,\theta$, where θ is the incidence angle to convert the slant plane range resolution to the ground plane resolution. A variety of waveforms are used but a linear FM chirp is common because its implementation is straightforward and it has good Doppler properties.

Angle measurement is directly related to antenna (aperture) size, for both real and synthetic apertures. Angle measurement is determined by beamwidth which is in turn a function of the antenna width. Synthetic apertures are formed by combining multiple physical apertures coherently (preserving phase). Synthetic apertures provide nice images because the target area is generally planar, so range-angle measurements may be transformed to Cartesian coordinate systems. When the topography is severe, this transformation is compromised.

4.4 Noise and dynamic range

4.4.1 Noise figure

An antenna with a single element has a straightforward calculation of noise figure. An ESA with multiple elements must sum noise contributions either coherently or noncoherently. In the SAR case, coherent noise may arise from spurious targets in sidelobes for which the SNR cannot be improved with longer integration times. Thermal noise is the classic incoherent source and the SNR improves with longer integration times.

The dynamic range calculation for an ESA in receive mode is complex because the signal is coherently summed through multiple stages. A very strong target return may not affect the many low-noise amplifiers (LNAs) but once their output is summed in the beamforming network, it may cause nonlinear distortion or saturation of subsequent amplifiers. This is a classic problem in radars with large variation in target range but it is slightly more complicated to handle in ESAs with their multiple gain stages. In transmit mode, the signal levels fall in a narrower range, so the amplifier effects are less of an issue.

4.4.2 Array noise figure model

This section follows the paper by Lee [10]. Figure 4.7 shows an equivalent circuit for the ESA receive circuit.

Total noise output is

$$n_0 = \sum_{k=1}^{N} n_k \tag{4.3}$$

where

$$\begin{aligned} n_k = &\frac{kT_iB_g}{L_fL_dL_k} + \frac{kT_0B(L_f - 1)g}{L_fL_dL_k} + \frac{kT_0B(F-1)g}{L_dL_k} \\ &+ \frac{kT_0B(L_d - 1)}{L_dL_k} + \frac{kT_0B(L_k - 1)}{L_k} \end{aligned} \tag{4.4}$$

Total signal output is

$$S_0 = \frac{s_ig}{L_fL_d}\left(\sum_{k=1}^{N}\sqrt{1/L_k}\right)^2 \tag{4.5}$$

where s_i is the input signal power from each element. Accordingly

$$\frac{S_0}{n_0} = \frac{Ns_i}{kTB}\frac{\left(\sum_{k=1}^{N}\sqrt{1/L_k}\right)^2}{N\sum_{k=1}^{N}(1/L_k)} \tag{4.6}$$

Define efficiency η:

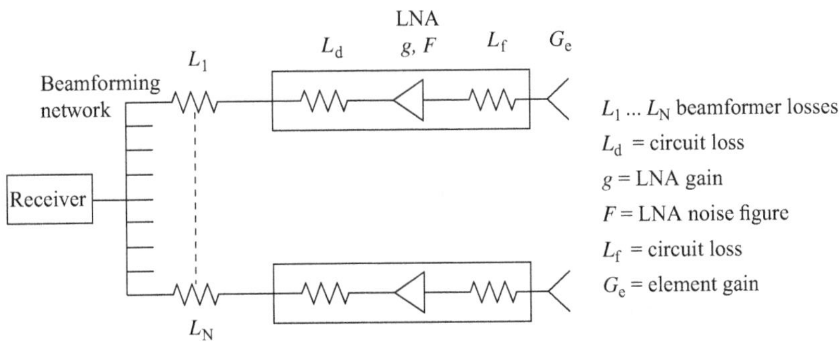

Figure 4.7 Noise figure model
After Lee [10]

$$\eta = \frac{\left(\sum_{k=1}^{N} \sqrt{1/L_k}\right)^2}{N \sum_{k=1}^{N} (1/L_k)} = \frac{\left(\sum_{k=1}^{N} a_k\right)^2}{N \sum_{k=1}^{N} a_k{}^2} \tag{4.7}$$

where a_k represents the amplitude weighting of each element.

Assuming an incident signal of power level P

$$S_{\mathrm{i}} = A_{\mathrm{e}} P \tag{4.8}$$

$$G_{\mathrm{e}} = \frac{4\pi A_{\mathrm{e}}}{\lambda^2} \tag{4.9}$$

$$s_{\mathrm{i}} = \frac{G_{\mathrm{e}} P \lambda^2}{4\pi} \tag{4.10}$$

$$G = \eta N G_{\mathrm{e}} \tag{4.11}$$

$$\frac{S_0}{n_0} = \frac{P\lambda^2 G}{4\pi kTB} \tag{4.12}$$

$$\frac{G}{T} = \frac{NG_{\mathrm{e}}\eta}{T_{\mathrm{i}} + T_0 \left(L_{\mathrm{f}}F - 1 - L_{\mathrm{f}}/g + NL_{\mathrm{f}}L_{\mathrm{d}}/\left(g\sum_{k=1}^{N} 1/L_k\right)\right)} \tag{4.13}$$

4.4.3 *Effect of aperture taper*

If L_k are identical, define

$$L_{\mathrm{c}} = N\left(\sum_{k=1}^{N} 1/L_k\right) \tag{4.14}$$

With no taper, $\eta = 1$ and G/T is

$$\frac{G}{T} = \frac{NG_{\mathrm{e}}}{T_{\mathrm{i}} + T_0(L_{\mathrm{f}}F - 1 - L_{\mathrm{f}}/g + L_{\mathrm{f}}L_{\mathrm{d}}L_{\mathrm{c}}/g)} \tag{4.15}$$

If L_k are equal to 1 (lossless taper), G/T is

$$\frac{G}{T} = \frac{NG_{\mathrm{e}}\eta}{T_{\mathrm{i}} + T_0(L_{\mathrm{f}}F - 1 - L_{\mathrm{f}}/g + L_{\mathrm{f}}L_{\mathrm{d}}/g)} \tag{4.16}$$

and the net effect on noise figure is confined to the aperture efficiency η.

4.4.4 *Dynamic range (third-order intercept)*

The dynamic range is simply the ratio of the maximum signal power to the minimum signal power and is represented in decibels. The minimum signal power will usually be the device noise. The maximum signal power is one of several things. All amplifiers will saturate with a peak-to-peak signal approximating the supply voltage, creating many spurious signals, so a lower signal level will usually be used. One metric is the harmonic distortion which results from the nonlinear transfer curve of the device. A second metric is the third-order intercept (TOI) which represents intermodulation products of two signals. It is called third order because it is the result of three terms. With two input signals, the intermodulation signal represents the difference of one frequency doubled and the other frequency: this combination represents an in-band signal. The second-order term which is simply the sum or difference of the two signals is generally guaranteed to be out of band and not of interest. The third-order intermodulation product has a magnitude proportional to the sum of the signal levels and so it increases three times as fast as any individual signal. These values are device specific and may be improved by operating the devices at higher voltages and powers, which unfortunately is a problem for power and heat. System needs will establish the limits on these parameters.

Holzman [11] discusses an example of this calculation.

4.5 System performance equations

4.5.1 SAR equation (NESZ)

Noise-equivalent sigma zero (NESZ) is a SAR figure of merit and represents the noise floor of the SAR image. Its calculation is shown in (4.17). The equation is based on Curlander and McDonough [12] with the substitution $L = 2V/\mathrm{PRF}$, i.e., two samples per antenna length. NESZ is sometimes termed sigma-naught or σ_0.

It represents the "darkest" area discernible in an image. NESZ is inversely proportional to the PG-squared term as well as the cube of the range. The range relationship differs from the radar range equation because the required integration time in SAR is proportional to range.

A water surface generally functions as a mirror for radio frequency so it has a very low monostatic cross section. Accordingly, water areas in the scene will display the noise floor directly. NESZ sets the lower bound on the dynamic range of the image and the upper bound is the radar cross section (RCS) of the largest object in the scene. The difference of the two scene contrast which determines image interpretability.

Performance of some SAR systems is shown in Table 4.4.

$$\mathrm{NESZ}(\sigma_0) = \left(\frac{4\pi r}{\lambda}\right)^3 \frac{2L\sin\theta_i}{PG_\mathrm{t}G_\mathrm{r}c\tau_\mathrm{pd}} \frac{k_B TB}{\eta_\mathrm{prop}\eta_\mathrm{sys}} \tag{4.17}$$

where

σ_0 = backscattering cross section
P = (peak) transmitted power
G_t and G_r = transmit and receive gains
c = speed of light
τ_pd = pulse width
λ = radar wavelength
r_i = (slant) range
k_B = Boltzmann constant

Table 4.4 System examples of NESZ

	Exponent	Units	Value	COSMO–SkyMed	Value	TerraSAR–X	Value	TAS Micro–Satellite
Satellite mass		kg		1,740		1,230		120 – 150
Altitude		km		620		515		300
Satellite velocity		km/s		7.56		7.61		7.73
Nyquist sampling rate		Hz		2,651		3,183		7,734
Incidence angle (extended)		°		20 – 60		20 – 55		
Incidence angle (normal)		°		25 – 50		20 – 45		20 – 50
Range (extended)		km		656 – 1,095		545 – 838		
Range (normal)		km		677 – 910		545 – 703		318 – 453
Antenna size (length × width)		m		5.7 × 1.4		4.78 × 0.7		2 × 0.4
System PRF (min, max)		Hz		2,905 – 3,874		3,000 – 6,500		10,000
Frequency		GHz		9.6		9.65		9.6
Wavelength		m		0.031		0.031		0.031
Antenna gain		Dimensionless		102,829		43,852		10,309
Tx duty factor (min, max)		Dimensionless		15% – 30%		13% – 20%		10%
Pulse width		s		0.000052		0.000043		0.00001
Number of T/R modules				1,280		384		TWT
Peak power per T/R module		W		~6 (~7)		~6		
Transmit power		kW		5		2.26		2.5
Bandwidth		MHz		300		300		300
	Exponent	**Units**	**Value**	**dB**	**Value**	**dB**	**Value**	**dB**
4π	3	Dimensionless	12.57	33	12.57	33	12.57	33
Boresight range (r)	3	m	784,900	177	618,912	174	379,596	167
Wavelength (λ)	–3	m	0.03	45	0.03	45	0.03	45
2	1	Dimensionless	2	3	2	3	2	3
Antenna length (L)	1	m	5.7	8	4.78	7	2	3
Sine of incidence angle $\sin(\theta_i)$	1	Dimensionless	0.58	–2	0.5	–3	0.52	–3
Boltzmann constant (k_B)	1	$m^2\ kg/s^{-2}\ K^{-1}$	1.38E–23	–229	1.38E–23	–229	1.38E–23	–229
Temperature (T)	1	K	300	25	300	25	300	25
Signal bandwidth (B)	1	MHz	300	85	300	85	300	85
Peak transmit power (P)	–1	W (kg m^2/s^{-3})	5,000	–37	2,260	–34	2,500	–34
Antenna transmit gain (G_t)	–1	Dimensionless	102,829	–50	43,852	–46	10,309	–40
Antenna receive gain (G_r)	–1	Dimensionless	102,829	–50	43,852	–46	10,309	–40
Speed of light (c)	–1	m/s	299,792,458	–85	299,792,458	–85	299,792,458	–85
Transmit pulse width (τ)	–1	s	0.000052	43	0.000043	44	0.00001	50
System NESZ (does not include system losses)				–35		–27.8		–19.3

B = bandwidth
θ_i = incidence angle
η_s (<1) = propagation and system losses
L = antenna length (= 2*Vs/PRF)

4.5.2 Radar range equation

The radar range equation, shown in (4.18), is well known. SNR is proportional to the inverse of the fourth power of range. Radar range is determined by the antenna size (area), transmit power, receive noise figure, and bandwidth.

One of the benefits of the ESA is the antenna gain, and pattern can be reconfigured between transmit and receive. This enables use of techniques which may prefer one or the other such as beam spoiling on transmit and beam scanning on receive or uniform weighting for transmit and tapered weighting for receive:

$$\mathrm{SNR} = \frac{P_t G^2 \lambda^2 \sigma}{(4\pi)^3 k_B T_e BFLR^4} \tag{4.18}$$

where

P_t = transmit power
G = antenna gain
λ = wavelength
σ= target cross section
k_B = Boltzmann constant
T = system temperature
B = system bandwidth
F = system noise figure
L = system losses
R = range to target

4.5.3 Friis transmission equation

The Friis equation shown in (4.19) is the fundamental communications metric. It is proportional to the inverse of the range squared because it is a one-way path length. It uses separate transmit and receive antennas and shows the gain of each explicitly:

$$\frac{P_r}{P_t} = G_t G_r \left(\frac{\lambda}{4\pi R}\right)^2 \tag{4.19}$$

where

P_r = received power
P_t = transmitted power
G_t = transmit antenna gain
G_r = receive antenna gain

References

[1] Doerry A.W. *Catalog of Window Taper Functions for Sidelobe Control*. Tech. Rep. SAND2017-4042. Prepared for the Advanced Research Projects Agency. Albuquerque, NM, USA: Sandia National Laboratories; 2017. Available from https://prod-ng.sandia.gov/techlib-noauth/accesscontrol.cgi/2017/174042.pdf.

[2] Lynch Jr D. *Introduction to RF Stealth*. Edison, NJ, USA: SciTech Publishing, an imprint of the IET; 2004.

[3] Melvin W.L., Scheer J.A. *Principles of Modern Radar*. Edison, NJ, USA: SciTech Publishing, an imprint of the IET; 2013.

[4] Martinez A., Marchand J.L. *SAR Image Quality Assessment*. Revista de Teledeteccion. 2 November 1993. Available from http://www.aet.org.es/revistas/revista2/AET22.pdf.

[5] Hannan P.W. 'Proof that a phased-array antenna can be impedance matched for all scan angles'. *Radio Science*. 1967, vol. 2(3), pp. 361–9.

[6] Pozar D. 'Scattered and absorbed powers in receiving antennas'. *IEEE Antennas and Propagation Magazine*. 2004, vol. 46(1), pp. 144–5.

[7] Kwon D.H., Pozar D.M. 'Analysis of maximum received power by arbitrary lossless arrays'. *2009 IEEE Antennas and Propagation Society International Symposium*. 2009, pp. 1–4.

[8] Best S.R., Kaanta B.C. 'A tutorial on the receiving and scattering properties of antennas'. *IEEE Antennas and Propagation Magazine*. 2009, vol. 51(5), pp. 26–37.

[9] Hopkins G.D., *et al*. 'Aperture efficiency of amplitude weighting distributions for array antennas'. *2007 IEEE Aerospace Conference*. 2007, pp. 1–9.

[10] Lee J.J. 'G/T and noise figure of active array antennas'. *IEEE Transactions on Antennas and Propagation*. 1993, vol. 41(2), pp. 241–4.

[11] Holzman E.L. 'Intercept points of active phased array antennas'. *1996 IEEE MTT-S International Microwave Symposium Digest*. 1996, vol. 2, pp. 999–1002.

[12] Curlander J.C., McDonough R.N. *Synthetic Aperture Radar: Systems and Signal Processing*. Hoboken, NJ, USA: Wiley; 1991.

Chapter 5
Mutual coupling effects

The analysis in Chapter 2 assumes that the element pattern and the array pattern may be separated and that the element pattern is independent of the array. This assumption is not correct, in general, and the interaction may be significant when elements are closely spaced. The interaction is termed mutual coupling or mutual impedance. There is a large body of research on this subject. The ESA design process must assess the interaction and mitigate or compensate it as necessary.

Mutual coupling is not an independent phenomenon. It describes the effects of incomplete boundary condition specification in the original problem. It operates at short-range and is not always apparent. In a very large array, the effect of the coupling for most elements is identical and its effect is incorporated into the element pattern leaving the array factor unchanged.

The treatment of mutual coupling in this section is qualitative as analytic techniques have not been very effective. It is difficult to measure in physical arrays. Its practical consequences have diminished as improved electromagnetic models and solvers deal with its effects directly with suitable boundary conditions.

Mutual coupling may introduce significant errors and uncertainties in the analysis. Of most concern, it may create unexpected nulls in the antenna pattern (scan blindness), but it also affects the element pattern and the array pattern. The following circumstances generally induce significant mutual coupling:

- Close lattice spacing
- Wide scan angle
- Wide bandwidth

At half wavelength spacing, it is problematic.

It is important to evaluate mutual coupling by various means including simulation and subscale tests. Frequently it is necessary to build test articles before accepting the design. The array size is a consideration: small arrays may be modeled in their entirety with numerical methods; large arrays may be partitioned or analyzed as infinite in extent.

5.1 Definition

Figure 5.1a illustrates the concept for a linear array. This arrangement is the same as that used for measurement of the embedded element pattern. A single element E_1 is provided with a transmit signal which it radiates. Some of the radiated power is coupled into adjacent elements. The amount of power coupled diminishes significantly with distance. In practice, all elements would be excited, and superposition provides the answer so long as the arrays operate in a linear fashion. There is no analogous test case for receive mode and it is generally assumed that reciprocity will suffice. The implicit assumption is that the results will be valid for the receive case. The principal difficulty in this illustration is that there is no insight into the ultimate disposition of scattered power, whether to free space or to the beam port of the antenna. Recent work has developed new models to treat receive as a separate case. Hui [1] identified some of the differences in 2004. Henault and Antar [2] provide a general framework for analysis.

Figure 5.1b and c represents a more realistic case of a square array which proliferates the coupling pairs. In this case, it is expedient to limit analysis to nearby elements as the more distant elements do not contribute measurably to the solution.

Mutual coupling may be characterized by a scattering matrix which is the ratio of scattered voltage to incident voltage; accordingly, it is a dimensionless quantity. The scattering parameter is sometimes reported as the squared magnitude, i.e., $|S_{12}|^2$, which approximates the power ratio. This is the approach used by Pozar [3] who considers mutual coupling in terms of scattering parameters.

Another characterization is mutual impedance representing induced voltage as a function of the transmitting element current. Its units are ohms. There is a direct relationship between mutual impedance and mutual coupling shown in (5.1). Orfanidis [4] shows the derivation in (14.3.5). The calculation of the mutual coupling requires knowledge of the impedance of both elements:

$$S_{12} = \sqrt{\frac{Z_2}{Z_1}} \frac{2\, Z_{12}\, Z_1}{(Z_{11} + Z_1)(Z_{22} + Z_2) - Z_{12}\, Z_{21}} \tag{5.1}$$

where

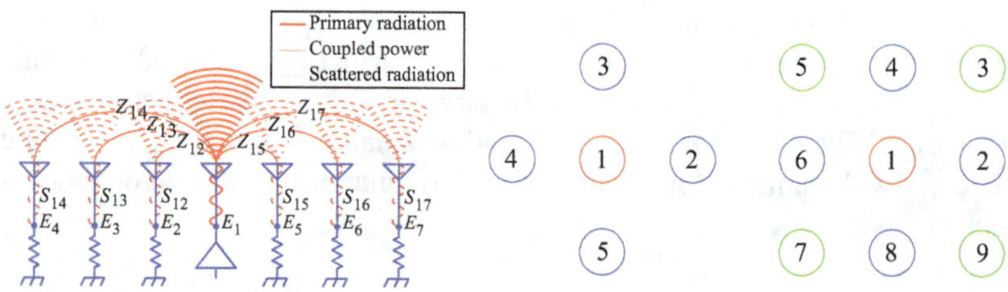

(a) Mutual coupling and mutual impedance (b) Array with four nearest neighbours (c) Array with eight nearest neighbours

Figure 5.1 Subarray illustration

$$
\begin{aligned}
Z_1 &= \text{load impedance of antenna-1} \\
Z_2 &= \text{load impedance of antenna-2} \\
Z_{11} &= \text{self-impedance of antenna-1} \\
Z_{22} &= \text{self-impedance of antenna-2} \\
Z_{12} = Z_{21} &= \text{mutual impedance between antennas 1 and 2}
\end{aligned}
\tag{5.2}
$$

Orfanidis [5] provides the following definition in Section 25.3.1:

> If antenna-1 is driven and antenna-2 is open-circuited, the near field generated by the current on antenna-1 will cause an open-circuit voltage, say $V_{21,\,oc}$ on antenna-2. The mutual impedance of antenna-2 due to antenna-1 is defined to be:
>
> $$Z_{21} = \frac{V_{21,\,oc}}{I_1} \tag{5.3}$$
>
> where I_1 is the input current on antenna-1.

Both approaches represent the pair of elements as a two-port network, much like a communications link characterized by the Friis equation (4.19). This analytical construct is incomplete for our purposes. In the transmit instance, the transferred power will be partitioned between the load and free space radiation. The power absorbed by the antenna element load is the quantity measured and represented by S_{12} or Z_{12}, but the scattered (reradiated) radiation is not commonly or conveniently measured. In the receive instance, the source is external to the array, so there is no convenient measurement port to measure the source. In consequence, modeling and simulation are the best approaches.

5.2 Problem

Mutual coupling can lead to undesirable effects such as "scan blindness" where the antenna pattern has an unexpected null. It also alters the load impedance presented to the amplifier by the element which will affect transmitted or received power. The effect is scan angle dependent and difficult to compensate with normal circuit methods.

Mutual coupling effects may perturb array patterns significantly when element spacing is less than one wavelength and can be dominant for element spacing less than one-half wavelength. The effect is to remove power from each element and reradiate it with some phase delay from adjacent elements. This is analogous to an overlapped subarray which is discussed in Section 2.7.2. As the gain and phase relationship is uncontrolled, in general, the effects are not likely to be advantageous. This is problematic because the effect is scan dependent and may reduce gain unexpectedly.

A useful technique to assess the interaction is to evaluate the element pattern in the array environment. This takes the form of simulating or measuring the element in an array in which the other elements are terminated by the characteristic impedance of the element. This methodology is inaccurate to the extent that the elements in the desired array are not, in general, terminated perfectly. They are connected to active circuitry which is not resistive and there will be significant reflections from the impedance mismatch. Pozar [3, 6] discusses this approach.

5.3 Parametric analysis

Mutual coupling is described as a single parameter relating the signal at the central element and at each of the surrounding elements. It may be understood as the power originating at the central element and traveling to the surrounding element with some attenuation and phase shift as a function of distance from the central element, with element details as secondary considerations. Considerable early work was published on determining the attenuation factor.

Several authors have tried to derive parametric relationships between element separation and mutual coupling. Hannan [7] reports that "if the elements are electrically small dipoles, the decay of mutual impedance at large distances is known to be $1/r$ in the H-plane and $1/r^2$ in the E-plane." This parametric relationship is represented by (5.4):

$$Z_{\text{E-plane}} = \frac{\exp(-jkr)}{r^2}$$
$$Z_{\text{H-plane}} = \frac{\exp(-jkr)}{r} \tag{5.4}$$

Galindo and Wu [8], in "On the asymptotic decay of coupling for infinite phased arrays," report the coupling shown in (5.5):

$$S \propto \exp(-jkr)/r^2 \tag{5.5}$$

Scattering and mutual impedance are each a single number representing the interaction of element pairs. In an array with n elements, they are $n \times n$ matrices. We may evaluate arrays using a hypothetical formulation in order to assess possible performance. Sections 5.4–5.6 describe some experimental and theoretical data. In this section, we consider a simple mutual coupling model where the magnitude of the coupling is inversely proportional to distance squared and the phase of the coupling is proportional to the distance. The result is a subarray pattern, sometimes termed an embedded element pattern, which may be used for associated array analysis.

The difficulty is that some of the power radiated by an array element does not launch into free space but instead interacts with neighboring elements which intercept that power and either absorb or reradiate it. To illustrate this point, consider the 5 element array in Figure 5.1b with a square lattice and $\lambda/2$ spacing. For an accurate analysis, it is necessary to include additional neighboring elements but this simple example is sufficient to illustrate the concept.

We assume equal scattering values of $S_{12}, \ldots, S_{15}$. Accordingly, we may represent the scattered electric field:

$$E_n = S_{1n}\, E_1 \tag{5.6}$$

The case where $E_{2,\ldots,5} = E_1/4$ is illustrative. The mutual coupling corresponding to this case is determined as follows. We stipulate that the total radiated power of all five elements is equal to 1 (in arbitrary units). This provides two equations in two unknowns shown in (5.7):

$$\begin{aligned} &\sum_{n=1}^{5} E_n^2 = 1 \\ &E_n = \frac{E_1}{4}, \qquad \text{where } n = 2, \ldots, 5 \end{aligned} \tag{5.7}$$

The solution to (5.7) is as follows:

$$\begin{aligned} E_1 &= \sqrt{\frac{16}{20}} \\ E_n &= \sqrt{\frac{1}{20}}, \qquad \text{where } n = 2, \ldots, 5 \end{aligned} \tag{5.8}$$

and the corresponding powers, in dB, are as follows:

$$\begin{aligned} &10\log(E_1^2) = -0.969 \text{ dB} \\ &10\log(E_n^2) = -13.01 \text{ dB} \end{aligned} \tag{5.9}$$

There is also a phase delay associated with the scattering. Assuming an element spacing of $\lambda/2$, the phase delay is on the order of 180° between the center element and each of the four other elements.

We may use these data to compute the pattern of the 5 element subarray and compare it to the pattern of the single central element without interaction. We assume that for each element, its radiated power angular dependence corresponds to the element pattern. Figures 5.2 and 5.3 illustrate the results. The blue line represents the single-element pattern; the red line represents the array factor resulting from the redistribution of power from the central element to the four adjacent elements; and the green line represents the resulting pattern which is the product of the element factor and the array factor.

One-half of the total power is radiated by the central element and the remainder by the four nearest neighbors. The scattered half of the power is exactly out of phase with the central element which results in a null in the gain pattern at boresight and peaks at about ±22° off boresight as illustrated in Figure 5.2a. Figure 5.2b with an element spacing of 0.51λ shows that the null fills in with small changes to element spacing. This 2% change in the element spacing also occurs with a 2% change in frequency, so the effect of mutual coupling is sensitive to frequency.

The boresight null has disappeared in Figure 5.3a and b with element spacings of 0.75λ and λ, respectively, primarily because the scattered contribution is more

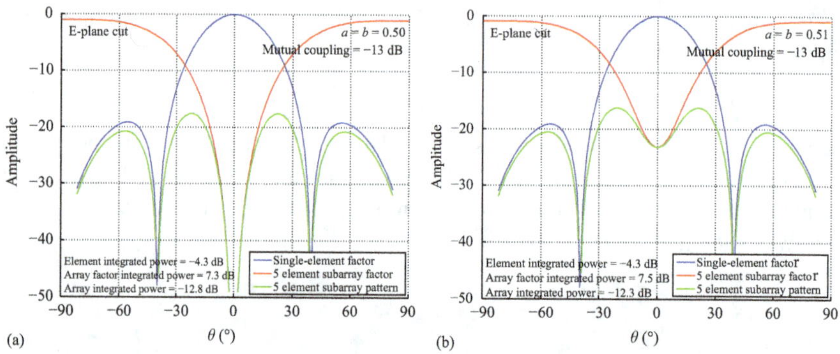

Figure 5.2 Cardinal plane cut of patterns with element spacing close to one-half wavelength: (a) spacing equals λ/2 and (b) spacing equals 0.51λ
SubArray.m.

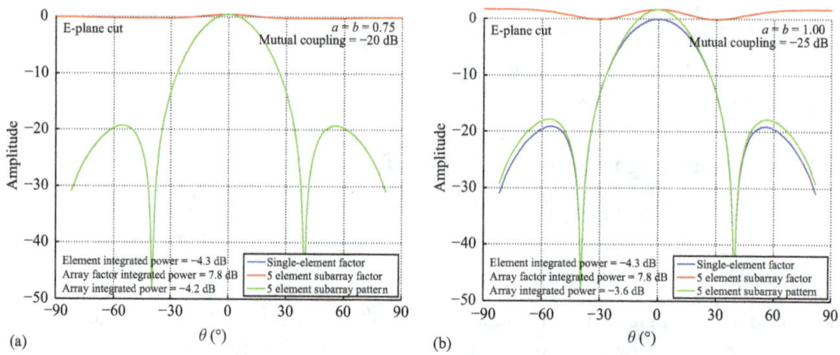

Figure 5.3 Larger element spacings: (a) spacing equals 0.75λ and (b) spacing equals λ
SubArray.m

in phase with the primary element but also because the magnitude of the scattered radiation decreases with distance.

Figure 5.4 represents a hypothetical mutual coupling relationship. Figure 5.4a shows the parametric relationship for coupling magnitude and phase assumed in this discussion. While actual values are design specific, the values assumed are plausible as will be seen in Sections 5.4–5.6.

Figure 5.4b shows the variation of boresight gain with lattice spacing. The gain variation is driven by the phase shift imposed on the scattered radiation and is most severe for element spacing close to λ/2, i.e., phase shifts close to 180°. The other three quantities are the integral of the power distribution over a hemisphere, representing the total power associated with each parameter. The element pattern integral is close to unity and does not change as it is the same for all cases. The subarray pattern is fairly constant and close to 8 dB ($10\log_{10}(2\pi)$), the expected

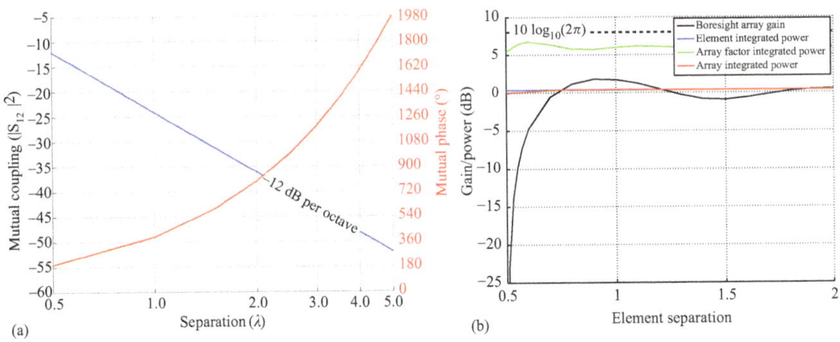

Figure 5.4 Functional relationship: (a) mutual coupling and phase as a function of element spacing and (b) boresight gain as a function of element spacing
SubArray.m

value for a uniform pattern. The total power in the array pattern oscillates around the value of the element pattern, as they involve the same input power, with slight enhancement at multiples of λ and slight decreases at half-integral multiples of λ element spacing.

In practice, the effects are somewhat less dramatic but still potentially problematic, notably as the design point of many ESAs preferentially involves lattice spacings between λ/2 and λ. Wideband arrays, particularly those with greater than a 2:1 bandwidth, may find that the λ/2 condition is unavoidable within their operating bandwidth.

The foregoing analysis has many limitations. The source element interacts with more than four elements, and they interact with each other. The mutual coupling values are more complicated than assumed. Reradiation of the coupled power is 100%. However, the example illustrates how easy it is to perturb the antenna pattern unknowingly.

Mutual coupling has been studied analytically for a small number of configurations and experimentally for more. With the advent of advanced modeling software and improved hardware (speed and RAM), it is not always necessary to quantify mutual coupling directly as its effects are embedded in the numerical model. It is useful to have a sense of its magnitude so as to avoid expending too much effort on problematic designs.

5.4 Patch element example

This section describes measurement and modeling results of a patch radiator, whose attributes are discussed further in Section 13.5.

Jedlicka *et al.* [9] constructed a test article and made a number of measurements. Subsequently, Pozar [10] analyzed the test configuration to validate an analytic model. The data from these two references is reproduced in Figure 5.5. Figure 5.5a illustrates the test set and Figure 5.5b compares the measured data reported by Jedlicka *et al.* [9] with model predictions by Pozar. Unfortunately, no phase data was reported.

Figure 5.5b replots the figure from the reference with three changes:

1. log scale for the x-axis,
2. x-axis recomputed to show scattering as a function of the distance between element centers rather than as a function of the separation of the elements, and
3. linear regression has been added to the data points to estimate the log–log relationship of scattering with distance.

The data is not apparently a simple function of distance. H-plane coupling decreases substantially faster than E-plane with a slope on the order of 16 dB per octave versus 6 dB per octave. Coupling at $\lambda/2$ is –10 dB in the H-plane and almost –20 dB in the E-plane.

The S-parameter test set assesses the coupling between the two antenna connectors which is not exactly the information we are looking for. The scattered radiation is certainly related to the power at the connector but this subject is not covered in the literature.

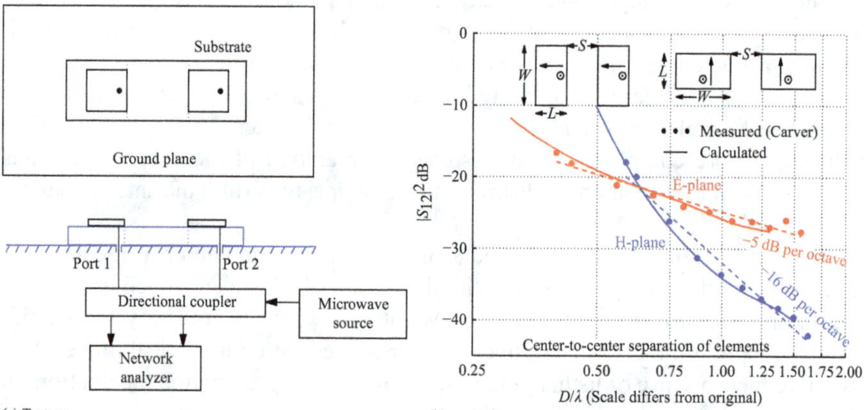

(a) Test set

(b) scattering

Symbol	Value
f	1.41 GHz
λ	21.3 cm
W	10.57 cm
L	6.55 cm
d	0.1575 cm
ε_r	2.55

(c) Element parameters

Figure 5.5 Calculated and measured mutual coupling

Source: (a) Jedlicka et al. [9] and (b) Pozar [10]

5.5 Waveguide element example

This section describes measurement and modeling results of a horn element, whose attributes are discussed further in Section 13.3.

Two types of radiating elements are termed waveguide elements: open-ended waveguides and slotted waveguides. Open-ended waveguides constitute a single radiating element, while slotted waveguides provide several radiating locations fed by a single waveguide. The latter incorporate the feed structure with the radiating element and their analysis is more complicated.

Steyskal [11] provides a comparison of simulated and measured data in "Mutual coupling analysis of a finite planar waveguide array." His data offers a basis for some observations. Unlike the example in Section 5.4, this example is explicitly an array and the author reports phase as well as magnitude of the coupling. Figure 5.6 reproduces his data and replots it for E and H planes.

Figure 5.6a reproduces the data and figure from the reference paper. The circles represent individual radiating elements and the data inside the circle represent the

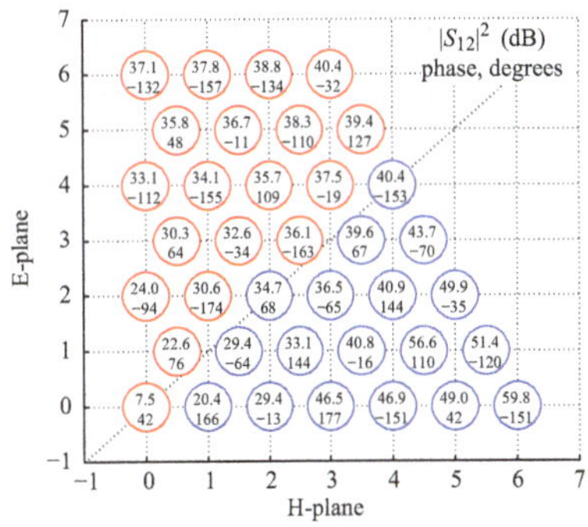

(a) One-quarter of hexagonal array

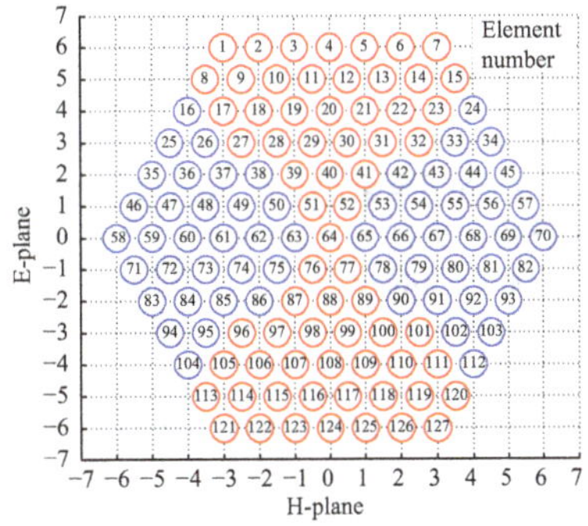

(b) Complete hexagonal array

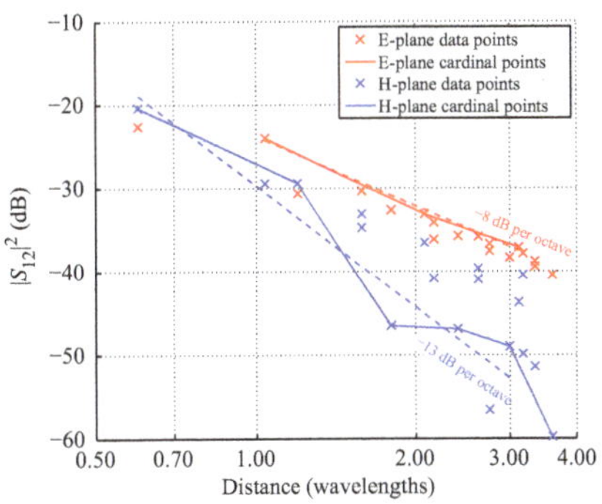

(c) Coupling coefficient as a function of distance from the center element

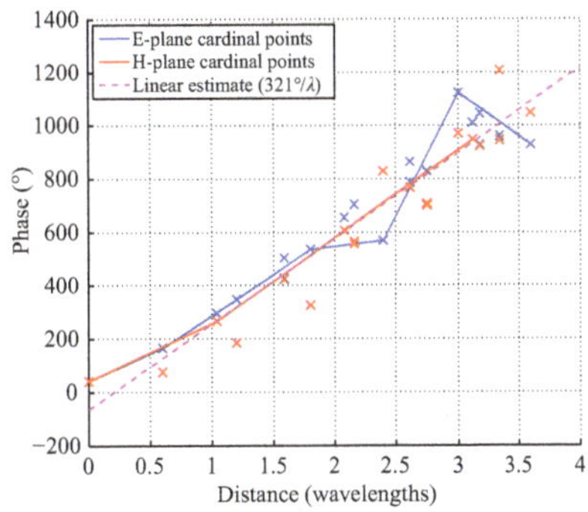

(d) Phase (unwrapped) as a function of distance from the center element

Figure 5.6 Steyskal analysis case

After Steyskal [11]

magnitude and phase of the coupled power. The phase is modulo 360° and must be unwrapped for our purposes. The elements in red are predominantly coupled in the E-plane and the elements in blue are predominantly coupled in the H-plane. Figure 5.6b depicts the full array; the paper relies on symmetry to analyze the single quadrant in Figure 5.6b.

Figure 5.6c plots the coupling coefficients as a function of distance from the array center. The data is shown as x's and the color of the data points corresponds to the color of the radiating elements. Solid lines connect data points which represent elements on the cardinal planes. The red line selects the elements along the cardinal plane in the E-field direction and shows a decrease in coupling from −25 to −38 dB as the distance increases from about 1 wavelength to about 3 wavelengths. The dashed red line represents a least-squares fit to the cardinal plane points and has a slope of −8 dB per octave. The blue line (H-field cardinal plane) shows a much steeper decrease from −20 to −60 dB as the distance increases from 0.6 wavelength to 3.6 wavelengths and the least-squares fit has a slope of −13 dB per octave. There is considerable dispersion in the data.

Figure 5.6d shows the phase relationship with distance which requires unwrapping the phase as the range exceeds 360°. The red line plots the elements on the cardinal plane and the line is very close to a linear estimate derived from the full data set. The blue line depicting the H-plane cardinal points shows more dispersion. The linear estimate is approximately 320° per wavelength. While 360° per wavelength might be expected assuming free-space wave velocity, the actual path of the EM power probably exceeds the distance between elements, so a value less than 360° seems plausible.

Another interesting calculation is the total power balance. The coupling coefficients represent the power absorbed/intercepted by each element. The summation of the power for each of the 36 elements totals 0.08 W or about −11 dBW. Subtracting this power from the central element shows that its power reduction is about −0.4 dB. The scattered radiation from the other 126 elements may add or subtract to the mainbeam gain.

The mutual coupling data may be used to construct an array factor similar to the process in Section 5.3. This is shown in Figure 5.7. It shows some reduction of boresight gain and enhancement of gain at scan angles between 60° and 90° which would be attenuated by the element pattern which has not been considered in this discussion.

5.6 Dipole example

This section describes measurement and modeling results of a dipole radiator, whose attributes are discussed further in Section 13.2.

While the preceding examples dealt with measurements, this example is purely analytic. It is a two-dipole configuration, in which the dipoles are parallel but separated in both the x-dimension and the y-dimension. This configuration has been analyzed by numerous authors (see Ref. [12], for example). Katagi *et al.* [13] provides a recent assessment. For our purpose, we follow the formulation of Balanis [14] in his (7-38–7-42). Figure 5.8 illustrates the dipole coordinates.

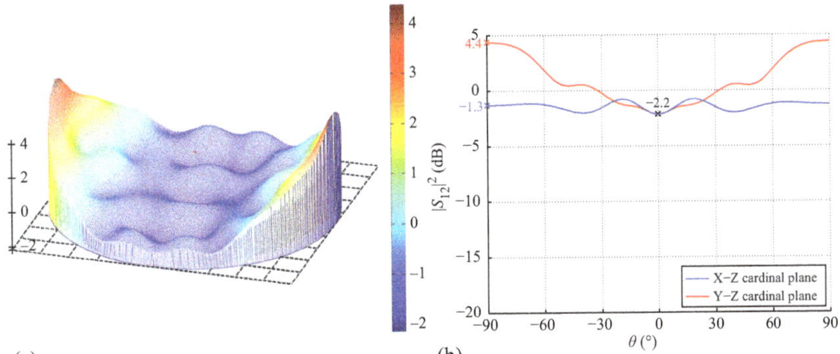

(a) (b)

Figure 5.7 Array factor of 127 element hexagonal array: (a) surface plot and (b) cardinal plane cuts
MutualCouplingWaveguide.m

It comprises a single-dipole source and a second dipole scatterer as indicated in Figure 5.8b. The second dipole is offset by a lateral distance of d and a vertical distance of s, as indicated in green in Figure 5.8b.

This example differs from an array in several respects. It deals with only two elements. The passive element does not include a feed port as would be required for an ESA. A feed port would provide an additional destination for the coupled power and reduce the reradiated power accordingly. The configuration also omits a ground plane which would be standard practice for a phased array.

The resultant mutual impedance is plotted in Figure 5.9. Figure 5.9a shows contours of constant mutual impedance for all pairs of x and y. The x-axis represents the

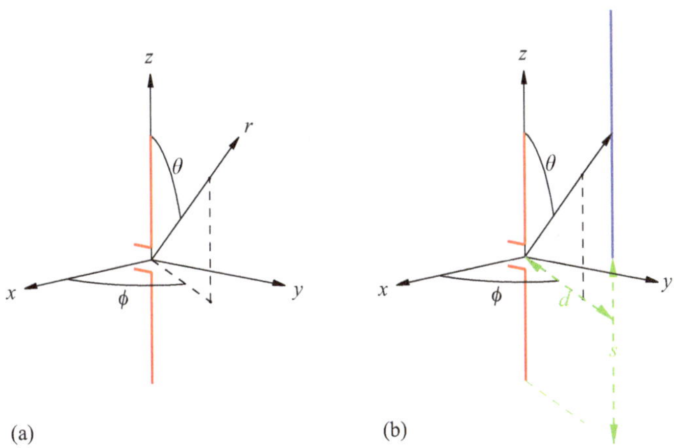

(a) (b)

Figure 5.8 Dipole geometry: (a) single dipole in nearfield and (b) two dipoles in nearfield
DipolePicture.m

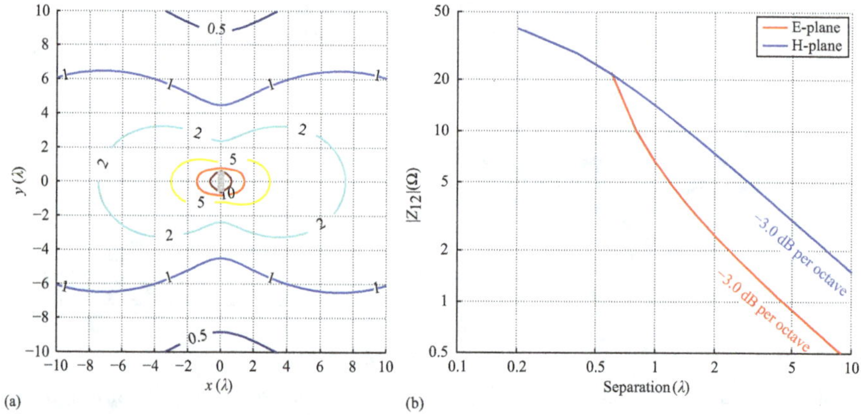

Figure 5.9 Mutual impedance (length = λ/2): (a) contours and (b) principal planes

CoupledDipoles.m

horizontal separation (d) of the dipole centers and the y-axis represents the vertical separation (s) of the dipole centers. Obviously, when $d = 0$, s must exceed the length of the dipole as they would otherwise occupy the same location, so this portion of the figures is not applicable.

Figure 5.9b shows the values of mutual impedance in the x–z (H)-plane and the y–z (E)-plane. The mutual impedance is greater in the X-plane (H-plane).

The phase is linear with distance between the sample points:

$$
\begin{aligned}
S_{12} &= \frac{2\,Z_0\,Z_{12}}{(Z_{11}+Z_0)(Z_{22}+Z_0)-Z_{12}Z_{21}} \\
&= \frac{2\,Z_0\,Z_{12}}{4\,Z_0^{\,4}-Z_{12}^{\,2}}
\end{aligned}
\tag{5.10}
$$

Mutual coupling is more interesting for our purposes, so we convert the mutual impedance value to its mutual coupling equivalent using (5.10) (see Chapter 14 in Ref. [4] for matrix transformations).

Figure 5.10 shows the mutual coupling values computed from the mutual impedance values portrayed in Figure 5.9 using (5.10). Figure 5.11 illustrates the value of mutual coupling based on the separation in the x-axis and y-axis. Dipoles are resonant structures, so the mutual impedance and mutual coupling will vary with frequency which is an additional layer of complexity and requires analysis over the planned bandwidth of operation.

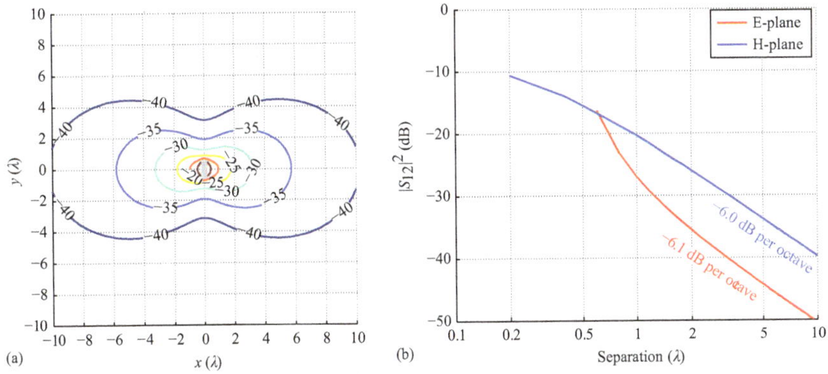

Figure 5.10 Dipole coupling (length = λ/2): (a) mutual coupling (dimensionless) and (b) mutual coupling (principal planes)
CoupledDipoles.m

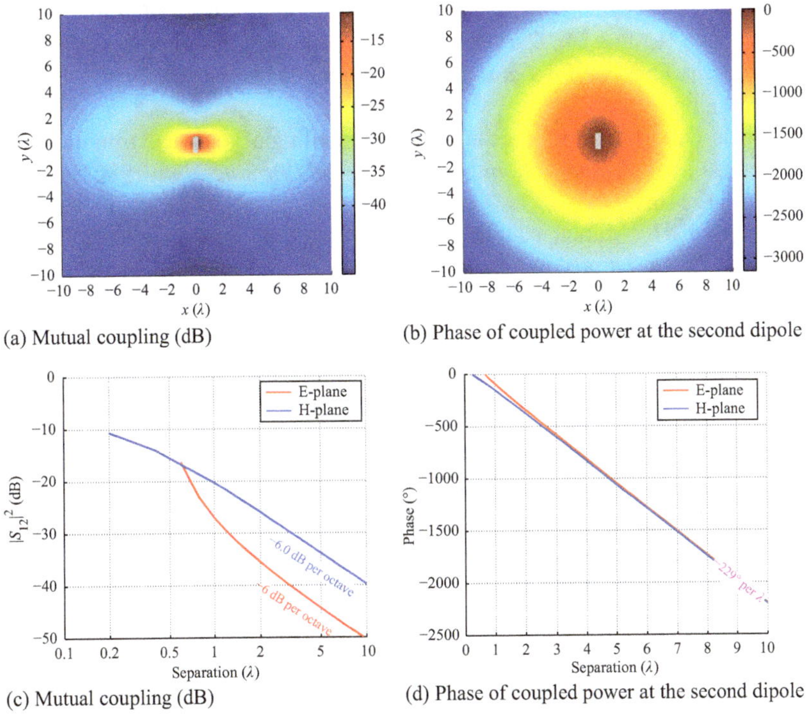

Figure 5.11 Mutual coupling (length = λ/2)
CoupledDipoles.m

References

[1] Hui H.T. 'A new definition of mutual impedance for application in dipole receiving antenna arrays'. *IEEE Antennas and Wireless Propagation Letters*. 2004, vol. 3, pp. 364–7.

[2] Henault S., Antar Y. 'Unifying the theory of mutual coupling compensation in antenna arrays'. *IEEE Antennas and Propagation Magazine*. 2015, vol. 57(2), pp. 104–22.

[3] Pozar D.M. 'The active element pattern'. *IEEE Transactions on Antennas and Propagation*. 1994, vol. 42(8), pp. 1176–8.

[4] Orfanidis S.J. *Electromagnetic Waves and Antennas*. Rutgers University. August 2016. Available from http://eceweb1.rutgers.edu/Orfanidi/ewa/ewa-1up.pdf.

[5] Orfanidis S.J. *Electromagnetic Waves and Antennas*. 2018. Available from http://www.ece.rutgers.edu/Orfanidi/ewa.

[6] Pozar D. 'Analysis of finite phased arrays of printed dipoles'. *IEEE Transactions on Antennas and Propagation*. 1985, vol. 33(10), pp. 1045–53.

[7] Hannan P. 'The ultimate decay of mutual coupling in a planar array antenna'. *IEEE Transactions on Antennas and Propagation*. 1966, vol. 14(2), pp. 246–8.

[8] Galindo V., Wu C.P. 'On the asymptotic decay of coupling for infinite phased arrays'. *Proceedings of the IEEE*. 1968, vol. 56(11), pp. 1872–80.

[9] Jedlicka R., Poe M., Carver K. 'Measured mutual coupling between microstrip antennas'. *IEEE Transactions on Antennas and Propagation*. 1981, vol. 29(1), pp. 147–9.

[10] Pozar D. 'Input impedance and mutual coupling of rectangular microstrip antennas'. *IEEE Transactions on Antennas and Propagation*. 1982, vol. 30(6), pp. 1191–6.

[11] Steyskal H. 'Mutual coupling analysis of a finite planar waveguide array'. *IEEE Transactions on Antennas and Propagation*. 1974, vol. 22(4), pp. 594–7.

[12] Frid H. 'Efficient methods to calculate mutual coupling between generic antennas on large platforms'. Licentiate dissertation. KTH Royal Institute of Technology. June 2014. Available from http://urn.kb.se/resolve? urn=urn:nbn :se:kth:diva-146903.

[13] Katagi T., Ohmine H., Miyashita H., Nishimoto K. 'Analysis of mutual coupling between dipole antennas using simultaneous integral equations with exact kernels and finite gap feeds'. *IEEE Transactions on Antennas and Propagation*. 2016, vol. 64(5), pp. 1979–84.

[14] Balanis C.A. *Antenna Theory: Analysis and Design*. 4th edn. New York, NY, USA: Wiley-Interscience; 2016.

Chapter 6
Errors and tolerances

In converting a design into hardware, there is seldom an exact match between as-designed and as-built, so performance will deviate from prediction. System requirements must be decomposed into subsystem specifications and tolerances compared with accuracy achievable with available components. Design and simulation with software tools utilize parameters that are exact to many decimal places. An essential step in the design process is to assess the sensitivity of predicted performance to inaccuracies, including temporal changes.

This chapter describes common error sources and assesses their effects. The effect of errors is to degrade antenna performance metrics that include beam pointing, beamwidth, sidelobe magnitude, and location. Correlated errors such as phase and gain quantization tend to be more noticeable but their average effect is comparable to random errors.

Errors may arise from component or fabrication variation or from partial or complete component failure. Intentional deactivation of elements, also described as a thinned array discussed in Section 8.3, may also be analyzed as a failure whose effects vary. For instance, turning off one-half of the array has greatly different effects depending on whether those elements are on one side of the array, alternate rows or columns of the array, or randomly selected. It is necessary to assess the specific design options to determine the importance of these errors.

Ruze was an early researcher in this field. His 1951 PhD thesis "Physical limitations on antennas" [1, 2] made the following observation:

> The antenna designer can readily compute by means of standard synthesis methods the aperture excitation necessary for a desired radiation pattern. However, when he constructs his antenna and measures its performance he finds that his experimental pattern only approximates the theoretical one.

6.1 Random and deterministic errors

Random errors include module and component failures and parametric variation, gain, phase, and physical position of each element. Arrays tolerate random errors because the multiple elements tend to reduce the effect of any single error.

Physical components approximate desired parameters to some degree of accuracy. Electronics advances have enabled good hardware accuracy: gain, match, and stability. The accuracy and repeatability are sufficiently good that sometimes their errors may be accommodated within reasonable error budgets which is another way of saying that device yield is much improved. Accordingly, theoretical performance may be closely realized.

Manufacturing variations affecting electrical length of multiple paths as well as gain and phase quantization require initial calibration and possibly recalibration. The array physical flatness is a concern as it is generally required to be lightweight, especially if deployment to final configuration is required. The manufacturing errors may have a random distribution but, being invariant, they are potentially remediable through calibration.

Deterministic errors include quantization of gain and phase and also invariant parametric values resulting from manufacturing tolerances. Examples of deliberate errors are array thinning and array clocking. The first reduces the number of elements for cost saving or other reasons. The second reorients subarray axes so that grating lobes will be smaller (but more numerous). Quantization errors may be taken as deliberate since the phase and gain steps are a design decision.

Quantization error is unavoidable and usually mitigated by selecting a sufficiently large number of bits in the phase and gain control. The required number of bits depends on the number of control elements (T/R modules) and decreases as the number of elements increases. Quantization results in a flat-topped approximation to the desired phase setting as shown in Figure 6.1.

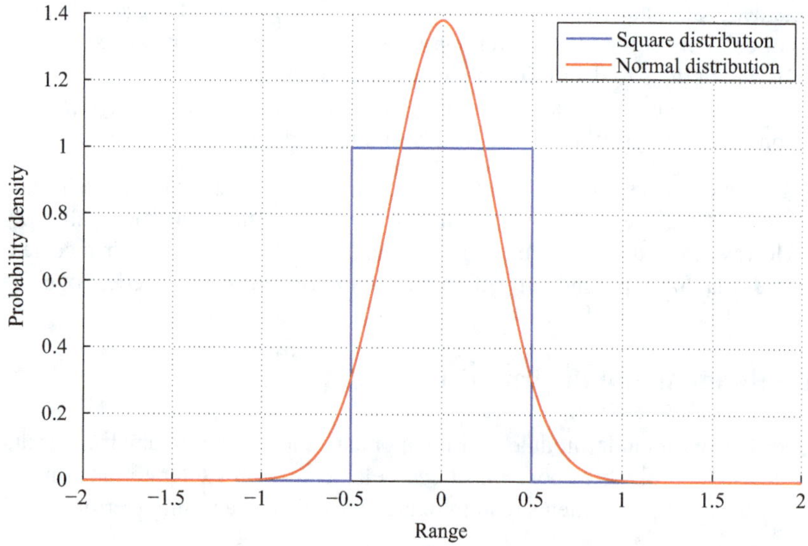

Figure 6.1 Error distribution functions

- Error analysis is similar for both continuous apertures and sampled apertures.
- Separate the desired term and the error term in the aperture distribution.
- Perform the farfield transform separately to obtain the desired beam pattern and an error beam pattern.
- Sum the two and square to assess the power.
- The squared error term will be negligible, so the cross-term represents the observed error.

6.2 Amplitude error

In transmit, the elements typically operate at full power, and the consistency of element power is determined by the components. There is little potential to compensate for element variability as the amplifiers are operated in saturation for best *DC-to-RF* conversion efficiency and gain control is relatively ineffective. However, the uniform power per element produces a pattern which is not very sensitive to amplitude variation.

In receive, the gain control is exercised to generate various patterns, typically to reduce sidelobes. Gain is usually set using switched attenuators. These components are subject to the ordinary variability resulting from fabrication inaccuracies but they also contribute a deterministic error due to their limited precision. The least significant bit determines the step size in selectable gain. Amplitude control is generally between 0 and 32 dB, so the least significant bit is $32/2^n$. The gain error term depends on the amplitude weighting scheme selected for the array but will not in general be a simple function.

Quantization is an issue in this case and can degrade the PSLR. Random errors have less effect on the PSLR but will increase the ISLR. Both errors reduce the boresight gain (by diverting power to and from sidelobes) and may also effect beamwidth. PSLR and ISLR are relatively more important with low sidelobe tapers, such as Taylor.

Ruze [3] published an early analysis deriving the effect of phase and amplitude errors on the average sidelobe error according to (6.1):

$$\begin{aligned} E &= \sum_m A_{\text{error}} A_m \, \mathrm{e}^{ik(x\,(u_1-u_0)+y\,(v_1-v_0))} \\ A_{\text{error}} &= (1+\Delta)\mathrm{e}^{ik\delta} \\ &\approx (1+\Delta)(1+\mathrm{i}\delta) \end{aligned} \tag{6.1}$$

Figure 6.2a is a graphical representation of (6.1). Ruze also calculated the effect of random failure (thinning) on average sidelobe level shown in Figure 6.2b.

Brookner [4] in chapter 3 summarizes the effect of random errors on peak sidelobe error in his equation (3.5) reproduced in (6.2) and (6.3):

$$P(r^2 \le r_{\text{p}}^2) = \int_0^{r_p} \frac{2r}{\overline{r_N^2}} \left[\exp\left(-\frac{r^2 + |E_0|^2}{\overline{r_N^2}} \right) \right] \boldsymbol{I}_0 \left(\frac{2r\,|E_0|}{\overline{r_N^2}} \right) \mathrm{d}rs \tag{6.2}$$

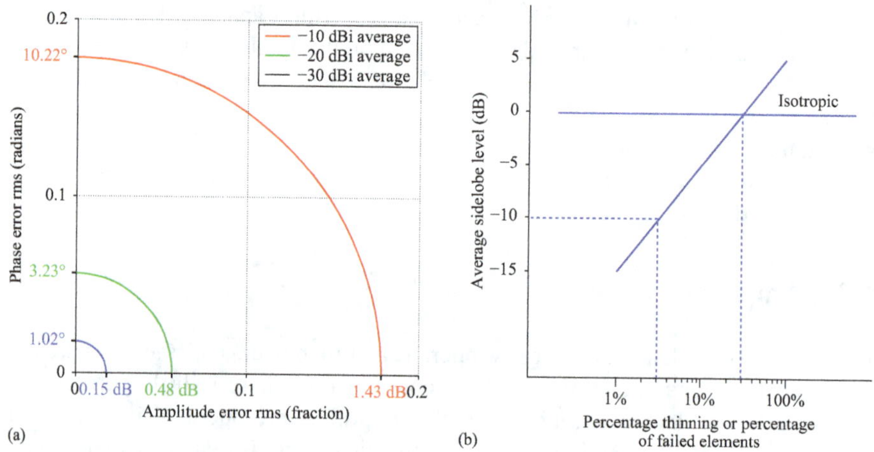

Figure 6.2 Parametric error relationships: (a) increase in average sidelobe level due to array errors and (b) increase in average sidelobe error due to failed elements
After figures 4 and 5 of Ruze [3]

where r is the actual sidelobe level, r_p is a parametric peak sidelobe level, and P is the probability that r_p exceeds r. $\boldsymbol{I_0}$ is the modified Bessel function of first kind. When the error-free voltage E_0 goes to zero in a null, (6.2) reduces to (6.3). This represents the null filling attribute of errors:

$$P(r^2 \leq r_p^2) = 1 - \exp\left(-\frac{r_p^2}{r_N^2}\right) \tag{6.3}$$

Figure 6.3 provides a graphical summary.

Mulcahey and Sarcione [5] describe an example showing uncalibrated and calibrated receive amplitudes as well as an array map of module performance for the THAAD UOES. The array map is discussed in Section 8.2. Figure 6.4a shows the probability distribution of element receive gain before and after calibration. Calibration reduces the gain standard deviation from 1.31 to 0.24 dB but also reduces the average gain from –3.74 to –7.04 dB. This gain reduction would be expected to increase the array noise figure but this information is not provided. The uncalibrated gain standard deviation reflects the cumulative effect of errors of six gain stages and other circuit elements shown in Figure 14.2.

McLachlan *et al.* [6] provide a somewhat more recent example of transmit power of groups of T/R modules. This information is reproduced in Figure 6.4. The calculated standard deviations are 0.64 and 0.26 dB, respectively. The Tx power amplitude variation across the aperture degrades the transmit pattern.

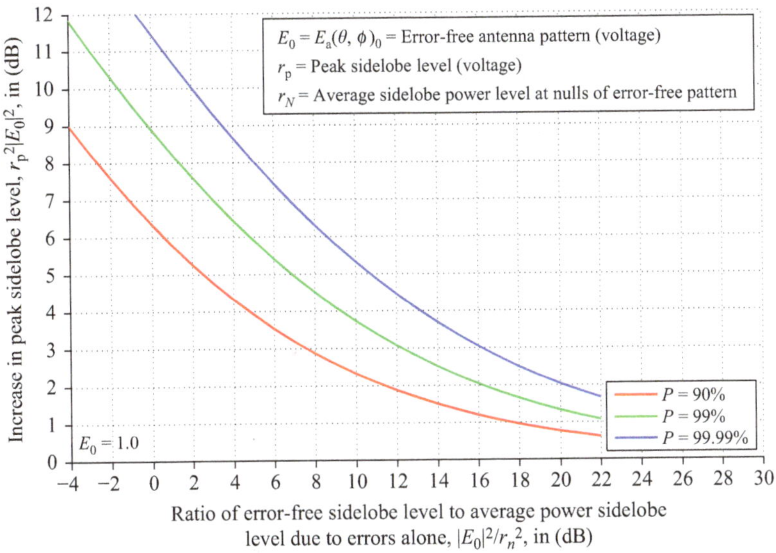

Figure 6.3 *Increase in peak sidelobe level due to array errors*
After figures 3–5 of Brookner [4]

6.3 Phase error

The beam pointing direction is determined by phase settings at each element which establish the appropriate phase slope across the array. Typically the required phase variation across the array exceeds 360°, but it is sufficient for most purposes to select the phase value modulo 360°.

Phase shifters are available with either continuous or with discrete values. The former are not widely used although there are some interesting examples, see for instance figure 2 of Corey [7], which describes a 720° ferrite phase shifter used with row/column array addressing. Ferrite phase shifters can handle considerable power (order of watts) so they are used to advantage with transmitters in passive ESAs. They are less suitable for T/R modules due to size.

Phase shifters with discrete values are produced as monolithic GaAs devices with high-pass/low-pass filters implemented in lumped inductances and capacitances providing a relatively constant phase shift over a reasonable bandwidth. This approach is discussed in detail in Section 14.5.3.

Phase is usually set using switched phase bits. These components are subject to the ordinary variability resulting from fabrication inaccuracies but they also contribute a deterministic error due to their limited precision. The least significant bit determines the step size in selectable phase. The phase control is generally between 0° and 360°, so if n is the number of bits, the least significant bit is $360/2^n$. Beam

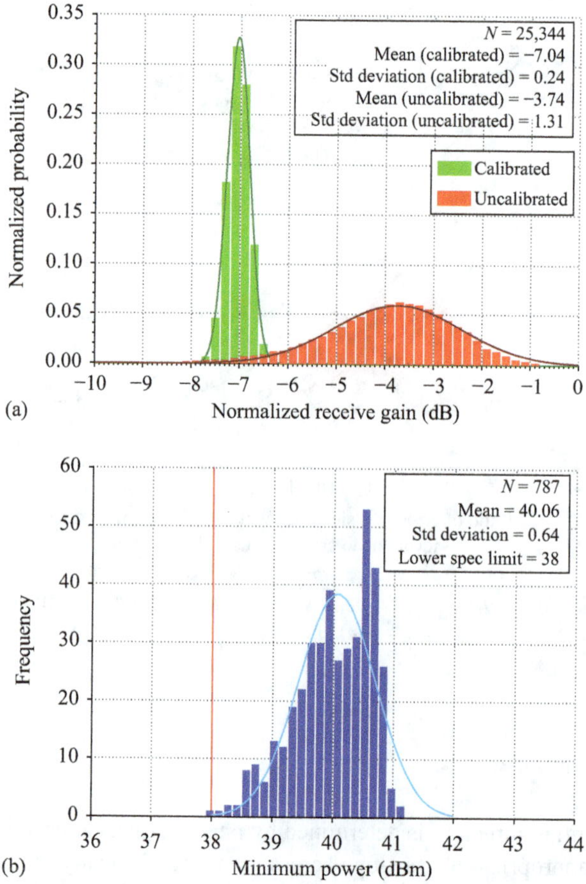

Figure 6.4 *Statistics for two different T/R module examples: (a) UOES #1 receive gain (*c. *1996; After Mulcahey and Sarcione [5], figure 2) and (b) SELEX transmit power (*c. *2007; After McLachlan* et al. *[6], figure 5)*

steering requires a uniform phase slope across the aperture. Accordingly, the phase error term is a sawtooth which may be readily assessed.

Figure 6.5 illustrates the case of a 64 element array with elements spaced at $\lambda/2$ and steered to 24°. The necessary phase slope spans several thousand degrees from one edge of the array to the other so the phase setting is "wrapped," i.e., computed modulo 2π radians. The wrapped phase is then approximated by the nearest available phase setting. The intermediate calculations and resulting error term are shown in Figure 6.6a. Note that while the data are plotted as continuous

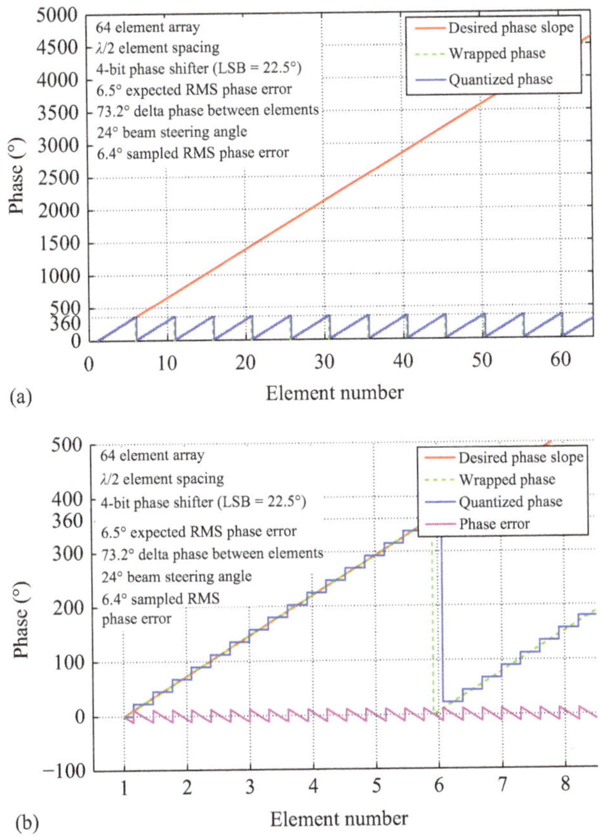

Figure 6.5 Phase approximation detail (24° scan): (a) quantized phase approximation and (b) resulting error term

functions of element number, only the phase values at integral values of element number are applicable.

This assessment is repeated for a beam steering angle of 30°. The result is illustrated in Figure 6.6. The phase increment is exactly 90° so the phase shifter settings incur no error. This is also the case at boresight (nominal zero phase setting) and for other pointing directions for which the phase increment between elements is a multiple of the LSB. The discrete phase settings are phase $= 0:2^{(n-1)}$. LSB and the corresponding pointing directions are beam steering directions = arcsin(phase/π). Thus there are $2^{(n-1)} + 1$ beam steering directions with no error.

Figure 6.7 summarizes phase error as a function of desired scan angle. Figure 6.7a shows the actual error for each element as a function of steering angle for a 4-bit phase shifter. The results for other numbers of bits are similar. The phase error is zero for element 1, since that is the reference element against which the other elements are measured. The error is also small for beam pointing

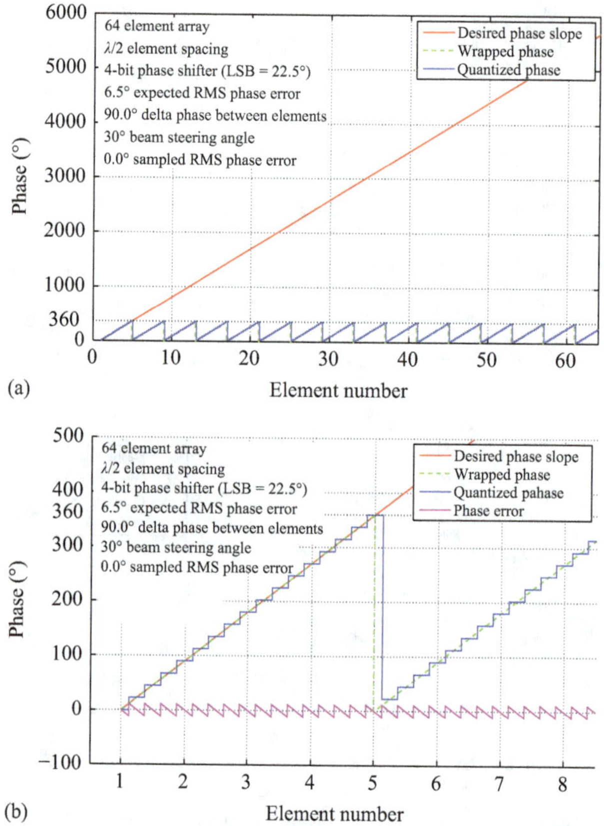

Figure 6.6 Phase approximation detail (30° scan): (a) quantized phase approximation and (b) resulting error term
PhaseError.m

directions close to 90° as the required phase shift for this case is 180°, which is available with all phase shifters. Intermediate steering angles of 14.5°, 22°, 30° 38.7°, 48.6°, and 61° also experience negligible errors.

Figure 6.7b shows the rms error for all elements as a function of steering angle for a 4-bit phase shifter. The rms error of a uniform distribution between limits of ±LSB is LSB/$\sqrt{12}$. This is a good approximation for quantization error except when the phase increment is a multiple of the LSB. Figure 6.7b represents these cases.

The quantized error case and the random error case have different effects on sidelobes. Since the quantized error is systematic, besides producing zero errors at some pointing directions, it can also generate discrete sidelobes at other pointing directions. This has the result of degrading the PSLR especially for antennas

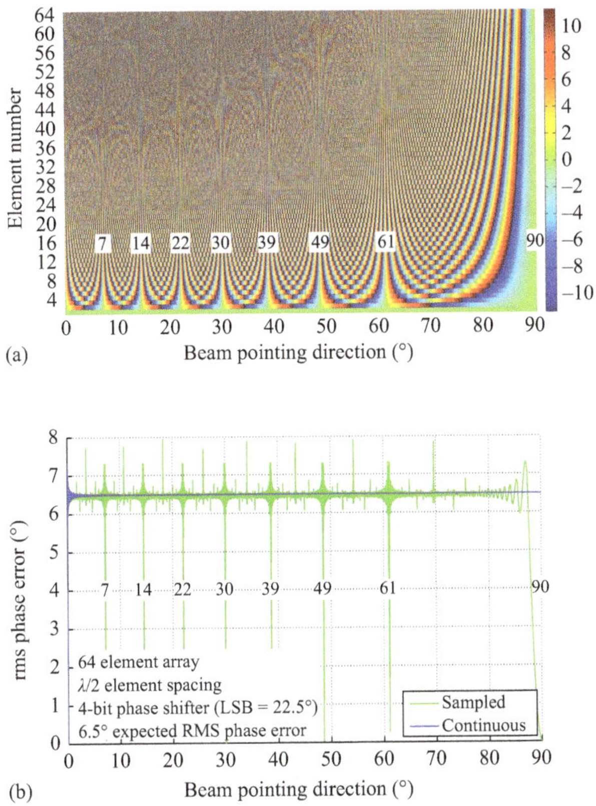

Figure 6.7 *Quantized phase error as function of scan angle: (a) element phase error (°) and (b) rms phase error (°)*
PhaseError.m

optimized for low sidelobes. Figure 6.8 shows two different phase slopes for a single set of phase settings..

The quantization errors are plotted for each element, for two cases, scanned to 24° and 30°. As previously noted, the second case is error free. The first case has periodicities, indicated by dotted and dashed lines, which would be expected to generate beams in corresponding directions: Δphase/π, 0.03°, and 1.8°. The magnitude of the sidelobe depends on the separation of the spurious parallel lines; it will be greatest when they are separated by a wavelength and minimized when the spacing approximates a half-integral wavelength.

The effect of the error may be determined by computing representative antenna patterns with and without the error. Periodic errors will produce antenna lobes at predictable directions. These examples are based on an array with elements spaced at $\lambda/2$ and scale with wavelength.

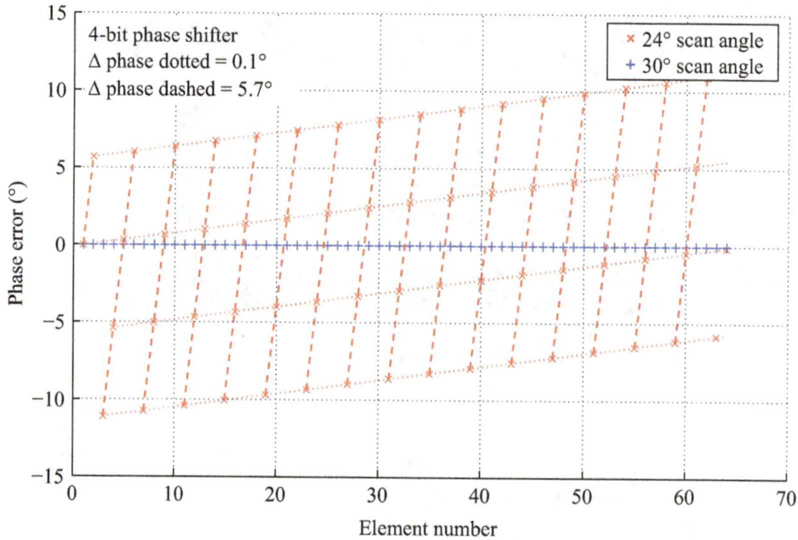

Figure 6.8 Quantization lobe origin
PhaseError.m

To assess the effect, we use an example array of 64 × 64 elements on boresight with phase errors corresponding to $\theta = 0°, 4°, 8°, 12°, 16°, 20°, 24°$, and $\phi = 0°$. The quantization error is based on beam pointing direction but it is assessed separately, corresponding to a boresight beam pointing direction with specified error term. The error term is applied only to the y axis (elevation direction). Amplitude weighting, either uniform or Taylor, is applied to both axes.

Figure 6.9 shows the calculation of PSLR and Figure 6.10 shows the calculation of ISLR for these cases. PSLR and ISLR values resulting from quantized phase error results are plotted in blue and those resulting from random phase error are plotted in red. The quantized phase error distributions differ in detail although their rms values are similar so the ISLR results are slightly different.

The rms phase value corresponding to the number of bits in the phase shifter is plotted in green. Each phase shifter bit provides 1–2 dB ISLR benefit up to a limit of about 6 bits where the ISLR value is approximately equal to the error-free case.

There is a noticeable difference in PSLR between the random phase case and the quantized phase case as illustrated in Figure 6.9. For the uniform illumination case, the best sidelobe ratio is 13.2 dB and the errors do not in general produce exceedances. For the Taylor case, the best sidelobe radio is 30 dB and errors produce sidelobes exceeding that number even with 5-bit phase shift control.

There is little difference in ISLR between the random and quantization error cases as illustrated in Figure 6.10. This illustrates the point that the random error distribution is the cause of increased integrated sidelobes while quantized phase errors produce discrete sidelobes.

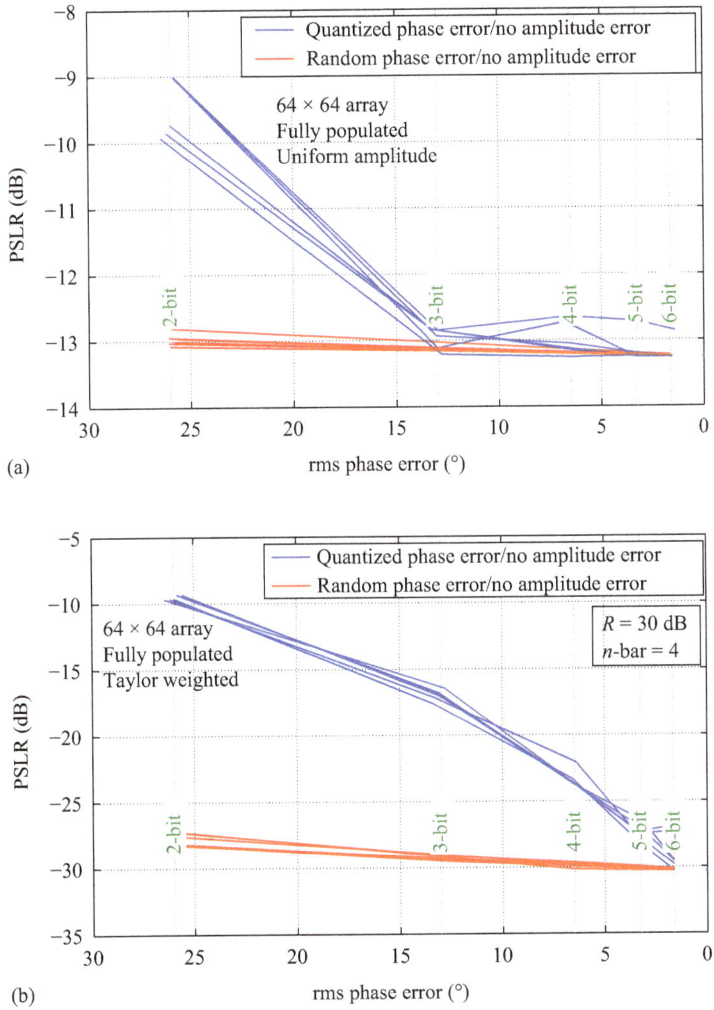

Figure 6.9 PSLR calculation as a function of phase shifter bits: (a) uniform case and (b) 30 dB Taylor-weighted case

We may examine representative cases of the beam pattern in elevation, the axis in which the errors are introduced. Figure 6.11 shows the reference, error-free pattern for uniform illumination on the left and Taylor weighting on the right. The expected values of 13.2 dB and 30 dB are observed in the first sidelobe.

Figure 6.12 shows the effect of a random error of 6.5° rms. There is no increase in the PSLR of the uniform illumination case, because the sidelobes induced by the phase error are lower than the first sidelobe. In contrast, the PSLR of the Taylor-weighted pattern degrades because the same sidelobes are competing with a 30

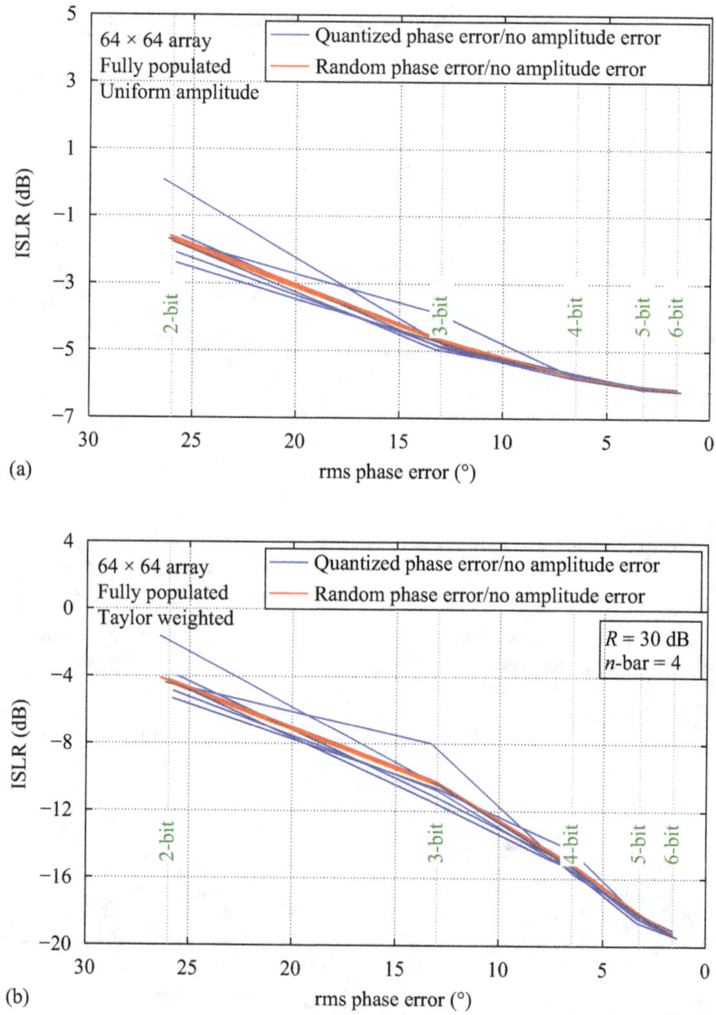

Figure 6.10 ISLR calculation as a function of phase shifter bits: (a) uniform case and (b) 30 dB Taylor-weighted case
ThinnedSquare.m

dB first sidelobe which is more easily exceeded. The ISLR of the uniform case is unchanged at –6 dB while the ISLR of the Taylor case worsens, from –20 to –16 dB. In both cases, the 3 dB and 10 dB mainlobe widths are unaffected.

Some figures of merit, notably peak gain, beamwidth, and peak sidelobes show little sensitivity to the number of bits of phase shift. The ISLR shows a discernable and direct relationship to the number of phase bits.

Figure 6.13 shows the effect of errors introduced by use of a 4-bit phase shifter. Its rms error is 6.5° matching the value of the previous random case. Both the

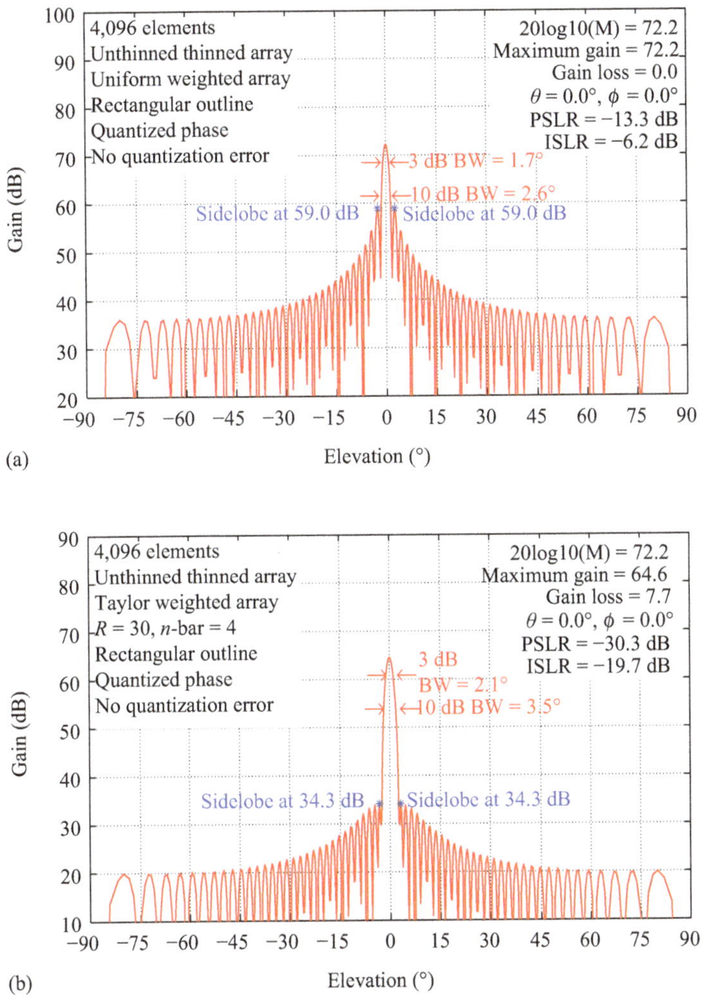

Figure 6.11 Elevation cut of boresight beam pattern error free: (a) uniform boresight case and (b) 30 dB Taylor-weighted boresight case
ThinnedSquare.m

uniform illuminated case and the Taylor case display noticeable sidelobes at about $\pm 45°$ whose magnitude is about 26 dB below the peak. The PSLR of the uniform case is unaffected while the PSLR of the Taylor-weighted cases increases to 26 dB. The ISLR of the quantized error case is the same as the random error case. In both cases, the 3 dB and 10 dB mainlobe widths are uneffected.

Quantization errors introduce beam pointing error, as for instance steering slightly off boresight will not effect any phase shift until the required phase shift is more than one-half LSB. We may rewrite (2.39) as (6.4) making the substitutions $\Delta x = \lambda/2$, N = number of elements, and $\Delta\phi = \text{LSB}/N$.

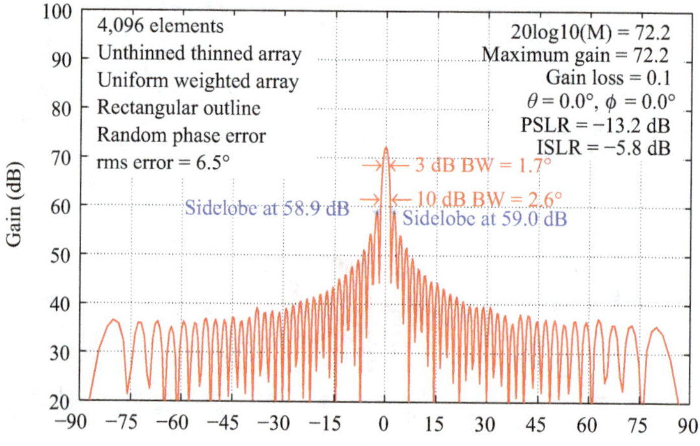

(a)

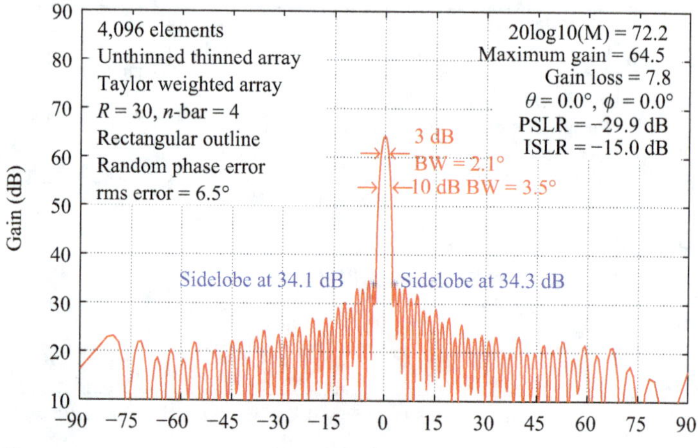

(b)

Figure 6.12 Elevation cut of boresight beam pattern 6.5° random error: (a) uniform case with error and (b) 30 dB Taylor-weighted case with error

ThinnedSquare.m

$$\Delta\theta = \arcsin \frac{\text{LSB}}{N\pi} \tag{6.4}$$

Table 6.1 shows a few examples for array size and number of phase bits. These minimum pointing step sizes are small fraction of the beamwidth. Figure 6.14 represents the actual beam pointing error as a function of commanded beam pointing direction for a 16 element array with $\lambda/2$ element spacing.

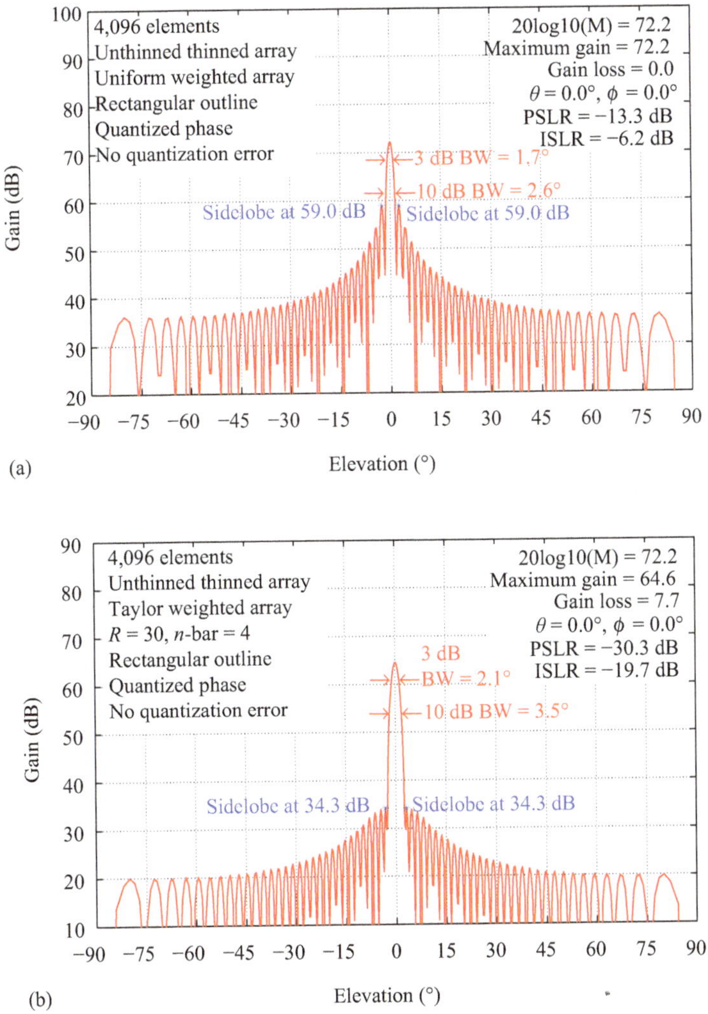

Figure 6.13 *Elevation cut of boresight beam pattern for 4-bit quantization error: (a) uniform case with error and (b) 30 dB Taylor-weighted case with error*

ThinnedSquare.m

6.4 Calibration

Known and stable errors may be mitigated by calibration. This is useful for manufacturing variation, for instance, differing receive gain among the ensemble of T/R modules or variation after manufacture due to environmental factors. Some of the phase and gain control is used to level the array elements to a constant gain and amplitude state. This assignment of control authority reduces the control

Table 6.1 Quantized phase scan increment for λ/2 lattice

		Minimum beam steering step (°)				
Array size	**3 dB beam width (°)**	**2-bit**	**3-bit**	**4-bit**	**5-bit**	**6-bit**
8	6.2	3.58	1.79	0.90	0.45	0.22
16	3.1	1.79	0.90	0.45	0.22	0.11
32	1.6	0.90	0.45	0.22	0.11	0.06
64	0.8	0.45	0.22	0.11	0.06	0.03

Assumes a square lattice and uniform illumination
QuantizedPhase.m

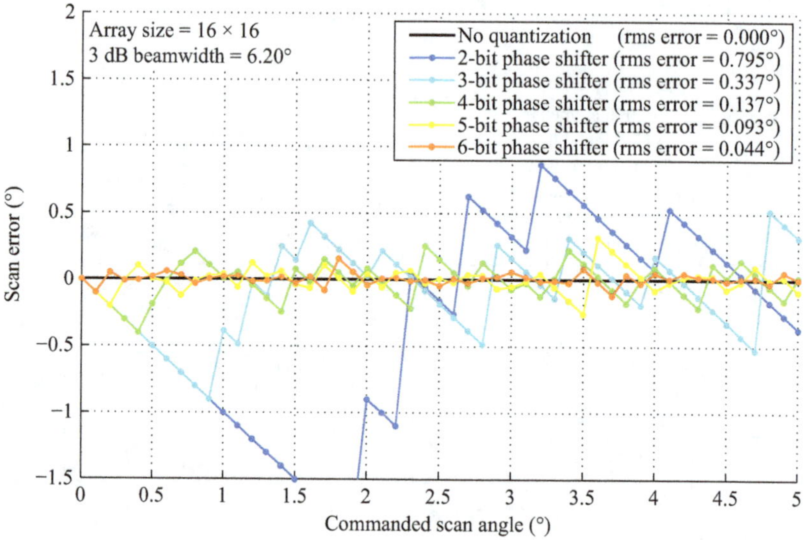

Figure 6.14 Beam pointing error
PhaseError.m

authority available to other purposes such as gain shaping and phase control for beam steering.

Corrections will be no better than allowed by the quantization error. Moreover, the errors may not be stable with time or constant over parametric variation such as temperature. It is possible to devise calibration circuits to allow recalibration during use which is useful when it is not possible to return to the lab for calibration.

Figure 6.4a shows an example of gain calibration. Gonzalez *et al.* [8] describe calibration experience with TerraSAR-X and Ahn *et al.* [9] describe calibration experience with an X-band T/R module.

References

[1] Ruze J. 'Physical limitations on antennas'. PhD thesis. Massachusetts Institute of Technology, May 1952. Available from http://hdl.handle.net/1721.1/16077.

[2] Ruze J. 'Physical limitations on antennas'. Prepared for the Army Signal Corps, the Navy Department (Office of Naval Research), and the Air Force (Office of Scientific Research, Air Research and Development Command). PhD thesis. Lexington, MA, USA: Research Laboratory of Electronics, MIT, October 1952. Available from http://hdl.handle.net/1721.1/4820.

[3] Ruze J. *Pattern Degradation of Space Fed Phased Arrays*. Tech. Rep. Project Report SBR-1. Air Force Contract F19628-80-C-0002 (ARPA order 3724). Lexington, MA, USA: Lincoln Laboratory, MIT, December 1979. Available from https://apps.dtic.mil/dtic/tr/fulltext/u2/a083264.pdf.

[4] Brookner E. 'Chap. 3 Antenna array fundamentals: Part 2' in Brookner E., Norwood M.A. (eds.). *Practical Phased Array Antenna Systems*. USA: Artech House; 1991. pp. 1–28.

[5] Mulcahey J.K., Sarcione M.G. 'Calibration and diagnostics of the THAAD solid state phased array in a planar nearfield facility'. *Proceedings of International Symposium on Phased Array Systems and Technology*. 1996, pp. 322–6.

[6] McLachlan A.D., Dunn M., Morrison G.D., *et al.* 'T/R module design and production processes for airborne radar systems'. *2007 IET International Conference on Radar Systems*. 2007, pp. 1–4.

[7] Corey L.E. 'A survey of Russian low cost phased-array technology'. *Proceedings of International Symposium on Phased Array Systems and Technology*. 1996, pp. 255–9.

[8] Gonzalez J.H., Bräutigam B., Schwerdt M., *et al.* 'TerraSAR X internal calibration experience and extension for TanDEM X'. *Committee on Earth Observation Satellites (CEOS) Workshop*. November 2008. Available from https://elib.dlr.de/56167/.

[9] Ahn C.S., Chon Sang-Mi., Kim Seon-Joo. 'Accurate characterization of T/R modules with consideration of amplitude/phase cross effect in AESA antenna unit'. *ETRI Journal*. 2016, vol. 38(3), pp. 417–24.

Chapter 7
Grating lobes

Grating lobes are a sampling phenomenon. They represent aliasing appearing from other Nyquist zones due to undersampling. They are detrimental because they divert power to unwanted directions in transmit and they introduce additional noise and spurious targets in receive.

It is both expensive and usually unnecessary to totally eliminate grating lobes. Scanning far off-boresight is not a common expectation of a planar antenna. Many applications require only 10°–30° scan capability. Instead, the optimum solution is usually a compromise of restricted scan volume and reduced grating lobes.

7.1 Lattice attributes

Rectangular and triangular lattices are typical for ESAs. Rectangular lattices in a rectangular aperture produce a separable array factor which is numerically equivalent to the product of two linear arrays at right angles.

Triangular lattices are slightly more complicated but they offer the interesting possibility of fewer T/R modules for a prescribed scan volume (see Sharp [1]). The triangular lattice may be created by shifting a copy of the original lattice as illustrated in Figure 7.1. This doubles the element density. Evidently, the triangular lattice may be analyzed as superposition of two rectangular arrays. As we will see, the shift adds sidelobe ridges besides those which occur on the cardinal planes.

Another technique to create a triangular lattice is to shift every other row (or column) of a rectangular lattice by one-half the lattice spacing in that direction leaving the element density unchanged. This is the case analyzed in Figures 7.2 and 7.3.

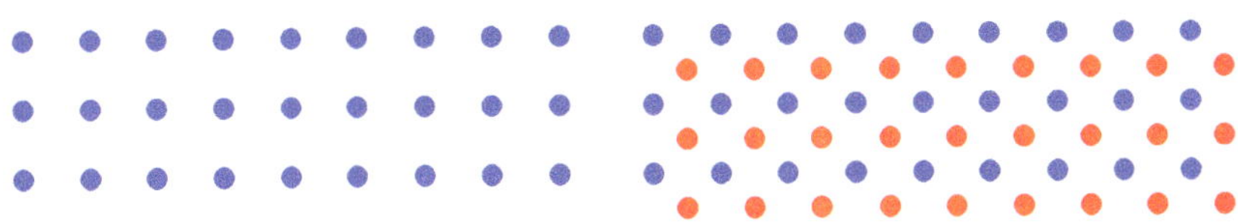

(a) (b)

Figure 7.1 Common lattice types: (a) rectangular array and (b) triangular array

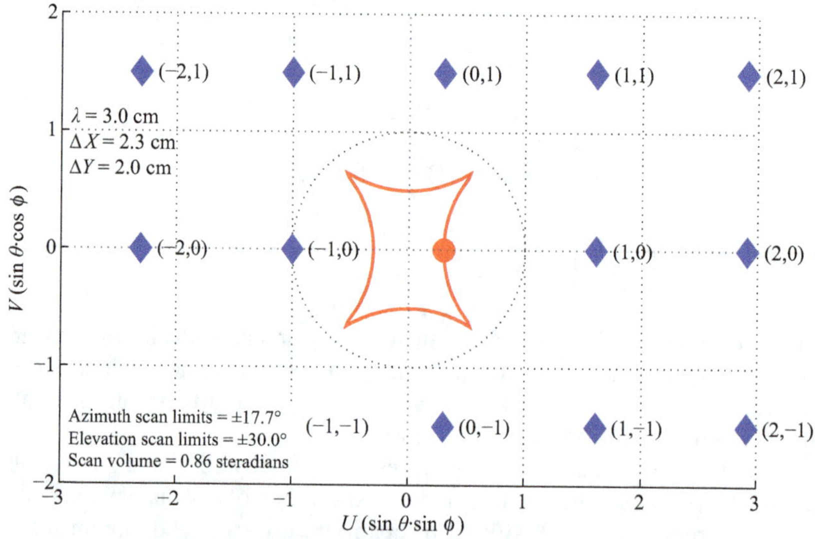

Figure 7.2 *Rectangular lattice grating lobes in (u, v) space*
GratingLobes.m

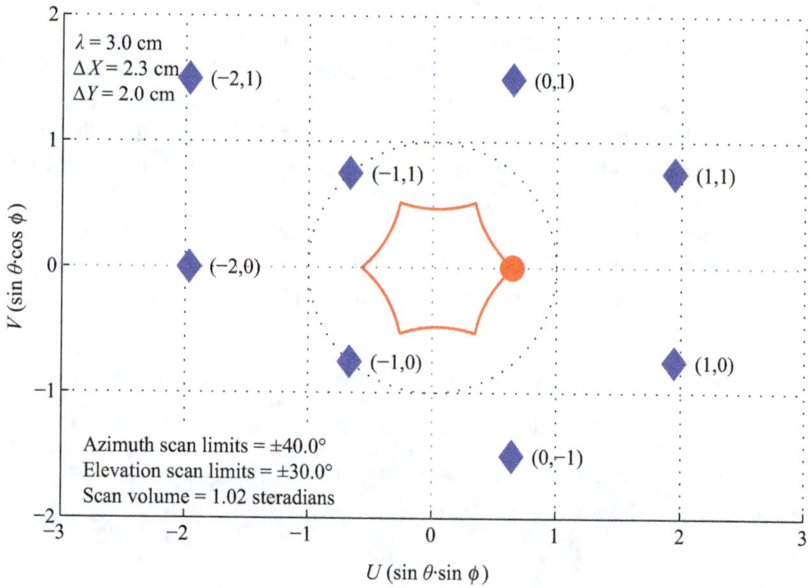

Figure 7.3 *Triangular lattice grating lobes in (u, v) space*
GratingLobes.m

7.2 Grating lobes location

Figure 7.2 shows grating lobes for a rectangular lattice computed using (2.37). They occur for pointing directions given by

$$\sin \theta_p = \sin \theta_0 + \frac{p \cdot \lambda}{d} \tag{7.1}$$

where

θ_p = grating lobe direction
θ_0 = beam direction
λ = wavelength
d = element spacing
$p = \pm(1, 2, 3, \ldots)$
The scan volume is free of grating lobes if θ satisfies

$$\theta \le \arcsin\left(\frac{\lambda}{d} - 1\right) \tag{7.2}$$

Conversely, if θ satisfies (7.3), then all beam steering directions experience grating lobes:

$$\frac{\lambda}{d} \le 1 \quad \text{(i.e., } d \ge \lambda) \tag{7.3}$$

The ultimate limit on beam scanning is $\theta_p = -\theta_0$ (equal and opposite) from which it follows that

$$\sin \theta_0 = \frac{p \cdot \lambda}{2 \cdot d} \tag{7.4}$$

This relationship is derived in Section 2.4.3.

Some of the solutions yield $\sin \theta_p$ greater than 1 or less than –1. Although this pointing direction does not exist (is "imaginary"), it possesses sidelobes which fall into the real radiation hemisphere.

Grating lobes in (u, v) space (rectangular lattice)

The (u, v) space representation yields insight into the behavior and effects of grating lobes. Figure 7.2 represents the mainlobe and possible grating lobes for an array with ΔX and ΔY shown in the upper left corner. The dashed circle represents real space or visible, i.e., pointing directions of 90° or less.

Pointing directions are plotted in (u, v) space and show the aliases of the mainlobe outside of visible space (radius ≤ 1). The mainlobe pointing direction is shown by the red dot. The aliases are shown by blue triangles and labeled with their (m,n) value. The initial pointing direction is chosen so that the (–1,0) grating lobe barely impinges on the visible space circle. Then the pointing direction is rotated counter-clockwise in ϕ, always keeping one grating lobe (whose identity changes) on the visible space circle. In this way, the red dot traces out the grating lobe-free region of this particular lattice (array factor). While the details vary, the grating lobe-free space is somewhat rectangular with sunken sides and has four prominent spurs at the corners. Evidently, the scan limits are most severe for the cardinal planes and there is some excess scan available at intercardinal positions.

The grating lobes are not infinitely sharp, even in the imaginary space. Grating lobes in the imaginary space which are adjacent to visible space generate sidelobes in visible space. We can generally rely on the element pattern to mitigate sidelobes which are just inside the circle.

Grating lobes in (u, v) space (triangular lattice)

Figure 7.3 represents a triangular lattice, achieved by shifting alternate rows of elements by one-half their spacing. Accordingly, the values of ΔX and ΔY are the same as in Figure 7.2 and the element density (elements per unit area) is the same. However, the grating lobes are positioned differently, with some benefit. As before, the initial beam position is chosen so that the grating lobe just impinges on the visible region circle. This time, there are two such grating lobes. As we rotate the pointing direction, the grating lobe free-region traced out is hexagonal with concave sides. The usable region is more difficult to describe. However, it is somewhat rotationally symmetric and might be advantageous for a nose-mounted aircraft radar in which the aircraft roll is variable as it would maintain track better as the aircraft rolled.

Scan volume comparison

Figure 7.4 compares the grating lobe-free regions of the two lattices.

The triangular lattice provides about 15% more accessible solid angle region with the same number of elements. Because the region shapes differ, there may still be application-specific reasons to prefer one or the other. Rectangular lattices remain more popular for most applications due to simpler mechanical arrangements. It is also true that there are several other considerations for operation at the scan limits, such as pattern degradation and scan loss.

7.3 Superresolution

The (u, v) representation provides some insight into oversampled lattices with spacing less than 0.5λ. It shows, for instance, that for some pointing directions (phase slopes) the mainbeam is outside real space although some sidelobe power may exist in real space. Accordingly, only a small fraction of the power radiates. The remainder of the power evidently reflects back to the source, except for a possible resistive loss in the array elements. This represents a reactive load with a very high voltage standing wave ratio (VSWR) or equivalently poor return loss. Apart from the fact that the array is not functioning in an efficient manner, the reflected power will stress and possibly damage the electronic components in the T/R module.

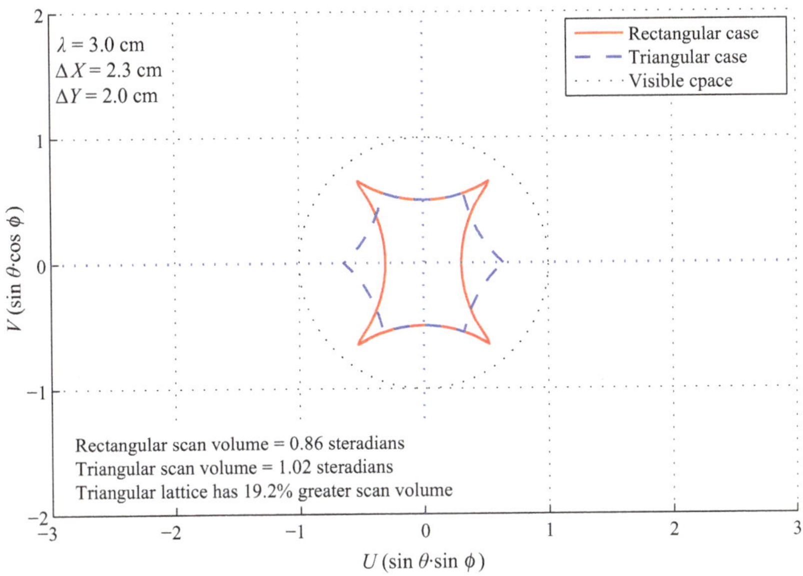

Figure 7.4 *Comparison of lattice grating lobes in (u, v) space*
GratingLobes.m

Superresolution uses the sidelobe power of imaginary beams to modify a real beam. This approach fails because of the excessive voltages or currents associated with the imaginary beams.

As a thought experiment, we may consider two arrays adjacent to each other, each with 0.5λ spacing and fed out of phase (180° phase difference). This represents a classic array configuration with a difference pattern, used for direction finding. If we superimpose the two arrays so that they are exactly interleaved, we have elements one-quarter wavelength separated, with opposite phases. This configuration obviously has no broadside gain. Not as obvious is the fact that it has no significant gain in any other direction. Effectively, the mainlobe is directed outside visible space and only the sidelobes appear.

Many radiating elements do not allow such close spacing as they tend to be larger than 0.5λ in size. Two types that could be used are wire loops and dipoles. Two adjacent loops constitute a transformer, and driving both sides of a transformer with opposite phases is usually destructive.

7.4 Linear array examples

As rectangular lattice two-dimensional patterns decompose into the product of two one-dimensional patterns, a one-dimensional lattice will illustrate the angular dependency of grating lobes.

Element spacing > λ/2 ⇒ grating lobes

ESAs are generally designed with lattice spacing greater than one-half wavelength which limits their grating lobe-free scan volume to less than one hemisphere. Figure 7.5 depicts two limits on scan (assuming a linear array). The more stringent limit is to exclude grating lobes from the visible region, which was the criterion used for the scan volume assessment in Figures 7.2–7.4. No grating lobes at the edge of the visible region is a fairly stringent criterion since the element pattern will do a good job suppressing them.

The less stringent criterion allows a grating lobe at the equal but opposite pointing direction. The element pattern (normally symmetric) does not reduce the grating lobe, so the antenna produces two equal magnitude beams in opposite direction. As a minimum, this would represent a 3 dB loss in gain in the desired direction. This loss would only be acceptable if there were small likelihood of interference through the unwanted lobe, for instance, if the unwanted lobe were pointing to space.

Element spacing ≤ λ/2 precludes grating lobes

Figure 7.6 represents scanning of an array of dipoles spaced at one-half wavelength. In this instance, the array generates two beams, one in the forward direction (up) and one in the backward direction (down). The two beams merge at ±90° scan direction (endfire).

The endfire configuration is widely used with fixed phase shifts between elements.

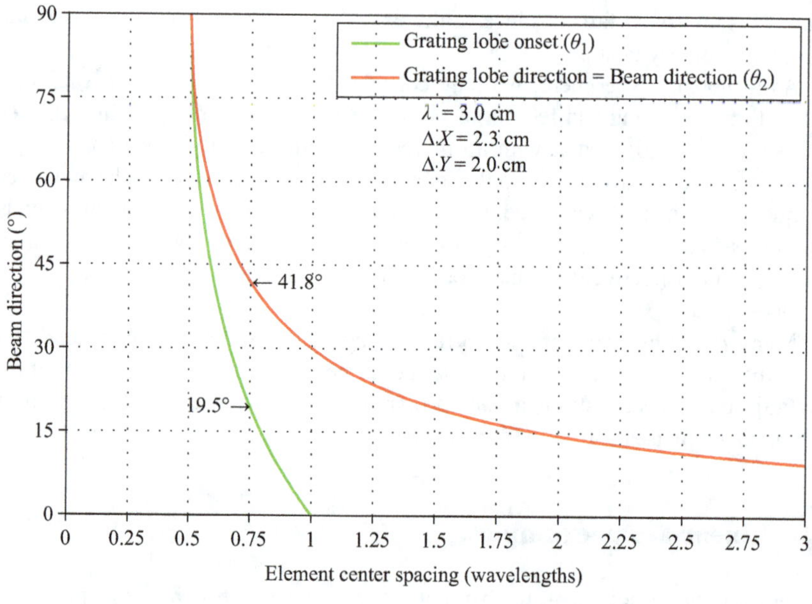

Figure 7.5 Triangular lattice grating lobes in (u, v) space
GratingLobes.m

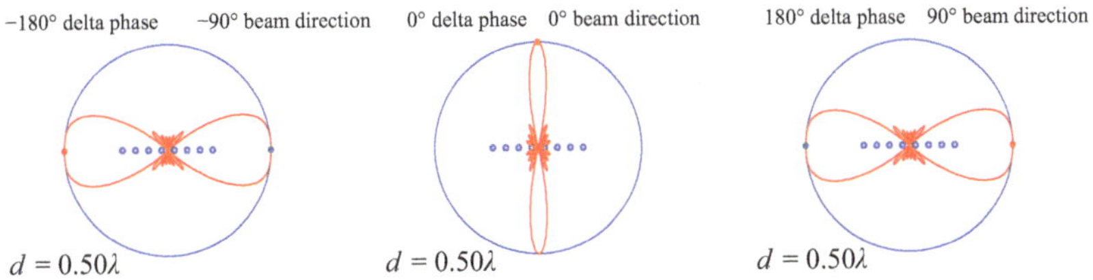

(a) –90° beam steering (b) 0° beam steering (c) 90° beam steering

Figure 7.6 λ/2 element spacing precludes grating lobes
DipoleArrayNoGroundPlane.m

Element spacing > λ/2 permits grating lobes

Figure 7.7 is similar to Figure 7.6 except that the elements are spaced at 3/4 λ and grating lobes appear at 19.5° as shown in Figure 7.5. The grating lobe is the same magnitude as the mainlobe in the pointing direction 41.8°.

Figure 7.8 illustrates the absence and presence of grating lobes for the two-element spacings, one-half and three-quarters wavelength.

7.5 Grating lobe suppression

Since some degree of grating lobe is almost always present, there is a considerable body of mitigation techniques summarized in Table 7.1.

Some systems use nonsteered subarrays several wavelengths in size. On boresight, the nonsteered subarray will have nulls matching the grating lobes of the array factor of the array. However, the grating lobes will emerge as soon as the beam is steered off boresight. Several of the satellite systems to be discussed use a single T/R module for 16 radiating elements (e.g., COSMO-SkyMed and TerraSAR). This constrains the azimuth scan capability, which is not an issue for side-looking SAR. It also enables use of a higher powered (more-efficient) T/R module and most importantly use of an existing T/R module. Alternatively, if the subarray beam may be

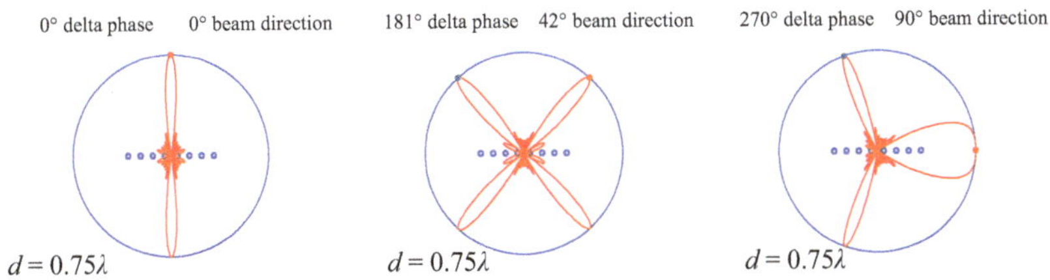

(a) 0° beam steering (b) 42° beam steering (c) 90° beam steering

Figure 7.7 ¾ λ element spacing produces grating lobes
DipoleArrayNoGroundPlane.m

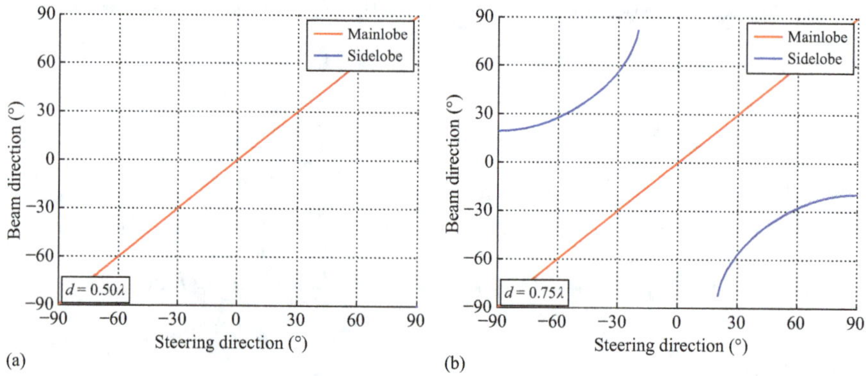

Figure 7.8 *Absence and presence of grating lobes: (a) no grating lobes and (b) grating lobes*
DipoleArrayNoGroundPlane.m

Table 7.1 *Grating lobe mitigation techniques*

Parameter	Comments
Restrict scan range	Primary approach
Restricted radiating element pattern will suppress grating lobes	This is almost always the case because elements larger than a wavelength become directional
Overlapped subarrays	Overlapped subarrays described in Section 2.7.2 provide additional control over the subarray pattern which assists in reducing grating lobes
Introduce uncorrelated errors	Redistributes grating lobe radiation so that the peaks are reduced although the total power is unaffected
Pointing them to empty space	Gain loss in the mainlobe will remain

steered, it is possible to keep its nulls (approximately) aligned with the grating lobes of the array factor of the thinned array.

Reference

[1] Sharp E. 'A triangular arrangement of planar-array elements that reduces the number needed'. *IRE Transactions on Antennas and Propagation*. 1961, vol. 9(2), pp. 126–9.

Chapter 8
Thinned arrays

The analysis in this book primarily concerns conventional arrays with elements spaced uniformly across a specified aperture to achieve specified performance: radiated power, sidelobes, etc. In this chapter, we consider thinned (or sparse) arrays by which we mean conventional arrays from which some number of elements have been deliberately removed. Rocca *et al.* [1] provide an overview of types of irregular arrays and associated design techniques.

Thinned arrays are useful in cases where the design complies with most performance requirements while exceeding others. It may be possible to reduce the excess performance (and cost) by selectively removing some elements.

One reason this circumstance might arise is that the beamwidth requires a certain aperture size while the corresponding aperture size provides more gain than necessary. Another possibility is that the element weighting is not uniform and some elements may contribute little or nothing in terms of transmitted/received power.

Failed elements may also be considered a thinned array. Although this case is unplanned and affects relatively few elements, if low sidelobes are a requirement, then element failure must be evaluated and specified. This approach was discussed in Section 6.2. Figure 5 in Ref. [2] reproduced in Figure 6.2a and b summarizes the analysis.

When the element count is reduced, the array factor gain is also reduced proportionately. The antenna gain may be preserved if the element is increased in area (and gain) maintaining aperture area. This design has the potential to reduce cost but the available scan volume will be reduced due to grating lobes (see Chapter 7). The radiating element, which is usually less than a wavelength in dimension, must be enlarged to efficiently capture incoming radiation.

One way to enlarge the element is to form a subarray from the smaller elements. In this case, the larger element will have its own beamwidth which will tend to suppress the far out grating lobes although still restricting the scan angle of the system.

Historically, the components and subassemblies of an ESA were relatively expensive, and considerable analysis was devoted to eliminating as many as possible. This is less of a concern now but it is still worthwhile to reduce weight and power whenever possible.

We consider the effects of thinning on the array factor of a 64×64 square array with $\lambda/2$ element spacing in Sections 8.1 and 8.3.

8.1 Random thinning

Random thinning may result from design or accident, e.g., failures. This section evaluates the reference array and summarizes relevant performance including gain, beamwidth, peak sidelobes (PSLR), and integrated sidelobes (ISLR).

The thinned array is created by randomly selecting elements to remove, based on the specified percentage of failures. The resulting percentage of vacancies is not exact. Performance is shown for 0% to 5% in 1% increments and 0% to 50% in 10% increments.

Figure 8.1 shows representative array patterns for no failures and 10% failures. The gain decreases by about 1.9 dB because the total elements are fewer. The expected reduction would be the ratio of elements or $10\log_{10}(922/1{,}024) = -2.1$ dB. The observed reduction of −1.9 dB is inexact because the deleted elements are random and there will be a spread of outcomes depending on the outcome of the *rand* function in MATLAB®.

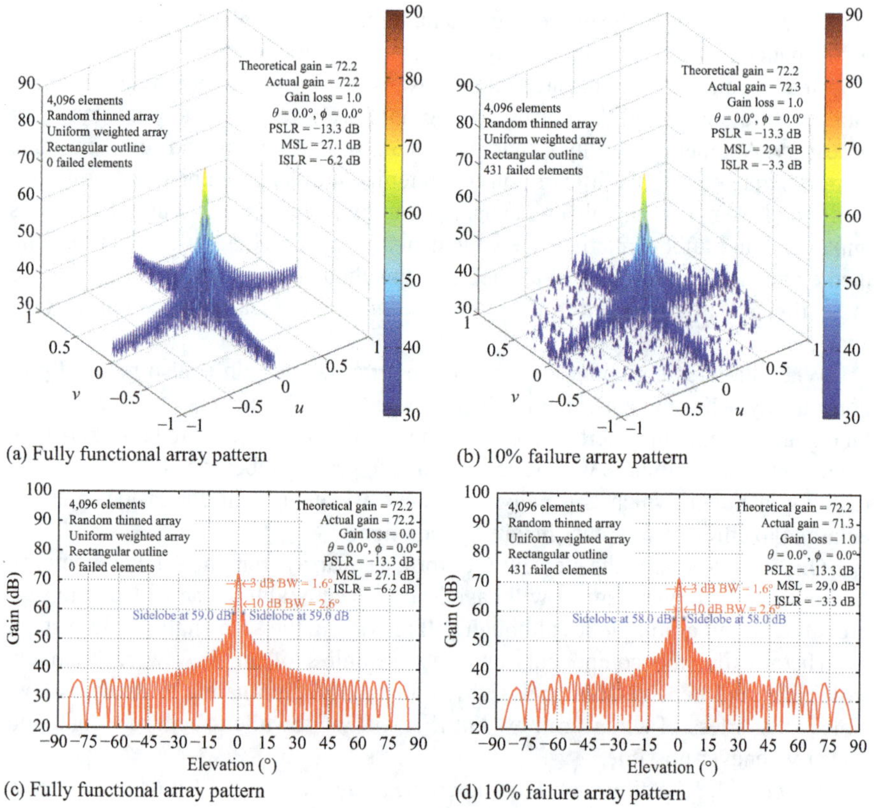

(a) Fully functional array pattern

(b) 10% failure array pattern

(c) Fully functional array pattern

(d) 10% failure array pattern

Figure 8.1 Thinned array analysis case
ThinnedSquare.m

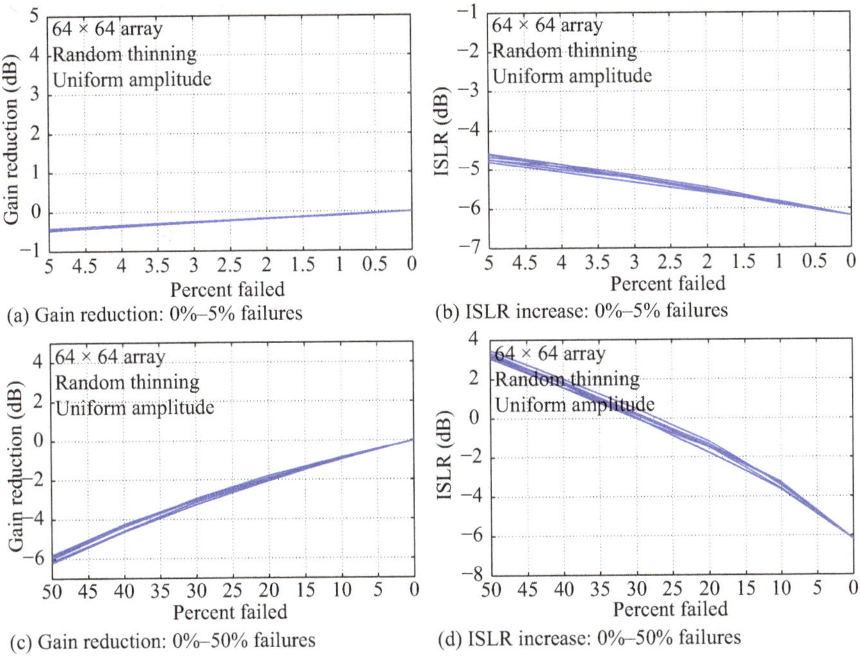

(a) Gain reduction: 0%–5% failures

(b) ISLR increase: 0%–5% failures

(c) Gain reduction: 0%–50% failures

(d) ISLR increase: 0%–50% failures

Figure 8.2 Thinned array analysis case
ThinnedSquare.m

The dominant effect of the thinning is to raise sidelobes in the off-axis regions which is the reason the ISLR increases from –6 to –3 dB. The beamwidth is unaffected, ±3.2° at –3 dB and ±5.3° at –10 dB.

The array is uniformly illuminated which means that the peak sidelobes and integrated sidelobes are already relatively high, so the degradation caused by element failure is less noticeable. Accordingly, the effect on performance may not be a problem. This is the reason that ESAs are considered tolerant of element failure. As we have seen in Chapter 6, low sidelobe patterns are considerably more sensitive to failures.

Figure 8.2 summarizes these parameters for element failures ranging from 0% to 50%. They generally apply across a broad range of failures.

8.2 Effects of amplitude errors

Mulcahey and Sarcione [3], also discussed in Section 6.2, describe an example of uncalibrated and calibrated receive amplitudes data on T/R module functionality after array installation and checkout. Figure 8.3 shows the module status for each position in the array. The data show that 96.33% of the modules met specification, while 3.67% were degraded in some respect, primarily low gain. This example demonstrates no significant degradation resulting from a few failed elements.

Array performance shown in Figure 8.4 illustrates the antenna pattern based on the worst-case assumption that degraded elements are nonfunctional. The figure shows the array elevation pattern for the error-free and actual case assuming uniform illumination. The peak gain and peak sidelobe ratios are virtually unchanged (0.1 and 0.3 dB, respectively). The integrated sidelobe ratio shows a small degradation of 0.7 dB.

Differences are more noticeable if we apply Taylor weighting (R = 30 dB, $\bar{n} = 4$). Figure 8.5 shows the resulting elevation pattern. Peak gain and peak sidelobes remain virtually unchanged while ISLR degrades by 6.5 dB.

For both the uniform illumination case and the Taylor weighting case, the principal plane patterns change very little due to missing elements. The primary effect is

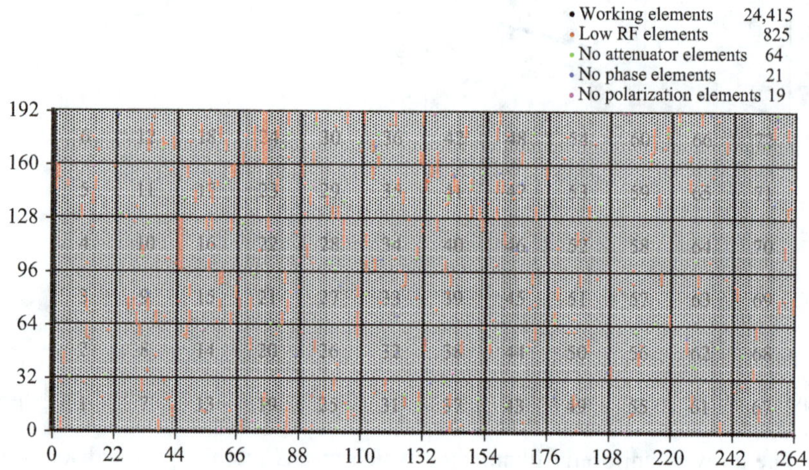

Figure 8.3 Receive check scan results for user operational evaluation system #1
After Mulcahey and Sarcione [3], Figure 1

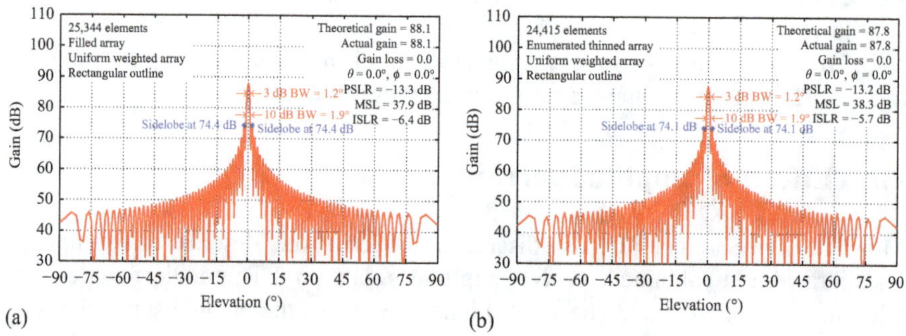

Figure 8.4 UOES #1 example with uniform weighting (transmit): (a) no failed elements and (b) 3.67% failed elements

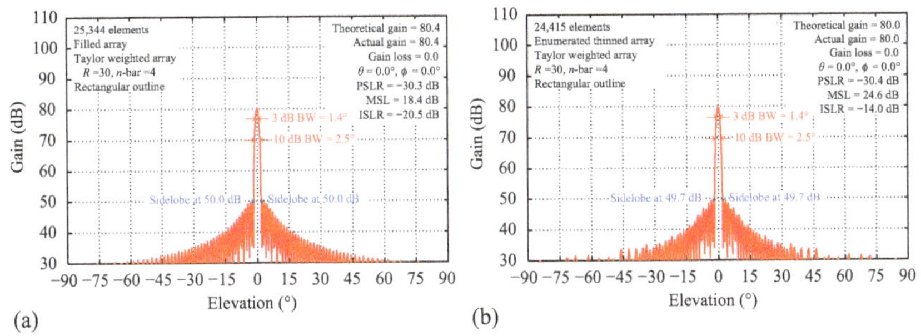

Figure 8.5 *UOES #1 example with Taylor weighting (receive): (a) no failed elements and (b) 3.67% failed elements*
ThinnedSquare.m

in the intercardinal planes where the filled array cases have extremely low sidelobes but missing elements bring the sidelobe level up.

8.3 Systematic thinning for amplitude taper

Taylor weighting, described in Section 9.6, is a popular technique for reduced sidelobes. It reduces the gain/power of array elements in a prescribed way but leaves the phase unchanged. The gain characteristic depends on the desired sidelobe level (R) and a prescribed number ($\bar{n}$) of sidelobes of constant level.

Taylor weighting is problematic in transmit because the transmit power amplifier achieves maximum efficiency when operating at saturation. Backing off the drive power does not greatly affect the power consumption of the amplifier, so the total power consumption does not decrease proportionate to the RF power reduction. This is a problem because available power and heat dissipation constrain performance. We may approximate the desired power distribution by operating some elements at full power and turning off others. As will be noted, thinning increases the sidelobe level and degrades the ISLR. While thinning provides more efficient utilization of the system power, it may not significantly decrease the cost of the array since the number of receive elements is typically unchanged.

Figure 8.6 illustrates the amplitude weights to achieve R = 30 dB and $\bar{n} = 4$ weighting. Figure 8.6a shows the map of element weights.

Figure 8.7 illustrates another technique to achieve the desired weighting. The thinned approximation to the Taylor weights is shown in Figure 8.7b where the Y-axis is simply the number of transmit elements in each row and column, respectively. The elements were selected with a probability equal to the desired amplitude weighting at that location. Hence, the center elements whose amplitude is near unity is densely populated. The amplitude weight near the edge of the array is small, so less than 10% of the elements are utilized. The array configuration may be determined randomly or by means of a goal-seeking algorithm; their performance varies

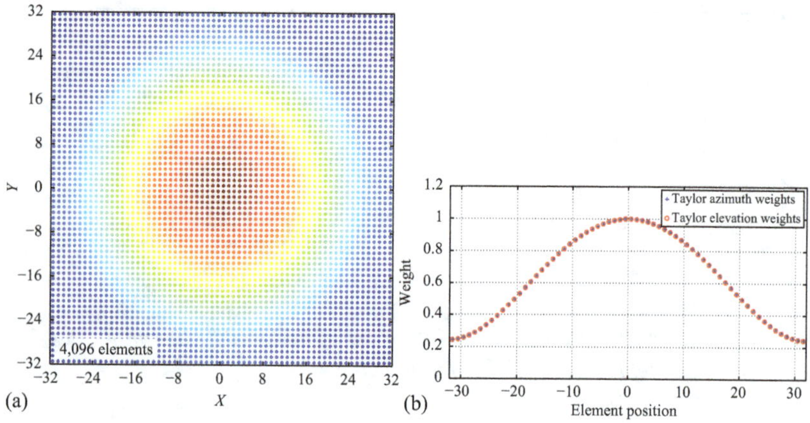

Figure 8.6 Taylor weighting for 64 elements: (a) array weighting and (b) Taylor weights for R = 30 dB and $\overline{n}$ = 4.
ThinnedSquare.m

slightly. Figure 8.7a shows the corresponding map of active elements, using fewer than 25% of the total elements.

Figure 8.6b shows computed amplitude weights for the azimuth and elevation axes of the array. The weight for each individual element is computed as the product of the azimuth and elevation weights; accordingly, the weight for edge elements is only about 6% and only the center elements operate at close to 100%.

We compute the antenna patterns for the two cases. The thinned transmit array has an inherent Taylor taper based on the criteria used to switch off specific elements.

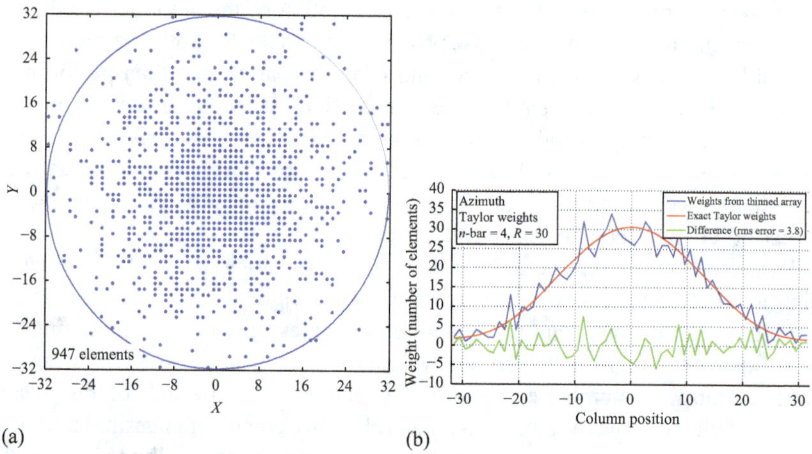

Figure 8.7 Thinned implementation of Taylor beam shaping: (a) array thinning and (b) thinned array equivalent weights
ThinnedSquare.m

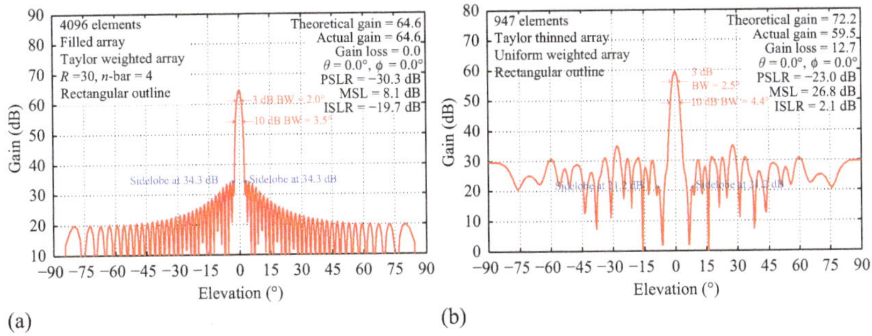

Figure 8.8 *Elevation pattern cuts for two cases: (a) Taylor weighting and (b) Taylor thinning*

ThinnedSquare.m

Figures 8.8 and 8.9 show the pattern for the thinned case (transmit) and the filled and Taylor-weighted case (receive). The thinned case incurs elevated sidelobes throughout the scan volume in comparison to the filled pattern whose sidelobes occur near the two principal planes. The overall sidelobe levels are comparable but the integrated sidelobes are much higher for the thinned array.

With the thinned case for transmit, it is still preferable to use the filled array for receive with the weighting, as shown in Figure 8.6b as it produces much better ISLR. Figure 8.10 illustrates the combined performance of the transmit and receive patterns. The combined transmit/receive performance is shown in Figure 8.10c and d and Table 8.1. The cost saving of omitting the unused elements in the thinned case is at least partially offset by the additional complexity of the thinned array.

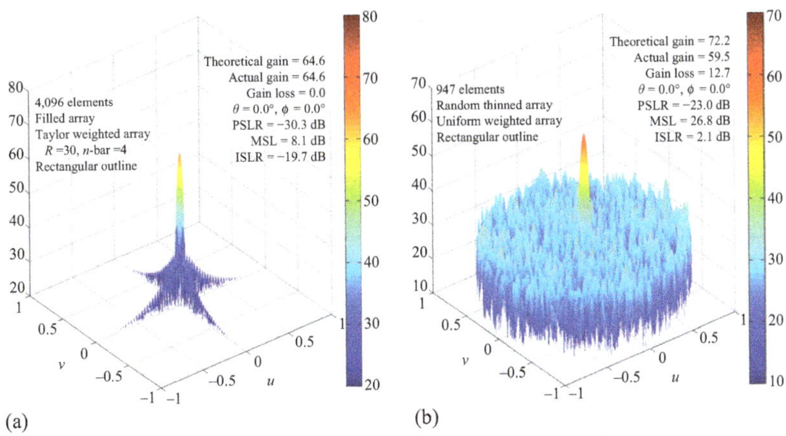

Figure 8.9 *Surface plots for two cases: (a) Taylor weighting and (b) Taylor thinning*

ThinnedSquare.m

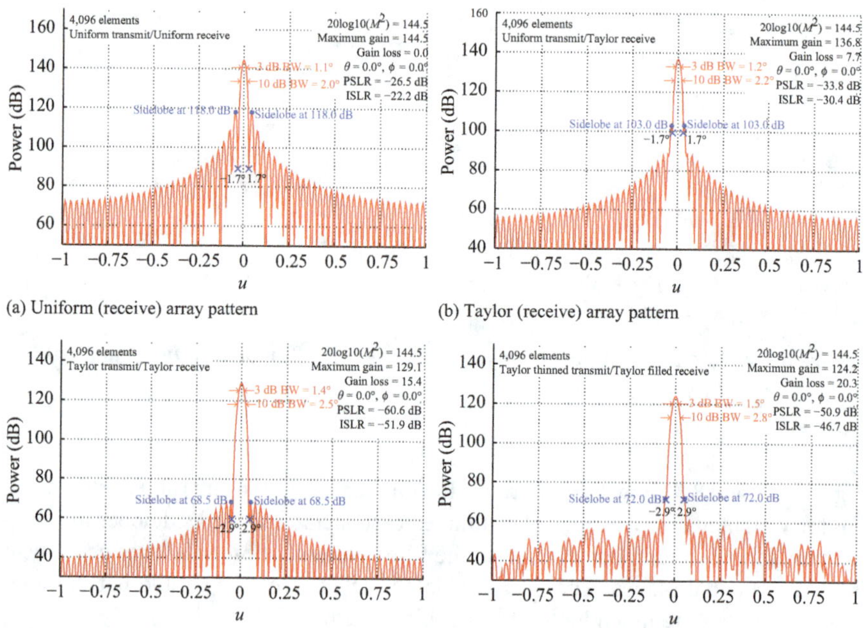

(a) Uniform (receive) array pattern

(b) Taylor (receive) array pattern

(c) Taylor transmit/receive array pattern

(d) Taylor thinned (receive) array pattern

Figure 8.10 Transmit receive product (two-way) pattern
ThinnedSquare.m

Planar arrays provide an elegant, but inflexible, solution to aperture tapering. The airborne warning and control system (AWACS) antenna shown in Figure 1.2 achieves sidelobes more than 50 dB down one way [4]. It uses unequal power dividers in its feed structure to allocate the power from the transmitter to each radiating element, so all of the transmit power is radiated. This technique is not applicable for ESAs as power is generated at each element and elements generating transmitting at less than peak power operate inefficiently.

Table 8.1 summarizes some relevant parameters for both approaches. Both power and area are reported in arbitrary units, so only the ratios are meaningful. Note that the computed ISLR is for the one-way pattern, whereas system requirements commonly specify the two-way ISLR. The two-way ISLR must be calculated using the product of the array factors and is not the product of two one-way values. Two-way ISLR and PSLR are described in Section 4.14.

The performance of the thinned array is similar to that of the fully filled array. The mainbeam power is comparable in the two cases. The sidelobe power is considerably higher for the thinned case, suggesting that the missing elements primarily controlled sidelobes and the thinned array ISLR is about 8 dB higher in consequence.

Leifer [5] has published a specific thinned design which is illustrative of the technique. He applied an optimization routine to the randomization procedure to improve the pattern. He also used the resulting design to examine a beam spoiling

Table 8.1 Comparison of taper techniques

Parameter	Thinned	Filled
Power mainbeam	1,994	3,788
Power sidelobe	2,907	41
Total power	4,901	3,828
Area mainbeam	0.012	0.007
Area sidelobe	6.272	6.276
L_final	35	34
Mainbeam peak power (dB)	59.7	64.6
Sidelobe peak power (dB)	35.6	34.3
Mainbeam average power (dB)	52.3	57.4
Sidelobe average power (dB)	26.7	8.1
Integrated mainbeam power (dB)	33.0	35.8
Integrated sidelobe power (dB)	34.6	16.1
ISLR (dB)	1.6	–19.7

technique. Section 10.3 will describe techniques for beam spoiling based on phase weighting.

References

[1] Rocca P., Oliveri G., Mailloux R.J., Massa A. 'Unconventional phased array architectures and design methodologies—a review'. *Proceedings of the IEEE.* 2016, vol. 104(3), pp. 544–60.

[2] Ruze J. *Pattern Degradation of Space Fed Phased Arrays*. Tech. Rep. Project report SBR-1. Air force contract F19628-80-C-0002 (ARPA order 3724). Lexington, MA, USA: Lincoln Laboratory, MIT; 1979. Available from https://apps.dtic.mil/dtic/tr/fulltext/u2/a083264.pdf.

[3] Mulcahey J.K., Sarcione M.G. 'Calibration and diagnostics of the THAAD solid state phased array in a planar nearfield facility'. *Proceedings of International Symposium on Phased Array Systems and Technology*. 1996, pp. 322–6.

[4] Schrank H.E. 'Low-sidelobe phased array and reflector antennas' in Brookner E. (ed.). *Aspects of Modern Radar*. Norwood, MA, USA: Artech House; 1988. pp. 337–68.

[5] Leifer M.C. 'An airborne radar phased array with low sidelobes and a spoiled beam'. *2018 IEEE Radar Conference (RadarConf18)*. 2018, pp. 0072–6.

Chapter 9

Beamwidth and sidelobes

The designer has almost unlimited freedom to specify the amplitude and phase distribution of the signal across the aperture. However, the range of beam shapes is limited. A uniform amplitude produces the maximum gain and minimum beamwidth but relatively high sidelobes. Any attempt to lower sidelobes generally reduces gain and increases beamwidth.

We may consider analysis and synthesis techniques for antenna optimization. In general, analysis produces a solution although perhaps not the best solution possible. Various optimization techniques have been used and are limited by computing resources, the existence of local optima, or both. Synthesis uses the desired solution and works backwards but does not necessarily produce a good result or any result at all. Stutzman and Licul [1] provide an overview of synthesis techniques.

This chapter considers analytic techniques. It describes a number of solutions and their beam characteristics. Chapter 10 will consider synthetic techniques.

Table 9.1 summarizes some classic design cases which will be discussed.

This section applies the Schelkunoff formulation to a variety of array weights designed to reduce sidelobes. It is a convenient formalism to describe the aperture efficiency, the mainbeam width, and peak sidelobe level. The examples in this section are presented with a common format. They are based on linear arrays with M elements, 11 in most examples. The elements are spaced at 0.5λ. The taper efficiencies in this chapter are one dimensional. If they are applied in both axes of a two-dimensional array, they must be squared.

Table 9.1 Classic antenna weighting

Analytic weighting	Comments
Uniform	Best gain and beamwidth
Triangular	Square of uniform pattern
Binomial	No sidelobes, restricted to small arrays
Dolph–Chebyshev	All sidelobes equal
Taylor	First n sidelobes equal
Synthetic	
Fourier	Unrestricted aperture function
Woodward–Lawson	
Quadratic	

9.1 Schelkunoff representation

Schelkunoff [2] developed an early method for understanding the relationship of the excitation function and the beam shape which he published in the *Bell System Technical Journal*. He rewrote the array factor (2.43) and with the use of a substitution of variable, transformed the sum to a product. This transformation is commonly termed the "z transform" but Schelkunoff avoided this terminology and derived his results directly from the representation:

$$AF = \sum_{m=0}^{M-1} a_m e^{jm(\psi - \psi_s)} \tag{9.1}$$

$$\psi = k\Delta x \sin\theta, \ \psi_s = k\Delta x \sin\theta_s \tag{9.2}$$

$$z = e^{j(\psi - \psi_s)} \tag{9.3}$$

$$AF = \sum_{m=0}^{M-1} a_m z^m = a_M \prod_{m=0}^{M-1} (z - z_m) \tag{9.4}$$

Schelkunoff theorems

Schelkunoff formulated several theorems based on this representation.

Theorem 1. Every linear array with commensurable separations between the elements can be represented by a polynomial and every polynomial can be interpreted as a linear array.

Theorem 2. There exists a linear array with a space factor equal to the product of the space factors of any two linear arrays.

Theorem 3. The space factor of a linear array of n apparent elements is the product of the space factors of $(n - 1)$ virtual couplets with their null points at the zeros of $\sqrt{\Phi}$: $t_1, t_2, \ldots, t_{n-1}$.

The formulation also provides additional insight into beam characteristics. Since z has unit magnitude, all the roots must also have unit magnitude and the product has a very simple form. The roots of the sum have a physical interpretation; they correspond to pattern nulls. Nulls alternate with peaks (sidelobes). The peak value is smaller when nulls are closer. Grouping the nulls away from the mainbeam direction reduces the sidelobes while broadening the peak.

For $0° \leq \theta \leq 180°$, ψ varies by $2k\Delta x$. Roots may fall inside or outside of this range representing nulls in real space or outside real space and corresponding to undersampling, Shannon sampling, and oversampling.

Single beam

Consider the uniform illumination case represented by (9.5)

$$AF = z^M + z^{M-1} + z^{M-2} + \cdots + z^2 + z + 1$$
$$AF = \sum_{m=0}^{M} z^m \prod_{m=0}^{M} (z - z_m) \tag{9.5}$$

whose roots are

$$z_m = e^{(2m-M)\pi/M} \quad \text{for } M \text{ even} \quad m = 1 : M, \quad M \neq \frac{M}{2}$$
$$z_m = e^{(2m-M-1)\pi/M} \quad \text{for } M \text{ odd} \quad m = 1 : M, \quad M \neq \frac{M+1}{2} \tag{9.6}$$

The resulting beam pattern is shown in Figure 9.1 with its associated element weights.

Addition of missing root

There is one missing root with value of 1. If we insert the missing root by adding a factor $(z - 1)$, we obtain (9.7) and Figure 9.2. The mainbeam disappears and is replaced by a sidelobe of one-half the width of the mainlobe. Additionally, 10 of the element weights become zero as seen in (9.7) so we are left with a two-element array, 180° out of phase:

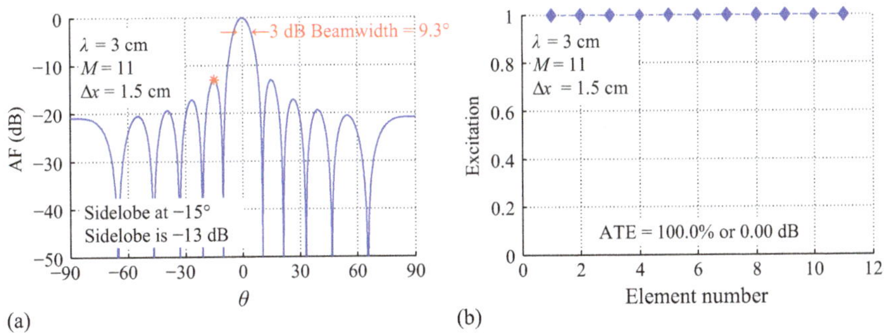

Figure 9.1 Original 11 element array: (a) equal weight subarray pattern and (b) equal weight subarray roots
Schelkunoff.m

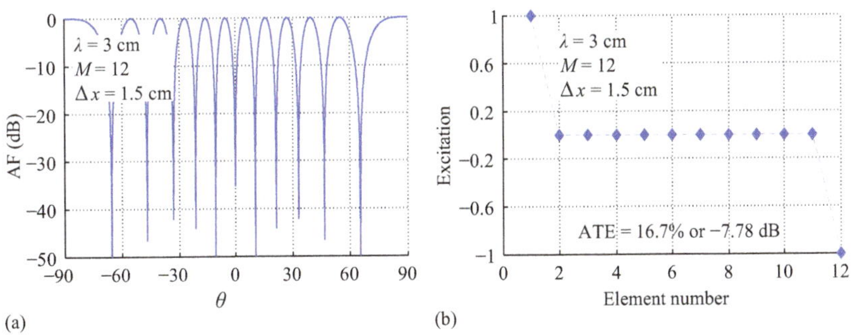

Figure 9.2 Two-element array resulting from addition of missing root: (a) added root subarray pattern and (b) added root subarray roots
Schelkunoff.m

$$\begin{aligned} AF &= (z-1)\left(z^M + z^{M-1} + z^{M-2} + \cdots + z^2 + z + 1\right) \\ AF &= z^{M+1} - 1 \end{aligned} \tag{9.7}$$

Also it is obvious that the roots of both polynomials have roots with magnitude 1.

Two slits separated by 5.5λ

Figure 9.3 shows the development of 10 maxima. It may be compared with Figure 2.2, which shows the pattern of an aperture of width 4λ. Two missing maxima at ±90° are suppressed by the element pattern of the slit. If we used dipoles for elements, i.e., omnidirectional, we would see the missing two maxima, similar to the end-fire condition in Figure 7.6c.

The previous example corresponds to the well-known double slit experiment in optics. The visible beams are not of equal amplitude because of the finite width slits which impose an element factor.

9.2 Uniform weighting (unweighted)

Uniform weighting is the default condition. It provides maximum gain and −13 dB peak sidelobes. It is useful for many cases as the two-way sidelobes are 26 dB. Figure 9.4 shows the analysis for uniform weighting.

Uniform example (M = 11)

Figure 9.4a plots the antenna pattern with several parameters as indicated. The pattern is boresight, centered on 0°. The half-power beamwidth is indicated and the red asterisk shows the first sidelobe location; sidelobe characteristics are indicated in the box in the lower left corner.

Figure 9.4b shows the excitation coefficient (magnitude). The taper efficiency is 100% and as indicated at the bottom of the figure.

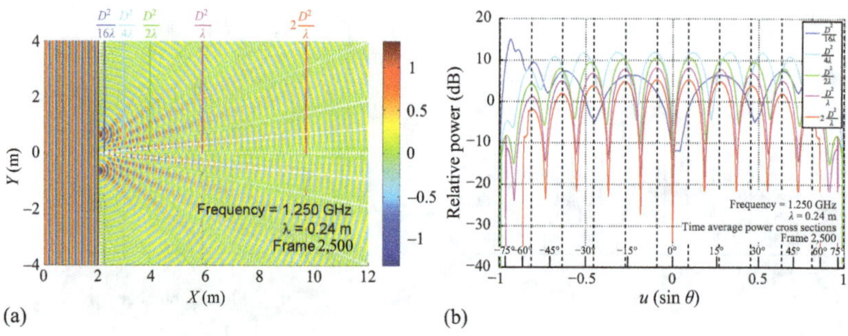

Figure 9.3 FDTD model of two slits separated by 5.5λ: (a) field strength and (b) cross sections
DoubleSlit.m

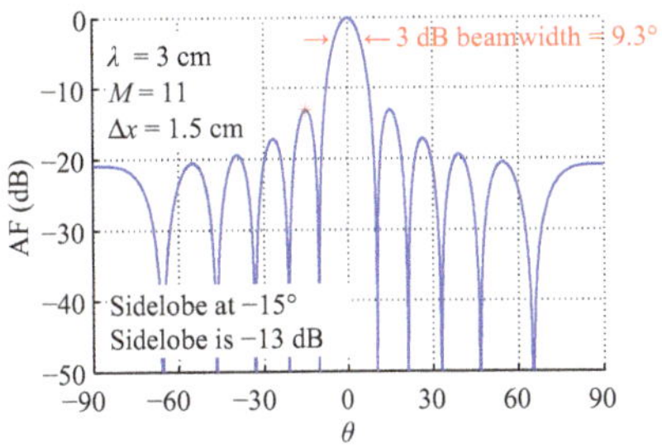

(a) Array pattern

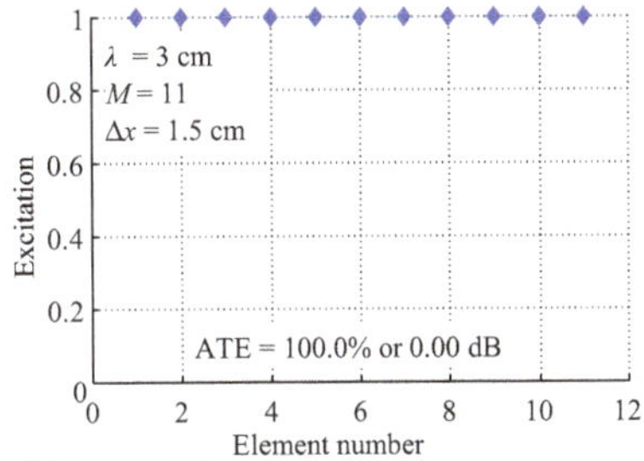

(b) Array weight

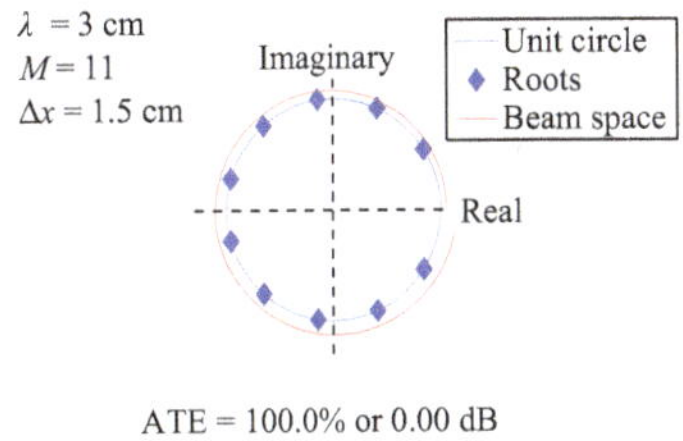

(c) Polynomial roots graph

Root	Real + imaginary	Magnitude	Angle
1	−0.959 + 0.282i	1.000	163.6°
2	−0.959 − 0.282i	1.000	−163.6°
3	−0.630 + 0.777i	1.000	129.0°
4	−0.630 - 0.777i	1.000	−129.0°
5	−0.090 + 0.996i	1.000	95.2°
6	−0.090 − 0.996i	1.000	−95.2°
7	0.785 + 0.619i	1.000	38.3°
8	0.785 − 0.619i	1.000	−38.3°
9	0.451 + 0.893i	1.000	63.2°
10	0.451 − 0.893i	1.000	−63.2°

(d) Polynomial roots table

Figure 9.4 Uniform example (M = 11)
Schelkunoff.m

Figure 9.4c and d plots and tabulates the roots of the excitation function. (Factorization is accomplished with MATLAB®.) The roots are equally spaced. The plot shows real space in terms of a red wrapper around the circle of roots. Since the element spacing is $\lambda/2$, the red wrapper spans exactly 360°. With undersampling ($> \lambda/2$), its span would be less than 360° and with oversampling ($< \lambda/2$), its span would be greater than 360°.

This weighting is most common for the transmit case due to its efficiency. It is much less common in receive where efficiency is less important than sidelobes. The cases that follow will show more points of interest.

9.3 Triangular weighting

Triangular weighting is interesting from an analytic point of view. It represents the autocorrelation function of the uniform illumination method, so we expect that the pattern is the square of the uniform pattern. The weights are zero at edges, unity in center, and linearly interpolated in-between. It is also a special case of binomial weighting for three elements. Figure 9.5 shows the analysis of triangular weighting.

Triangular example (M = 11)

The beamwidth is 12.3° as compared to 9.2° for the uniform case. We previously observed that lower sidelobes necessarily entail wider mainbeams. The sidelobes are

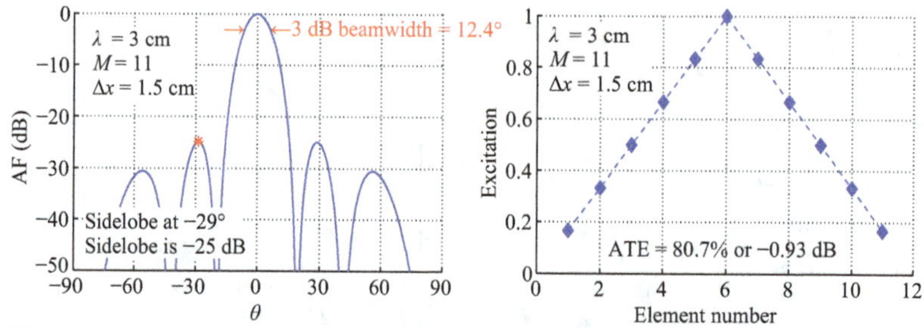

(a) Array pattern

(b) Array weight

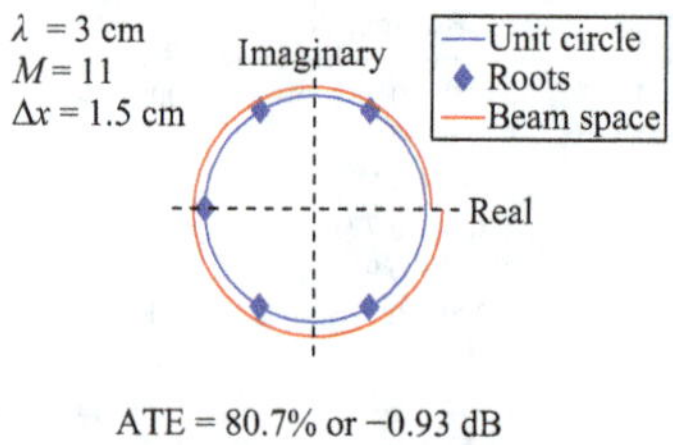

Root	Real + imaginary	Magnitude	Angle
1	0.500 + 0.866i	1.000	60.0°
2	0.500 − 0.866i	1.000	−60.0°
3	0.500 + 0.866i	1.000	60.0°
4	0.500 − 0.866i	1.000	−60.0°
5	−1.000 + 0.000i	1.000	180.0°
6	−1.000 + 0.000i	1.000	180.0°
7	−0.500 + 0.866i	1.000	120.0°
8	−0.500 − 0.866i	1.000	−120.0°
9	−0.500 + 0.866i	1.000	120.0°
10	−0.500 − 0.866i	1.000	−120.0°

(c) Polynomial roots graph

(d) Polynomial roots table

Figure 9.5 Triangular example (M = 11)
Schelkunoff.m

considerably lower, –25 versus –13 dB. The lower left shows the aperture taper and the taper efficiency is now 80.7%.

The right-hand side shows the roots and we see that they paired (consistent with squaring the illumination function). As there are only five distinct roots, the number of nulls in the pattern is fewer, three on each side versus five for the uniform case.

9.4 Binomial weighting

Binomial weighting is an extreme case with little practical consequence. It is, however, a good example of how to eliminate discontinuities at array edges by moving all of the zeros of the polynomial to the edge of the scan volume. Positioning all of the nulls at the edge of the scan volume, i.e., $\psi = 0$, so that $z_m = 1$ for all m creates a beam pattern with no sidelobes. This is termed the binomial array because the weighting function is the binomial function. This technique was first proposed by John Stone Stone in US Patents 1,643,323 and 1,715,433. Figure 9.6 shows the parameters for binomial weighting.

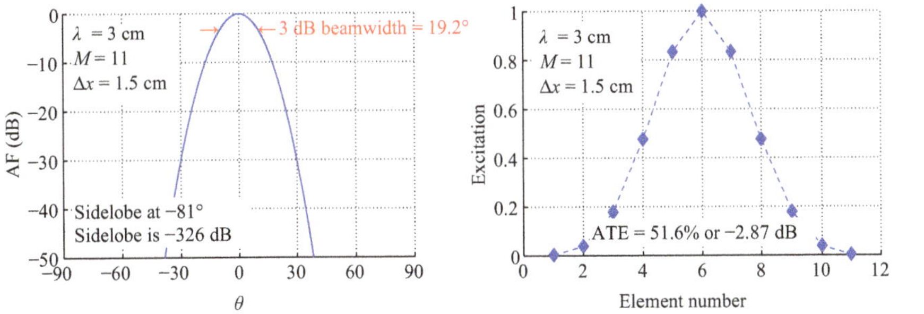

(a) Array pattern (b) Array weight

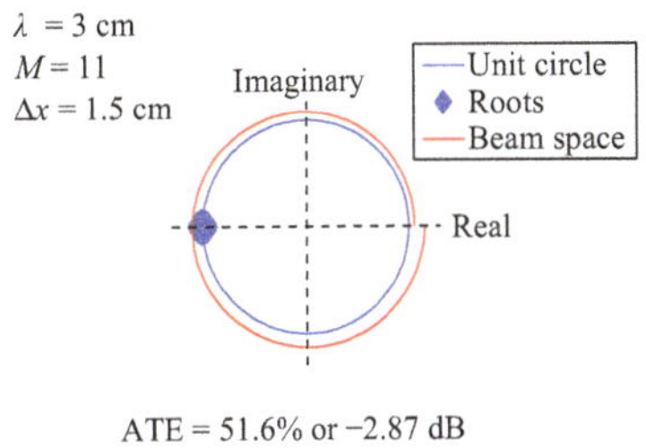

Root	Real + imaginary	Magnitude	Angle
1	−1.046 + 0.000i	1.046	180.0°
2	−1.038 + 0.027i	1.038	178.5°
3	−1.038 − 0.027i	1.038	−178.5°
4	−1.015 + 0.044i	1.016	177.5°
5	−1.015 − 0.044i	1.016	−177.5°
6	−0.986 + 0.045i	0.987	177.4°
7	−0.986 − 0.045i	0.987	−177.4°
8	−0.962 + 0.028i	0.963	178.4°
9	−0.962 − 0.028i	0.963	−178.4°
10	−0.953 + 0.000i	0.953	180.0°

(c) Polynomial roots graph (d) Polynomial roots table

Figure 9.6 Binomial example (M = 11)
Schelkunoff.m

Binomial example (M = 11)

The beamwidth of 19.1° is even wider than the triangular case. MATLAB® computed a first sidelobe level of –325 dB which is only a computational artifact.

The aperture efficiency has dropped to 51.6% which is problematic. Most of the array is severely underutilized. The problem only becomes worse for larger arrays, so this distribution is not useful for large arrays.

9.5 Dolph–Chebyshev

Dolph developed a weighting using Chebyshev polynomials which achieves a constant sidelobe level. This technique matches the roots of a Chebyshev polynomial with the roots of the aperture illumination function. Its applicability is limited because as the array becomes large and the mainbeam becomes very narrow, the sidelobes consume an increasing amount of the total power as their amplitude does not change. This weighting provides the narrowest beamwidth (at first null) for specified sidelobe level or lowest sidelobe level for specified beamwidth.

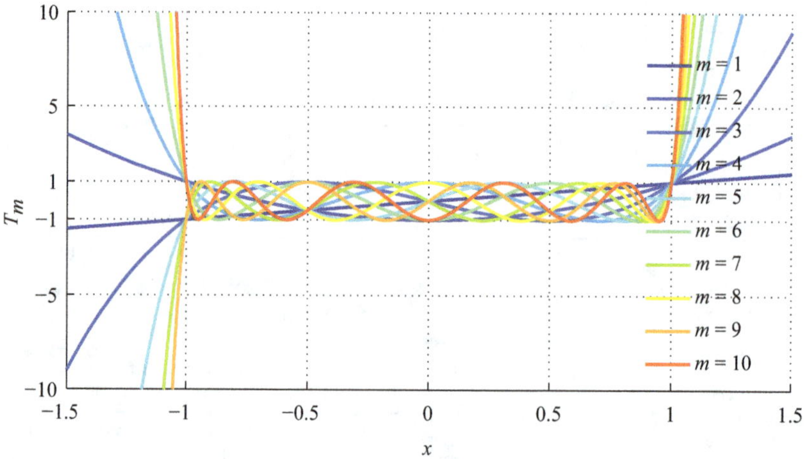

Figure 9.7 Chebyshev polynomials
Chebyshev.m

Chebyshev polynomials

Figure 9.7 graphs the first few Chebyshev polynomials.

The definition of the polynomials is shown in (9.6) and (9.7):

$$\begin{aligned} T_m(x) &= \cos(m \cdot \cos^{-1}x), & |x| \le 1 \\ T_m(x) &= \cosh(m \cdot \cosh^{-1}x), & x > 1 \\ T_m(x) &= (-1)^m \cosh(m \cdot \cosh^{-1}x), & x < -1 \end{aligned} \tag{9.8}$$

$$\begin{aligned} T_0(x) &= 1 \\ T_1(x) &= x \\ T_{n+1} &= 2xT_n(x) - T_{n-1}(x) \end{aligned} \tag{9.9}$$

Aperture weight derivation

The derivation of the coefficients is fairly elaborate:

$$\mathrm{AF} = \sum_{m=0}^{M-1} a_m \, e^{jkm\Delta x \sin\theta \, \phi} \tag{9.10}$$

$$\mathrm{AF}(\theta) = e^{jk_0(M+1)/2\Delta x \sin\theta} \sum_{-(M-1)/2}^{(M-1)/2} a_m \, e^{jk_0 m\Delta x \sin\theta} \tag{9.11}$$

$$\mathrm{AF}(\theta) = \sum_{-(M-1)/2}^{(M-1)/2} a_m \, e^{jk_0 m\Delta x \sin\theta} \tag{9.12}$$

$$\mathrm{AF}(\theta) = a_0 + \sum_{m=1}^{(M-1)/2} a_m \cos(2m \cos^{-1}x) \tag{9.13}$$

Result

For M odd:

$$a_m = \sum_{i=1}^{M} T_{M-1}\left(c\cos\frac{\psi_i}{2}\right)\ \cos\left(m\psi_i\right) \tag{9.14}$$

For M even:

$$a_m = \sum_{i=1}^{M} T_{M-1}\left(c\cos\frac{\psi_1}{2}\right)\ \cos\left(\left(m-\frac{1}{2}\right)\psi_i\right) \tag{9.15}$$

c is a function of the sidelobe ratio R:

$$c = \cosh\left(\frac{\cosh^{-1}(R)}{M-1}\right) \tag{9.16}$$

These are the desired weights which are different depending on whether the number of elements is even or odd.

Dolph–Chebyshev example (M = 11)

Figure 9.8 shows the analysis for the Dolph–Chebyshev example.

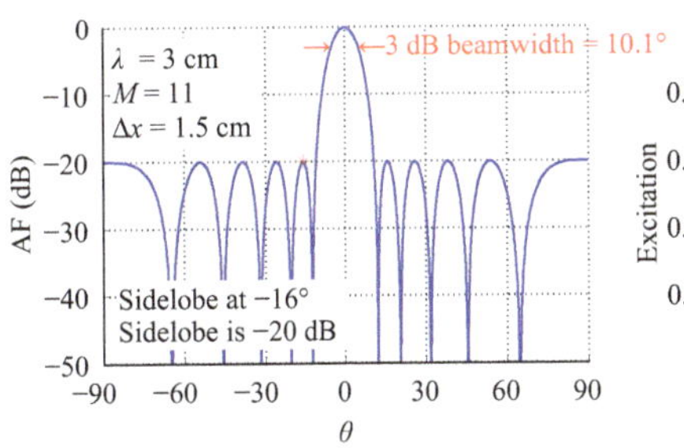

(a) Array pattern

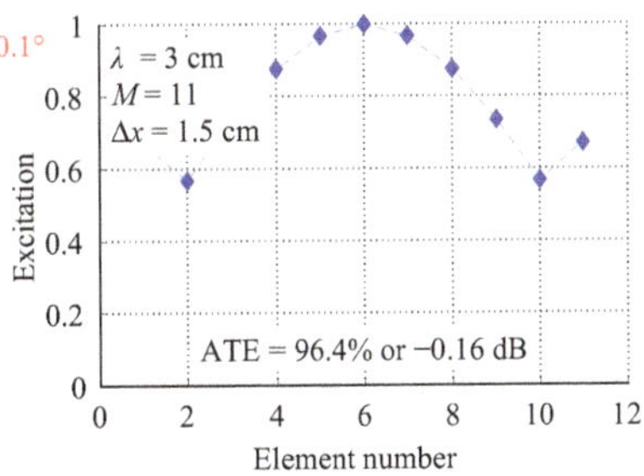

(b) Array weights

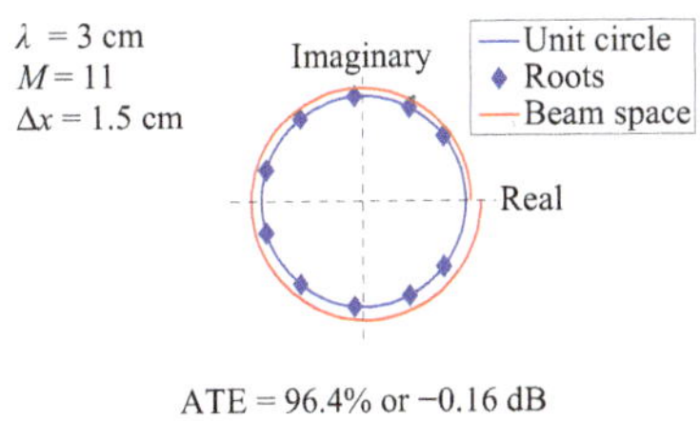

(c) Polynomial roots graph

Root	Real + imaginary	Magnitude	Angle
1	0.786 + 0.618i	1.000	38.2°
2	0.454 + 0.891i	1.000	63.0°
3	−0.085 + 0.996i	1.000	94.8°
4	−0.623 + 0.783i	1.000	128.5°
5	−0.955 + 0.296i	1.000	162.8°
6	−0.955 – 0.296i	1.000	−162.8°
7	−0.623 – 0.783i	1.000	−128.5°
8	0.786 – 0.618i	1.000	−38.2°
9	0.454 – 0.891i	1.000	−63.0°
10	−0.085 – 0.996i	1.000	−94.8°

(d) Polynomial roots table

Figure 9.8 Dolph–Chebyshev example (M = 11)
Schelkunoff.m

Figure 9.8 shows the case for 11 elements. As noted, the sidelobe level is constant at –20 dB. This is a parameter in the equation. The beamwidth is 10.1 which is only slightly greater than the uniform illumination case.

The aperture weighting is unusual in that the minima are not at the edge of the array. The uptick near the edge improves efficiency as well as raising the far out sidelobes, compared to the uniform case.

9.6 Taylor weighting

Taylor observed that it was possible to combine the best properties of the uniform and Dolph weightings. He modified the Dolph–Chebyshev, retaining the near sidelobe structure (and polynomial zeros) and modifying the far sidelobe structure (and polynomial zeros) to use the zeros of the sin x/x function which has lower far sidelobes. The transition between the two functions is based on two parameters σ and n-bar, where σ is the scale factor for the Dolph–Chebyshev function and n-bar is the number of Dolph–Chebyshev equal sidelobes. This approach may be viewed as reducing the near-in sidelobes of the uniform distribution by replacing them with Chebyshev roots. Figure 9.9 shows the analysis for the Taylor example.

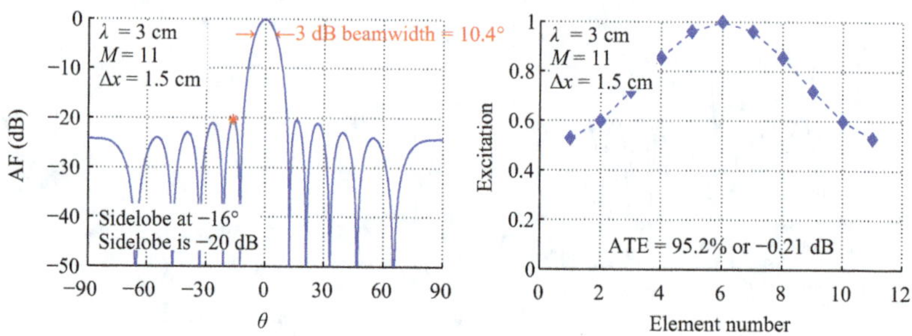

(a) Array pattern

(b) Array weights

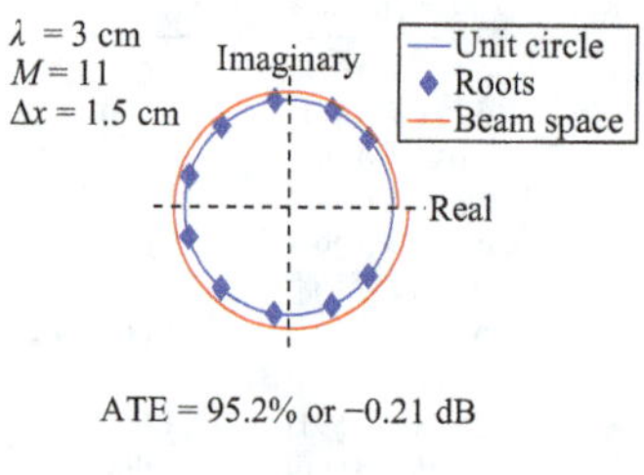

(c) Polynomial roots graph

Root	Real + imaginary	Magnitude	Angle
1	– 0.959 + 0.282i	1.000	163.6°
2	– 0.959 – 0.282i	1.000	−163.6°
3	– 0.630 + 0.777i	1.000	129.0°
4	– 0.630 – 0.777i	1.000	−129.0°
5	– 0.090 + 0.996i	1.000	95.2°
6	– 0.090 – 0.996i	1.000	−95.2°
7	0.785 + 0.619i	1.000	38.3°
8	0.785 – 0.619i	1.000	−38.3°
9	0.451 + 0.893i	1.000	63.2°
10	0.451 – 0.893i	1.000	−63.2°

(d) Polynomial roots table

Figure 9.9 Taylor example (M = 11)
Schelkunoff.m

Table 9.2 Beam shape comparisons for 11 element arrays

Method	**Beamwidth (°)**	**Efficiency (%)**	**First sidelobe (dB)**
Uniform	9.2	100	–13
Triangular	12.3	80.7	–25
Binomial	19.1	51.6	None
Dolph–Chebyshev	10.1	96.4	–20
Taylor (*n*-bar = 5)	10.1	96.3	–20
Fourier reconstruction to first null	13.6	67.3	–22
Fourier reconstruction to second null	16.9	52.5	–23
Fourier reconstruction to third null	17.8	43.5	–23
Woodward–Lawson	27.7	69.8	–15
Quadratic phase	26.7	100	–6

Fourier reconstructions required 14, 25, and 36 elements, respectively

9.7 Beam performance comparisons

Table 9.2 summarizes the performance of the weighting techniques described in Sections 9.2–9.6.

The preceding techniques achieve their stated objective with some adverse effect on other parameters. Taylor weighting on receive is the most common application in practice. Transmit beamshaping is not common as the aperture efficiency decreases the mainbeam effective isotropic radiated power (EIRP). The electrical efficiency will generally be lower because transmitters are most efficient at or near saturation. Since the system response is the product of the transmit and receive beams, it is more effective to shape the receive beam aggressively than to shape both transmit and receive equally.

References

[1] Stutzman W., Licul S. 'Synthesis methods for antennas' in Balanis P.C.A. (ed.). *Modern Antenna Handbook*. New York, NY, USA: John Wiley & Sons, Inc; 2008.

[2] Schelkunoff S.A. 'A mathematical theory of linear arrays'. *Bell System Technical Journal*. 1943, vol. 22(1), pp. 80–107.

Chapter 10
Beam shaping and spoiling

The techniques described in Chapter 9 dealt primarily with sidelobes without constraining the resulting mainbeam width and shape. It is sometimes desirable to form a broad beam, of constant or shaped amplitude. Examples include fan beams for increasing SAR swath coverage or cosecant beams for air surveillance where range varies with elevation angle. This chapter discusses the last three techniques listed in Table 9.1.

10.1 Fourier synthesis technique

This technique is very straightforward. Since the beam shape is the Fourier transform of the illumination function, take the inverse Fourier transform of the beam shape to obtain the required illumination function. The Fourier technique shows promise since it could, in principle, provide an arbitrary mainbeam shape.

The problem arises that the result is an illumination function infinite in extent. Accordingly, the computed illumination function is truncated, which of course distorts the desired antenna pattern. Moreover, even to achieve a close match to a desired shape many terms with small magnitudes are required (analogous to square wave synthesis) which requires a large antenna and provides low aperture efficiency. Booker and Clemov [1] provide a discussion of the Fourier transform as it relates to antenna pattern analysis.

Fourier transform synthesis

The Fourier transform function is shown in (10.1)

$$a = \frac{d_x}{\lambda} \int_{-\lambda/(2d_x)}^{\lambda/(2d_x)} F(u)\, \mathrm{e}^{-\mathrm{j}(2\pi/\lambda) u n d_x} \mathrm{du} \tag{10.1}$$

where $F(u)$ is the desired beam shape and a is the aperture weight function.

We consider the case of a flat-topped beam. This is a rectangular beam shape, and the resulting excitation is a sinc function reversing the roles of our uniform aperture and sinc-shaped beam. When we synthesize beam shape based on finite limits, we see ripple in the beam shape, termed Gibbs phenomena. The aperture needs to be long enough to encompass several zeros of the sinc in order to produce an approximately rectangular beam which causes poor efficiency. We will evaluate various aperture sizes which extend to the first, second, and third null of the aperture function.

Fourier transform – first null

Figure 10.1 illustrates the situation for the Fourier transform technique when the aperture is sized to include only the first null in the aperture distributions. This requires 14 elements and provides a somewhat broad (13.6°) beam which does not appear flat-topped. The first sidelobe is at 22 dB which is good. The amplitude taper includes the first null of the sinc function. The taper efficiency is only 67.3%, however. Most of the roots occur on the unit circle but roots 1 and 13 are real and complementary (product one) corresponding to complex values of psi and non-nulling.

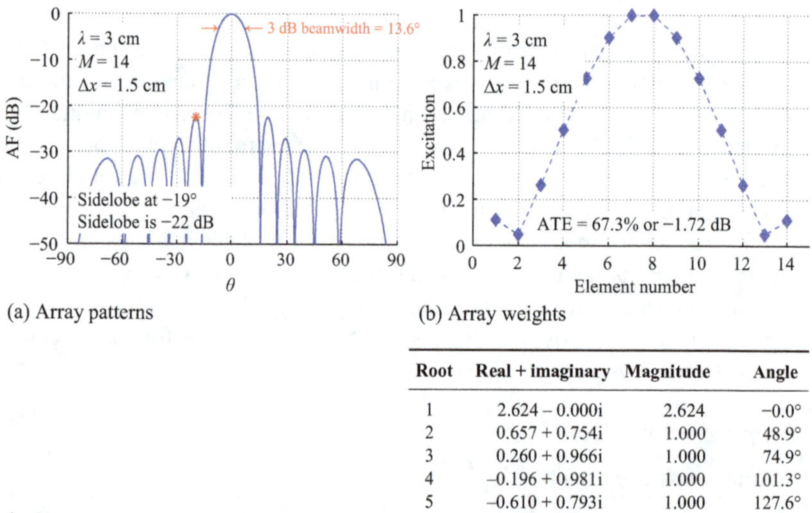

(a) Array patterns

(b) Array weights

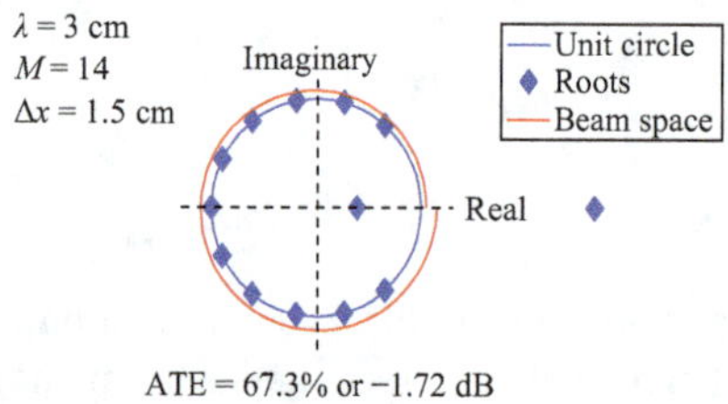

(c) Polynomial roots graph

Root	Real + imaginary	Magnitude	Angle
1	2.624 – 0.000i	2.624	–0.0°
2	0.657 + 0.754i	1.000	48.9°
3	0.260 + 0.966i	1.000	74.9°
4	–0.196 + 0.981i	1.000	101.3°
5	–0.610 + 0.793i	1.000	127.6°
6	–0.897 + 0.441i	1.000	153.8°
7	–1.000 – 0.000i	1.000	–180.0°
8	–0.897 – 0.441i	1.000	–153.8°
9	–0.610 – 0.793i	1.000	–127.6°
10	–0.196 – 0.981i	1.000	–101.3°
11	0.260 – 0.966i	1.000	–74.9°
12	0.657 – 0.754i	1.000	–48.9°
13	0.381 + 0.000i	0.381	0.0°

(d) Polynomial roots table

Figure 10.1 Fourier transform – first null
Schelkunoff.m

Fourier transform – second null

Figure 10.2 illustrates the situation for the Fourier transform technique when the aperture is sized to include the second null in the aperture distributions. This requires 25 elements and provides a beamwidth of 16.9° which is starting to look somewhat flat. The taper efficiency is only 52.5%. There are two pairs of complementary real roots with complex psi (and no nulls).

Fourier transform – third null

Figure 10.3 illustrates the situation for the Fourier transform technique when the aperture is sized to include the third null in the aperture distributions. This requires

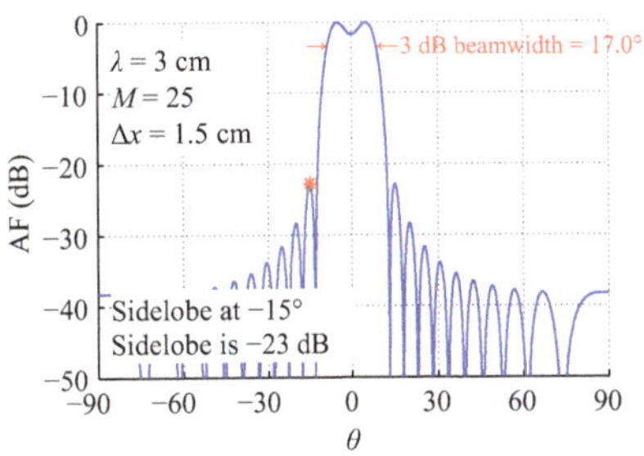

(a) Array patterns

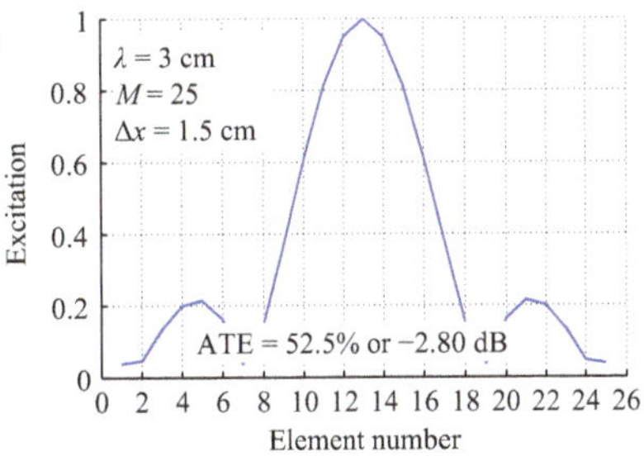

(b) Array weights

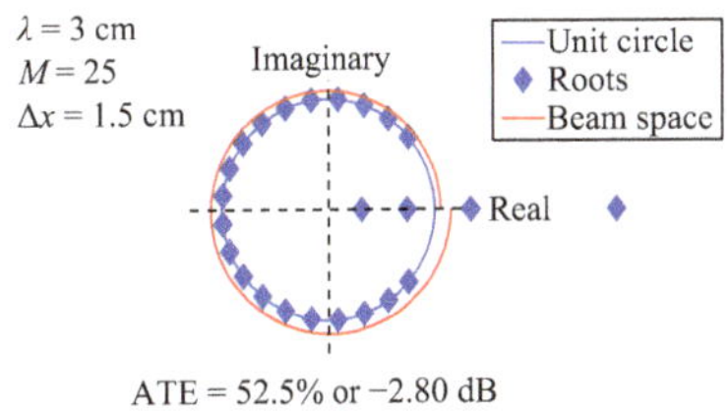

(c) Polynomial roots graph

Root	Real + imaginary	Magnitude	Angle
1	3.098 – 0.000i	3.098	–0.0°
2	1.332 + 0.000i	1.332	0.0°
3	0.758 + 0.653i	1.000	40.7°
4	0.573 + 0.820i	1.000	55.1°
5	0.346 + 0.938i	1.000	69.8°
6	0.095 + 0.995i	1.000	84.6°
7	–0.162 + 0.987i	1.000	99.3°
8	–0.407 + 0.913i	1.000	114.0°
9	–0.625 + 0.780i	1.000	128.7°
10	–0.803 + 0.597i	1.000	143.4°
11	–0.927 + 0.374i	1.000	158.0°
12	–0.992 + 0.127i	1.000	172.7°
13	–0.992 – 0.127i	1.000	–172.7°
14	–0.927 – 0.374i	1.000	–158.0°
15	–0.803 – 0.597i	1.000	–143.4°
16	–0.625 – 0.780i	1.000	–128.7°
17	–0.407 – 0.913i	1.000	–114.0°
18	–0.162 – 0.987i	1.000	–99.3°
19	0.095 – 0.995i	1.000	–84.6°
20	0.346 – 0.938i	1.000	–69.8°
21	0.573 – 0.820i	1.000	–55.1°
22	0.758 – 0.653i	1.000	–40.7°
23	0.751 + 0.000i	0.751	0.0°
24	0.323 – 0.000i	0.323	–0.0°

(d) Polynomial roots table

Figure 10.2 Fourier transform – second null
Schelkunoff.m

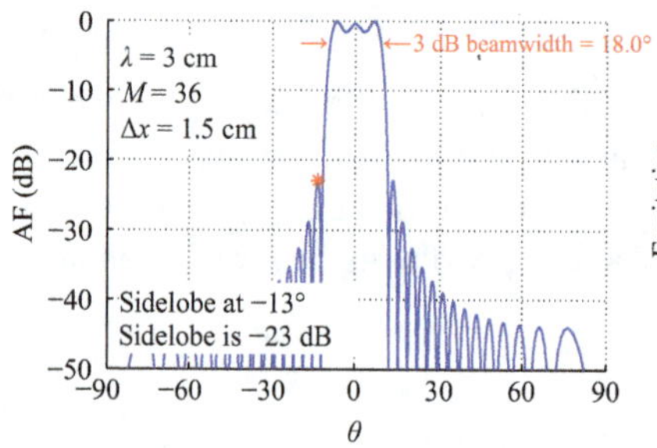

(a) Array patterns

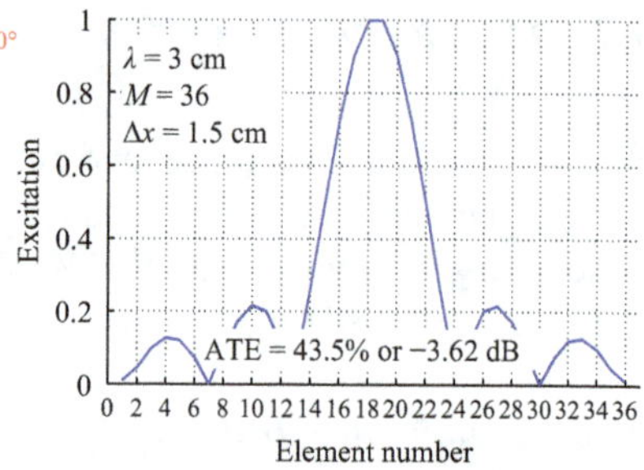

(b) Array weights

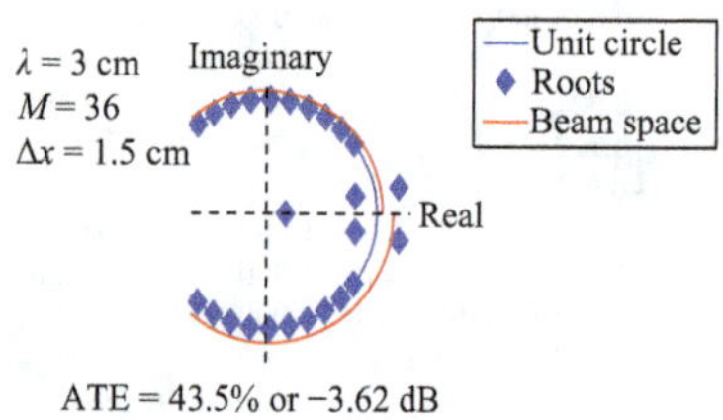

(c) Polynomial roots graph

Root	Real + imaginary	Magnitude	Angle
1	5.722 + 0.000i	5.722	0.0°
2	1.196 − 0.234i	1.219	−11.1°
3	1.196 + 0.234i	1.219	11.1°
4	0.792 − 0.611i	1.000	−37.7°
5	0.676 − 0.737i	1.000	−47.5°
6	0.536 − 0.844i	1.000	−57.6°
7	0.378 − 0.926i	1.000	−67.8°
8	0.208 − 0.978i	1.000	−78.0°
9	0.031 − 1.000i	1.000	−88.2°
10	−0.148 − 0.989i	1.000	−98.4°
11	−0.320 − 0.947i	1.000	−108.7°
12	−0.483 − 0.876i	1.000	−118.9°
13	−0.630 − 0.776i	1.000	−129.1°
14	−0.758 − 0.653i	1.000	−139.3°
15	−0.861 − 0.508i	1.000	−149.5°
16	−0.938 − 0.348i	1.000	−159.6°
17	−0.984 − 0.177i	1.000	−169.8°
18	−1.000 + 0.000i	1.000	180.0°
19	−0.984 + 0.177i	1.000	169.8°
20	−0.938 + 0.348i	1.000	159.6°
21	−0.861 + 0.508i	1.000	149.5°
22	−0.758 + 0.653i	1.000	139.3°
23	−0.630 + 0.776i	1.000	129.1°
24	−0.483 + 0.876i	1.000	118.9°
25	−0.320 + 0.947i	1.000	108.7°
26	−0.147 + 0.989i	1.000	98.4°
27	0.031 + 1.000i	1.000	88.2°
28	0.208 + 0.978i	1.000	78.0°
29	0.378 + 0.926i	1.000	67.8°
30	0.536 + 0.844i	1.000	57.6°
31	0.676 + 0.737i	1.000	47.5°
32	0.792 + 0.611i	1.000	37.7°
33	0.805 − 0.158i	0.820	−11.1°
34	0.805 + 0.158i	0.820	11.1°
35	0.175 + 0.000i	0.175	0.0°

(d) Polynomial roots table

Figure 10.3 Fourier transform – third null
Schelkunoff.m

36 elements and provides a beamwidth of 17.3° which is distinctly flat. However, the taper efficiency is only 43.5%. There are three pairs of real complementary roots.

This is about as far as we can reasonably go.

10.2 Woodward–Lawson synthesis

Woodward [2] developed a clever approach that constrains the aperture size and still provides considerable flexibility in beam shape. It is based on the observation that a finite array produces a set of beams equal to the number of elements whose nulls coincide with each other and with the single peak of each.

The synthesis procedure starts with basis functions for beam shape based on a finite aperture. These basis functions are uniformly weighted beams steered at increments of $2\pi/M$ with the result that nulls coincide. This allows a direct computation of weights to approximate any desired beam shape. Elliott [3] subsequently published a refinement of the technique.

Figure 10.4 shows 11 such beams generated with uniform illumination and a delta phase shift equal to $2 > \pi \cdot m/M$. The sum of the center three beams is also shown. The sum beam has a relatively flat mainlobe and the sidelobes are similar to those of a single beam (generated with uniform illumination).

Figure 10.5 shows Schulkunoff representation for the combination of three beams in Figure 10.4. The beamwidth is 27.7° and the first sidelobe at −15 dB. The taper efficiency is 69.8%. The illumination function resembles the Fourier reconstruction to the first null but the mainbeam is wider and the sidelobes are higher.

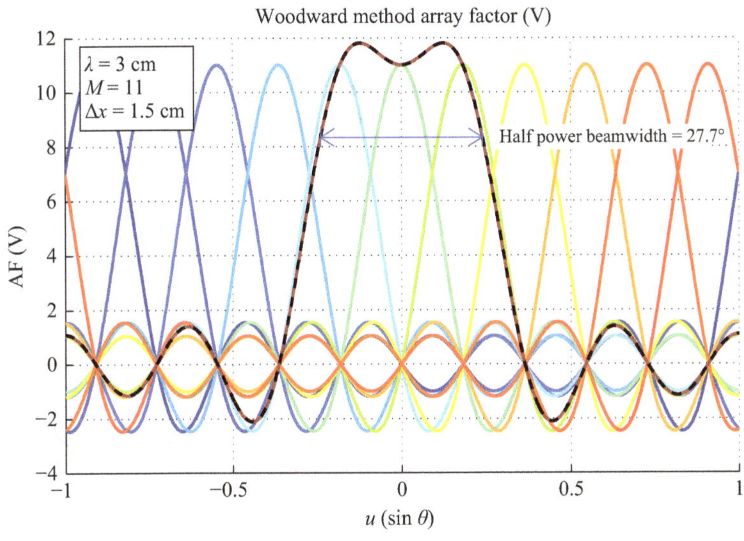

Figure 10.4 Center three beams combined
WoodwardSynthesis.m

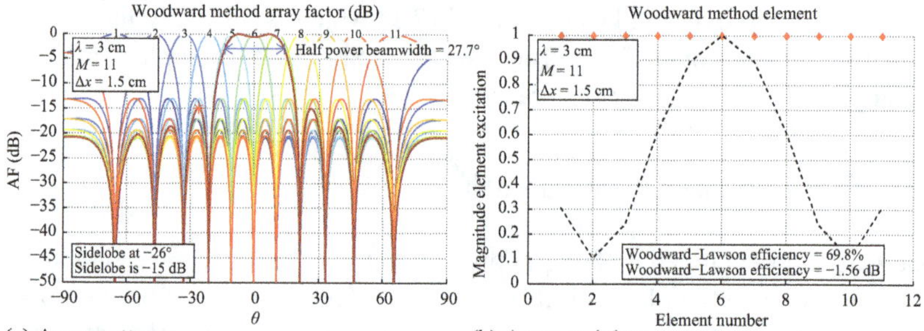

(a) Array patterns (b) Array weights

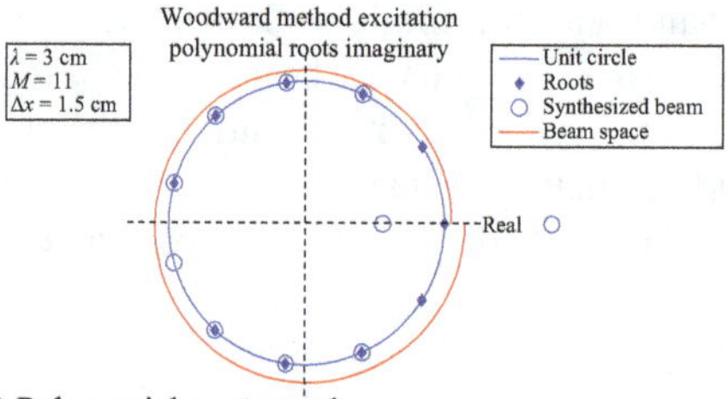

Root	Real + imaginary	Magnitude	Angle
1	1.785 + 0.000i	1.785	0.0°
2	−0.959 + 0.282i	1.000	163.6°
3	−0.959 − 0.282i	1.000	−163.6°
4	−0.655 + 0.756i	1.000	130.9°
5	−0.655 − 0.756i	1.000	−130.9°
6	−0.142 + 0.990i	1.000	98.2°
7	−0.142 − 0.990i	1.000	−98.2°
8	0.415 + 0.910i	1.000	65.5°
9	0.415 − 0.910i	1.000	−65.5°
10	0.560 + 0.000i	0.560	0.0°

(c) Polynomial roots graph (d) Polynomial roots table

Figure 10.5 Woodward–Lawson example
WoodwardSynthesis.m

10.3 Phase-only beam broadening

The preceding techniques alter both amplitude and phase to affect the beam pattern. As noted, the amplitude weighting reduces the aperture efficiency which is disadvantageous, especially in transmit. So, techniques which affect only phase would seem to offer an advantage with respect to aperture efficiency which would be unchanged from the uniform illumination case.

The restriction to phase weighting affects sidelobe levels, which tend to increase. A number of papers propose various techniques to choose the phase weights optimally.

An approach with some similarity to Woodward–Lawson beam synthesis is to divide the aperture and generate two overlapping beams, each half of the aperture generating one of the beams. Superficially, this produces the same result but without the necessity of amplitude weighting, thereby improving aperture efficiency. The split aperture will produce wider beams than the Woodward–Lawson approach which uses the full aperture but this is not necessarily disadvantageous if the end object is a broadened beam. Otherwise, the mainbeams will be similar and we may surmise that differences will primarily be observed in the sidelobes.

By analogy to the Woodward–Lawson synthesis, we may divide the aperture into equal segments, e.g., two, and apply varying phase slopes to each piece to generate multiple beams. The Woodward–Lawson technique uses the full aperture for each beam, with the well-known λ/D beamwidth. The alternative approach uses only the partial aperture for each beam, so the beamwidth is

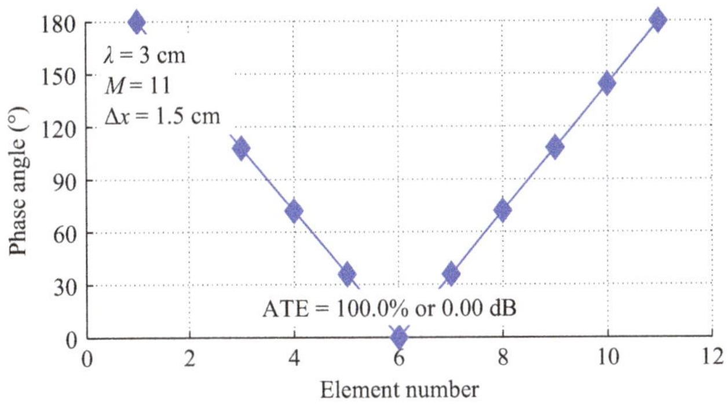

Figure 10.6 Bilinear phase term
Schelkunoff.m

$N \cdot \lambda / D$, where N is the number of segments. This beam broadening augments the end objective of a broader beam.

This provides some beam broadening but a dip appears in the center and grows as the beam directions diverge. A particular phase profile is shown in Figure 10.6. The example shown varies element phase linearly from the center of the array to the edge, from 0° to 180°. The corresponding array factor and related information is shown in Figure 10.7. We might expect two oppositely skewed beams from the bilinear case but by adjusting the phase slope and corresponding beam angle, the expected dip between the beams may be filled for a more uniform beam shape.

It might be beneficial to partition the weights into more and smaller facets. Figure 10.8 illustrates the result of several doublings of the number of facets and halving the size of facets. If each of the two aperture halves is divided, four overlapped beams will result. Continuing this process down to the element level, we observe that the phase weights approximate a quadratic distribution, which is a classic technique, as illustrated in Figure 10.8a. Figure 10.8b shows the resulting beam patterns overlaid on one another. In this example, the differences are small for more than four facets. The resulting beam shape is well behaved but represents an array with 64 elements in the spoiling axis. For smaller number of elements, the beam shape is more irregular.

Quadratic beam spoiling

Instead of the iterative approach described, which requires that the number of elements be a power of 2, we may simply vary the phase of each element continuously so that the phase offset is proportional to the square of the element distance to the center. This is explicitly quadratic phase spoiling. A particular phase profile is shown in Figure 10.9. The example shown varies element phase quadratically from the center of the array to the edge, from 0° to 180°. The corresponding array factor and associated information is shown in Figure 10.10.

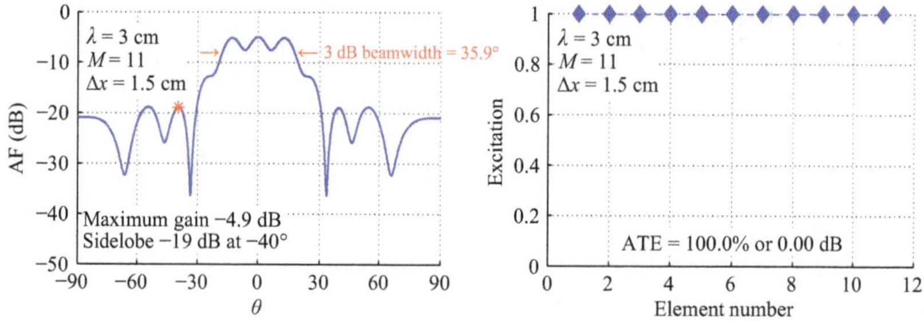

(a) Array patterns (b) Array weights

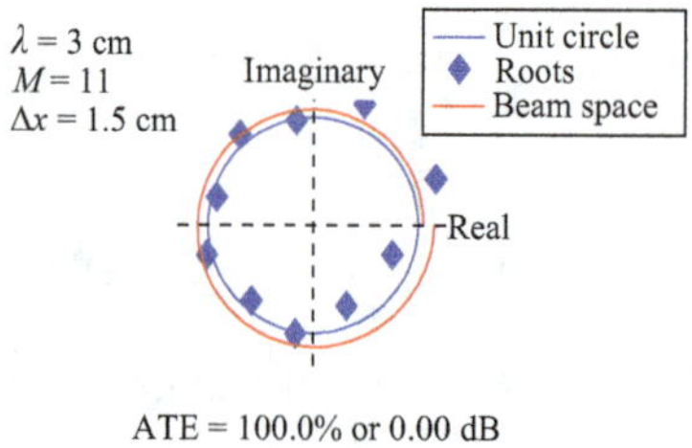

(c) Polynomial roots graph

Root	Real + imaginary	Magnitude	Angle
1	1.164 + 0.424i	1.239	20.0°
2	0.491 + 1.120i	1.222	66.3°
3	−0.160 + 0.967i	0.980	99.4°
4	−0.703 + 0.834i	1.091	130.1°
5	−0.921 + 0.255i	0.956	164.5°
6	−1.008 + −0.279i	1.046	−164.5°
7	−0.591 + −0.701i	0.917	−130.1°
8	−0.167 + −1.006i	1.020	−99.4°
9	0.758 + −0.276i	0.807	−20.0°
10	0.328 + −0.749i	0.818	−66.3°

(d) Polynomial roots table

Figure 10.7 Bilinear phase example
Schelkunoff.m

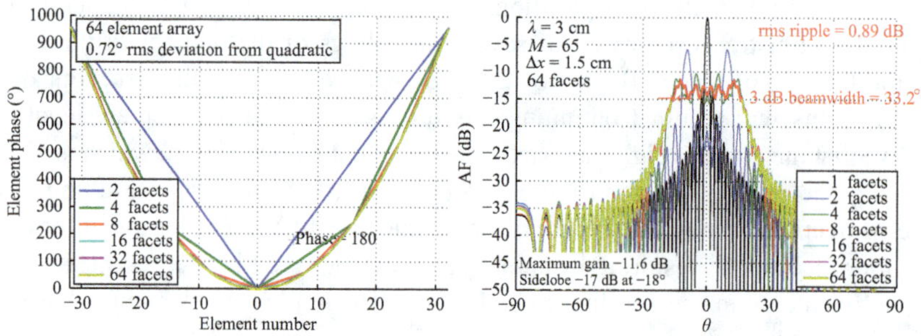

(a) (b)

Figure 10.8 Phase spoiled beam examples: (a) phase-only beam broadening and (b) phase spoiled beam examples
Quadratic2D.m

Quadratic phase example

To adapt this technique to a general case with a beamwidth of θ, we select the end element phase slope to produce a beam direction of $\theta/2$ and use (2.39) to compute the desired phase weighting in (10.2)

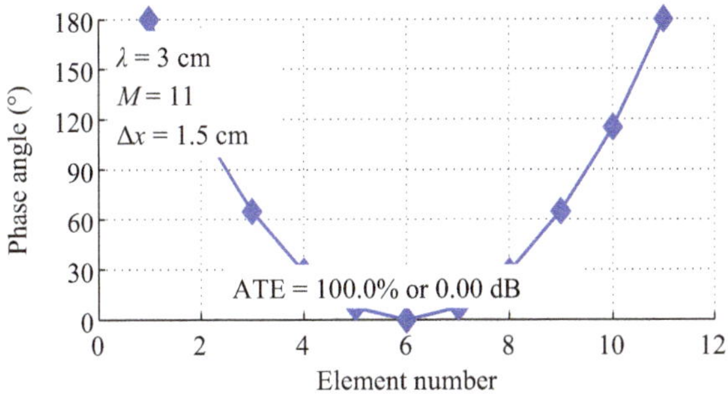

Figure 10.9 Additional phase term
Schelkunoff.m

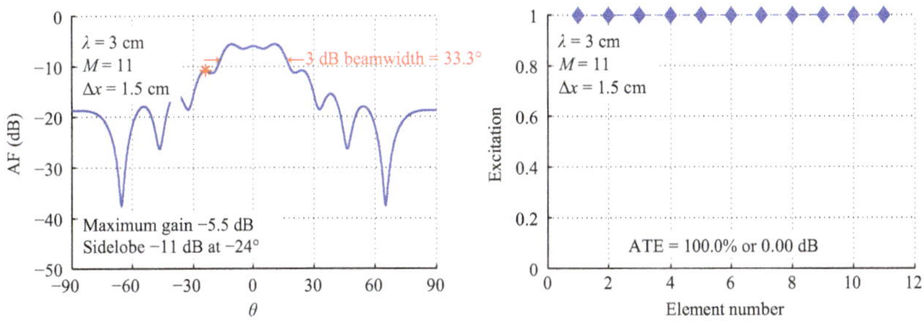

(a) Array patterns (b) Array weights

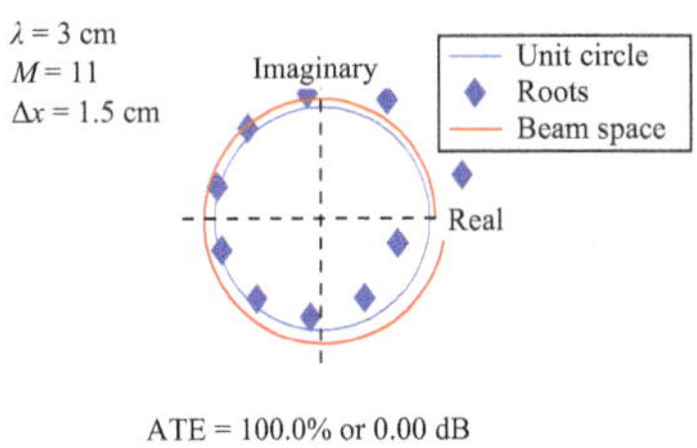

(c) Polynomial roots graph

Root	Real + imaginary	Magnitude	Angle
1	1.300 + 0.397i	1.360	17.0°
2	0.608 + 1.067i	1.229	60.3°
3	−0.127 + 1.126i	1.133	96.4°
4	−0.691 + 0.815i	1.069	130.3°
5	−0.979 + 0.290i	1.021	163.5°
6	−0.939 + −0.278i	0.979	−163.5°
7	−0.606 + −0.714i	0.936	−130.3°
8	−0.099 + −0.877i	0.882	−96.4°
9	0.403 + −0.707i	0.814	−60.3°
10	0.703 + −0.215i	0.735	−17.0°

(d) Polynomial roots table

Figure 10.10 Quadratic phase example
Schelkunoff.m

$$\Delta\phi = \frac{2\pi\Delta x}{\lambda \sin(\theta_{max})} \tag{10.2}$$

or

$$\frac{\Delta\phi}{\Delta x} = \frac{2\pi}{\lambda \sin(\theta_{max})} \tag{10.3}$$

Our quadratic phase relationship for an array with $2N + 1$ elements is shown in (10.4)

$$\phi = \frac{x}{x_{max}} \phi_{max} \tag{10.4}$$

where

$$\begin{aligned} y &= \frac{x^2}{4F} \\ x &= (0 : N)\,\Delta x \\ \Delta x &= \lambda/2 \end{aligned} \tag{10.5}$$

Accordingly,

$$\Delta\phi = \frac{\Delta x}{x_{max}} \phi_{max} \tag{10.6}$$

or

$$\phi_{max} = x_{max} \frac{\Delta\phi}{\Delta x} \tag{10.7}$$

Combining (10.3) and (10.7) gives us the relationship between beam steering angle θ_{max} and maximum phase shift ϕ_{max} in (10.8)

$$\begin{aligned} \phi_{max} &= \frac{2\pi x_{max}}{\lambda \sin(\theta_{max})} \\ &= \frac{2\pi N \Delta_x}{\lambda \sin(\theta_{max})} \end{aligned} \tag{10.8}$$

Beam spoiling and Taylor weighting

Leifer [4] describes a technique to add Taylor weighting to achieve better sidelobe control. The addition of the amplitude weighting requires a small adjustment in the phase weighting for lowest ripple. The aperture efficiency decreases because of the amplitude weighting but the sidelobes are much improved. A variation of this technique uses aperture thinning according to a Taylor distribution to regain much of the aperture efficiency but increases the ISLR.

References

[1] Booker H.G., Clemmow P.C. 'The concept of an angular spectrum of plane waves, and its relation to that of polar diagram and aperture distribution'. *Proceedings of the IEE–Part III: Radio and Communication Engineering.* 1950, vol. 97(45), pp. 11–17.

[2] Woodward P.M. 'A method of calculating the field over a plane aperture required to produce a given polar diagram'. *Journal of the Institution of Electrical Engineers - Part IIIA: Radiolocation.* 1946, vol. 93(10), pp. 1554–8.

[3] Elliott R. 'Criticisms of the Woodward-Lawson method'. *IEEE Antennas and Propagation Society Newsletter*. 1988, vol. 30(3), p. 43.
[4] Leifer M.C. 'Revisiting a method of beam shaping using phase weights'. *2016 IEEE International Symposium on Phased Array Systems and Technology (PAST)*. 2016, pp. 1–3.

Chapter 11
Reflector applications

Reflectors may be incorporated into an ESA design to increase the aperture area. By their nature, ESAs require considerable electronics with associated size, weight, power, and cost. High-gain, narrow-beam systems require many T/R modules. If other requirements such as electronic field of view allow, reflectors can provide increased effective area without additional electronics. The added passive mechanical structure is generally lower cost than the equivalent ESA area.

This approach combines some of the benefits (and some of the disadvantages) of ESAs and reflectors. ESA feeds may be used with both cylindrical (one-dimensional curvature) and rotationally symmetric reflectors (two-dimensional curvature). The basic trade-off is to exchange electronic field of regard (EFOR) for fewer T/R modules; this is exactly analogous to a thinned array described in Chapter 8.

There are many examples which place a multi-element feed at the focal point of a parabolic reflector. These are termed switched beam systems and have been in service for a long time, height-finding radars being an example.

As the feed becomes large and displaced from the optical axis, multiple elements may be combined with phase differences and so may be considered ESA-ish. Warnick *et al.* [1] provide a detailed analysis in their book. Cappellen *et al.* [2] describe the application of ESA feeds for the Square Kilometer Array.

Many authors cover this subject in great detail, among which are chapter 12 of Skolnik's *Radar Handbook* by Cooley and Davis [3] and the *Handbook of Reflector Antennas and Feed Systems* by Rao *et al.* [4]. Cooley [5] provides a detailed discussion for their use in space.

Section 2.6.2 described techniques to achieve multiple simultaneous beams with a beamformer. Reflector systems can provide multiple simultaneous beams with the addition of additional feed elements. The simplest reflector antenna uses a single feed at the focal point of the reflector, providing a single beam. For our purposes, a reflector may be fitted with additional feed elements corresponding to additional beams without reducing gain. This provides a means of rapid beam steering or multiple simultaneous beams.

Optical systems are analogous; microwave beams correspond to individual pixels in the focal plane. There is a large body of optical theory to apply to reflector design for microwave systems. The shorter wavelength of optical systems means that they may be analyzed with particle-like formalisms such as ray tracing, while wave-like formalisms for field calculations are more appropriate for microwave systems. Table 11.1 summarizes some of these differences.

Ray tracing, used in this chapter, provides qualitative insight, into microwave reflector systems. Field calculations require finite-element method (FEM) and finite-difference time-domain (FDTD) modeling, using software packages such as HFSS, GRASP, and FEKO. These tools are very accurate but very resource-intensive. They are ultimately required to complete a detailed design.

Reflector aperture efficiency is constrained by the conflicting objectives of full illumination of the reflector with no spill-over (illumination) outside the boundary of the reflector. This leads to feed design which tapers to near zero at the reflector edge, frequently of a Gaussian distribution. Efficiencies of 60%–70% are typical and the resulting farfield pattern has very low sidelobes. This idealized case is somewhat degraded if physical support structure obstructs the signal path.

An ESA feed will produce one or more beams and the reflector magnification produces narrower beams and higher gain. Mailloux [6] terms this a hybrid antenna in his book. Jensen [7] refers to it as an AESA fed reflector while Croswell et al [8] uses the term phased array-fed reflector.. This approach trades off electronic field of view with gain (analogous to a thinned array) and is not suitable for all applications.

The scan capabilities of reflector systems are constrained by inherent optical aberrations that cause incoming plane waves to be focused at differing points resulting in a blurred spot instead of a sharp point. These aberrations are at a minimum on boresight and become problematic at pointing directions a few degrees off boresight. The usability limit is determined by the system application. Ruze [9] discusses some of these issues.

A reflector is necessary for all these configurations. They have been demonstrated on a number of programs. Mobrem *et al.* [10] describe the reflector design process for a 6-m diameter Astromesh™ for the Soil Moisture Active Passive (SMAP) sensor.

11.1 Configurations

The telescope is the optical analog of most microwave reflector systems. One or two reflectors arranged on a common axis with radial symmetry are usually sufficient.

Table 11.1 Comparison of optical and microwave systems

Parameter	Optical system	Microwave system
Physical size	>1 cm	>10 cm
Wavelength	1 μm	1 mm to 10 m
Ratio (size/wavelength)	$>10^4$	10^3 to 10^{-2}
Sensors	Noncoherent	Coherent
Focal plane	Megapixels	Few pixels
Number of optical elements	Several	One or two
Element type	Aspheric	Conic (circle, parabola, etc.)

The underlying principle for reflectors is that they enlarge the feed, including the feed spacing in the case of an ESA feed, and the aperture area. The effect is to provide higher gain but reduced scan capability. These effects are similar to those of a thinned array.

A fundamental consideration is how many reflectors. A single parabolic antenna collimates a point source feed into a parallel beam of rays or incoming plane waves into a point. Two reflectors provide additional design freedom as well as shorter physical length compared to its electrical (optical) length. Both configurations may be constructed in the on-axis or offset geometries. Mailloux observes that each reflector may be associated with a Fourier transform and characterizes the configurations as one- and two-transform systems. An ESA feed provides opportunity for third transform.

The single reflector system produces a real image of the external environment on the focal plane. A two-reflector system may produce a real image or a smaller bundle of rays parallel to the incoming rays. A real image is produced by a parabolic primary reflector and a hyperbolic secondary reflector. This is an analog of a Newtonian telescope with two positive lenses.

Paired parabolic or hyperbolic reflectors produce a parallel bundle of rays. This is an analog of a Galilean telescope with one positive and one negative lens. It is termed a beam expander in optical applications and is used, for instance, to reduce the beamwidth of lasers. Fitzgerald [11, 12] termed this approach a "near-field Cassegrainian" system in his technical reports published by Lincoln Laboratory on axial and offset-limited electronic scanning systems, respectively.

11.2 Single reflector

Figure 11.1 illustrates a single-reflector configuration using a parabolic primary. Its performance is shown using ray tracing in two dimensions displaying the effect of the reflector on the incoming plane wave. The figures may be interpreted as a cylindrical reflector or a principal plane cut of a circularly symmetric reflector. Each figure includes boresight (nonscanned) and scanned cases. In the boresight case, all the rays converge on the focal point; in off boresight cases, the rays diverge because of various aberrations.

Figure 11.1 illustrates a multiple beam capability of a reflector system. Each feed creates a beam in a specified direction.

Figure 11.1b shows the single reflector scanned to 6°. Convergence is defocused and falls off the focal plane. This amount of defocus might be unusable in an optical system but is probably acceptable in a communications or SAR system whose resolution is not determined by the spot size.

This configuration places the feed array at the focal plane, so on axis, there is some blockage of the incoming signal. Two beams are shown, boresight in blue and one angled off axis in green.

Single-reflector configurations are used in several satellites, including TecSAR satellites. The single-reflector configuration has been selected for L-band SAR satellites including DESDyni, NISAR, and TanDEM-L.

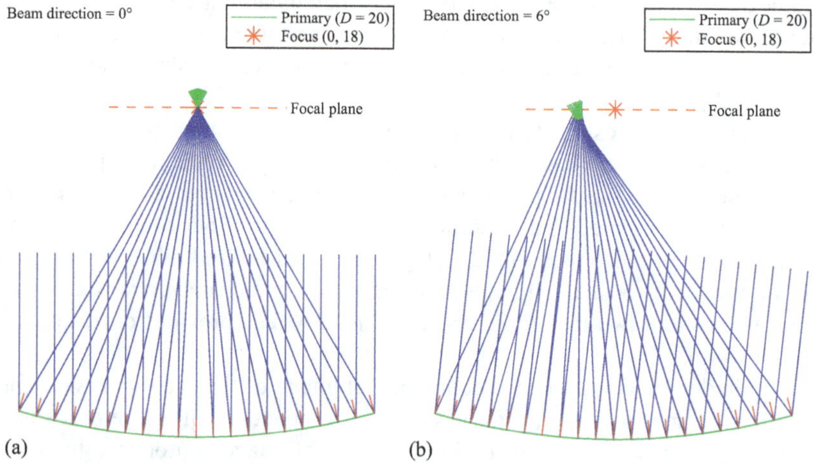

Figure 11.1 *No secondary: (a) ray tracing boresight case and (b) ray tracing beam scanned case*
TwoDimensionalRayTracing.m

Table 11.2 *Two-reflector system configurations*

Name	Primary	Secondary
Near-field Cassegrainian	Parabolic	Parabolic
Gregorian	Parabolic	Ellipsoidal
Ritchey–Chrétien	Hyperbolic	Hyperbolic

11.3 Dual reflector

Table 11.2 summarizes some configurations. The first configuration on the list is the near-field Cassegrainian which differs from the others in that it does not provide a focal point. In optics, it is termed a "beam expander" and is analogous to the Newtonian optical telescope with two lenses, the first positive focal length and the second a negative focal length. Fitzgerald [11] analyses this configuration in detail in his Lincoln Laboratory technical report, *Limited Electronic Scanning with a Near-Field Cassegrainian System*. Dual-reflector systems are not represented by any fielded systems but Airbus has proposed a system called "HybridSAR" for a radar system with 25 cm resolution.

Figure 11.2 shows the near-field Cassegrain and Figure 11.3 shows the Gregorian dual-reflector configurations. Both use a parabolic primary: Figure 11.2 illustrates a parabolic secondary and Figure 11.3 illustrates a hyperbolic secondary. Their performance is shown using ray tracing in two dimensions displaying the effect of the reflector on the incoming plane wave. The figures may be interpreted

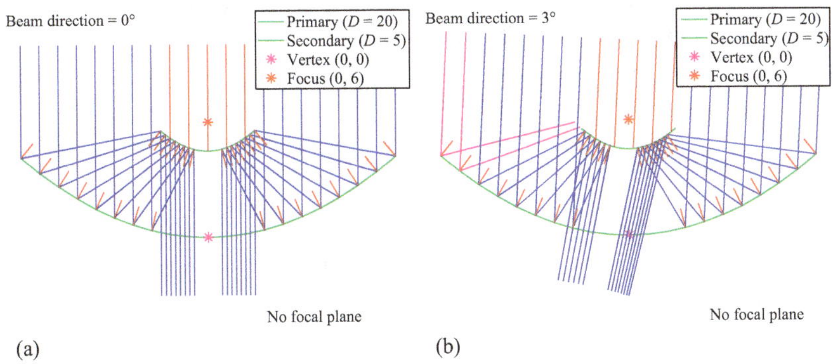

Figure 11.2 Parabolic secondary: (a) ray tracing boresight case and (b) ray tracing beam scanned case
TwoDimensionalRayTracing.m

as a cylindrical reflector or a principal plane cut of a circularly symmetric reflector. Each figure includes boresight (nonscanned) and scanned cases. In the boresight case, all the rays converge on the focal point; in off boresight cases, the rays diverge because of various aberrations.

No provision is shown for mechanical support of the reflectors or any associated electronics. This structure degrades performance by blocking or scattering some of the incoming signal, reducing gain and increasing sidelobes. The "offset" configuration moves the focal point and secondary reflector or feed out of the path of the incoming signal to avoid this problem but this technique reduces symmetry and increases aberrations.

Most antennas focus in both planes for which three-dimensional ray tracing is necessary. It will be illustrated in Section 11.4 and utilized in Chapter 18.

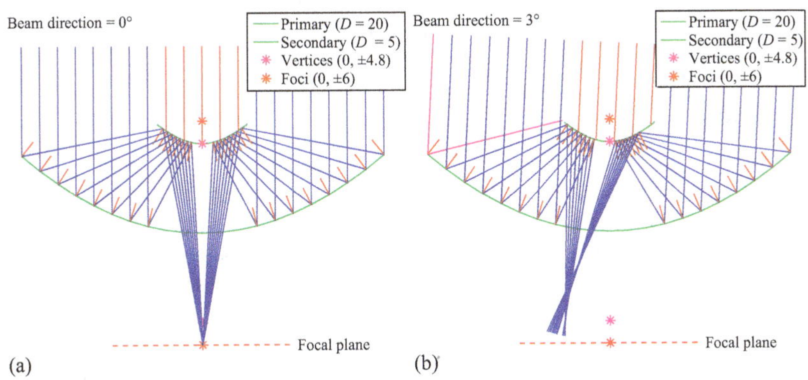

Figure 11.3 Hyperbolic secondary: (a) ray tracing boresight case and (b) ray tracing beam scanned case
TwoDimensionalRayTracing.m

Figures 11.2 and 11.3 display the boresight performance of two configurations: (1) parabolic primary reflector with parabolic secondary and (2) parabolic primary reflector with hyperbolic secondary reflector. The light green line represents the primary reflector and the dark green line represents the secondary reflector. The focus and vertex of the reflectors are identified in the legend. Blue lines represent incoming rays and converge on the focal point. The nonscanned case shows perfect convergence of the rays.

Each figure displays the performance of the same reflector configurations with a scanned beam. Figures 11.2b and 11.3b display the performance of two-reflector configurations scanned to 3°. The green lines represent the primary (light green) and secondary (dark green) reflectors. The foci and vertices of the parabola and hyperbola are shown as indicated in the legend. Blue, magenta, and red lines represent incoming rays. Red lines are blocked by the secondary, magenta lines reflected from the primary miss the secondary, while blue lines are unobstructed.

For the parabolic secondary, the primary and secondary reflectors have coincident foci. For the hyperbolic secondary, one focus of the hyperbola coincides with the focus of the parabolic primary, while the second hyperbolic secondary focus constitutes the focal point of the system. The hyperbolic secondary provides a long effective focal length in a short package. A disadvantage is that the focus degrades and moves off the focal plane as the beam scans.

All configurations behave well on boresight. Performance degrades off boresight, rather more in the case of the hyperbolic secondary. The received field in the parabolic secondary case moves laterally with beam scan, requiring an oversized feed to support beam scanning. In the case illustrated, the ESA feed is a sizeable fraction of the primary reflector diameter diminishing the benefit of this configuration.

All of the configurations discussed incur blockage of the incoming rays, either by the secondary reflector or by the feed in the case of the single reflector as well as mechanical structure. Blockage is detrimental for two reasons. It reduces the signal strength and scatters some power into sidelobes. If we use only a portion of the optical system, for instance, the left or right half, we can avoid this blockage but will sacrifice symmetry which mitigates some distortions.

11.4 Offset reflector

Each approach may be configured with the feed location offset from the incoming optical path offset to utilize a unobstructed portion of the optical path. This eliminates scattering by the feed and its structure which would otherwise raise sidelobe levels. The antenna feed is offset to the side of the reflector; it is still located at the focal point of the reflector, but the focus is located to the side as the reflector is an asymmetric segment of a paraboloid. This configuration is shown in Figure 11.4. The system utilizes a portion of the full paraboloid which is shown in outline. The green dashed line is the axis of symmetry of the paraboloid and the feed is centered on the focal point of the paraboloid, indicated by a green dot. Incoming rays are shown in blue, while the ray which reflects from the center of

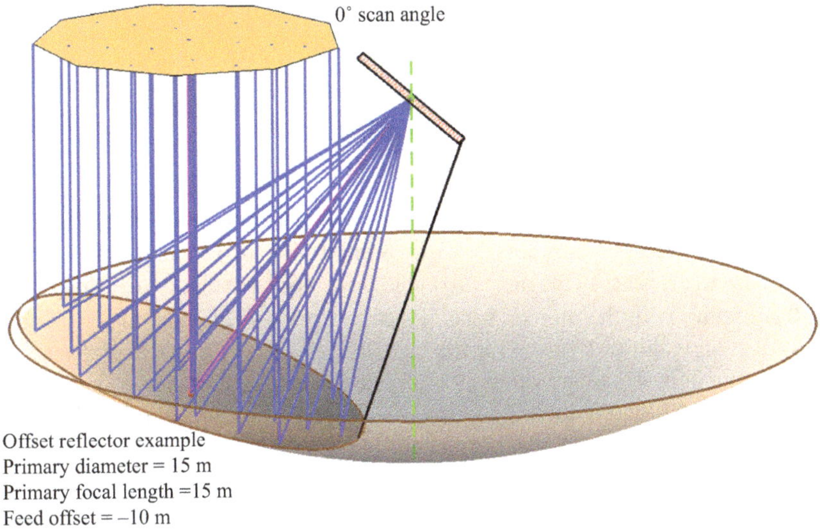

Figure 11.4 Offset-fed parabolic reflector
ThreeDimensionalRayTracing.m

the paraboloid is shown in magenta. In all cases, the ray which reflects from the center of the paraboloid will be termed the central ray. This configuration will be used in Chapter 18.

11.5 Feed design

The feed should provide the desired illumination of the reflector while minimizing the power that misses the reflector. In general, the reflector is in the farfield of the feed. To compute the feed size, we consider the necessary beamwidth which corresponds to the angle (θ_r) subtended by the rim of the reflector from the point of view of the feed. This angle is a function of reflector diameter (D_r) and focal length (F_r) as indicated in (11.1) for a nonoffset design:

$$\begin{aligned} \frac{\theta_r}{2} &= \arctan\left(\frac{D_r/2}{F_r - \delta_y}\right) \\ \delta_y &= \frac{(D_r\ r/2)^2}{4F_r} \end{aligned} \tag{11.1}$$

The situation is slightly more complicated for an offset feed because we have to compute the difference between two equations of the form of (11.1) as shown in (11.2):

$$\begin{aligned} \theta_r &= \theta_2 - \theta_1 \\ \theta_r &= \arctan\left(\frac{\text{Offset} + D_r/2}{F_r - \delta_1}\right) - \arctan\left(\frac{\text{Offset} - D_r/2}{F_r - \delta_2}\right) \end{aligned} \tag{11.2}$$

where

$$\delta_1 = \frac{(\text{Offset} + D_r/2)^2}{4F_r}$$
$$\delta_2 = \frac{(\text{Offset} - D_r/2)^2}{4F_r}$$

The practical effect of the offset case is that the reflector is more distant from the feed and the feed accordingly must possess a narrower beamwidth, i.e., it must be larger. So, the on-axis case is actually the most favorable from the standpoint of feed size.

Continuing with the on-axis case, we compute the feed beam angle using the first null beamwidth as a proxy for the customary –12 dB of a Gaussian-weighted aperture function. Th beam width (θ) for a uniformly illuminated round aperture is given by (11.3) (based upon the formula in Table 2.2 for uniform illumination):

$$\begin{aligned}\frac{\theta_b}{2} &= \frac{1.22\lambda}{D_{\text{feed}}}\\ D_{\text{feed}} &= 1.22\frac{\lambda}{\theta_b/2}\end{aligned} \tag{11.3}$$

Substituting θ_r for θ_b in (11.3), we obtain (11.4):

$$\begin{aligned}D_{\text{feed}} &= \frac{1.22\lambda}{\arctan\left(\dfrac{D_r}{F_r - \delta_y}\right)}\\ D_{\text{feed}} &\approx 1.22\lambda\,\frac{F_r - \delta_y}{D_r}\end{aligned} \tag{11.4}$$

The feed diameter is proportional to F/D, so obscuration becomes an increasing issue for increasing focal lengths. In these cases, an offset feed may also be preferred.

More generally, however, if we wish to form contiguous beams with our feed array, the feed element size must be compatible with required feed element spacing. The selected beam pointing angle (θ) results in a displacement (δ_x) of the central ray in the feed plane away from boresight according to (11.5):

$$\begin{aligned}\theta &= \tan\left(\frac{\delta_x}{F_r}\right)\\ \delta_x &= F_r \arctan\theta\\ \delta_x &\approx F_r\,\theta\end{aligned} \tag{11.5}$$

The feed element beamwidth (full width radians at –3 dB power level) using the data in Table 2.2 is represented by (11.6). This relationship assumes uniform illumination of the reflector:

$$\Delta\theta = 1.03\,\frac{\lambda}{D_r} \tag{11.6}$$

Combining (11.5) and (11.6), we have (11.7):

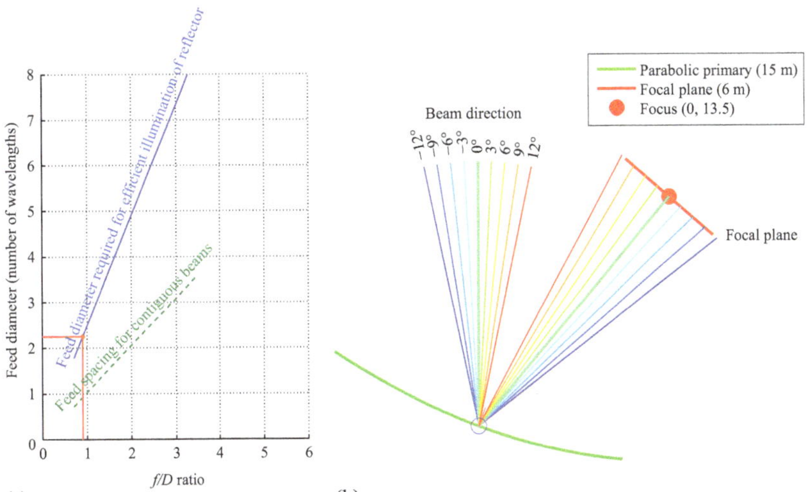

(a) (b)

Figure 11.5 Feed element size and spacing: (a) parametric horn size and (b) central ray geometry
ReflectorFeedSize.m

$$\delta_x \approx 1.03\,\frac{\lambda F_\mathrm{r}}{D_\mathrm{r}} \tag{11.7}$$

If we take the ratio of (11.4) and (11.7), we obtain (11.8), indicating that the required element spacing is about 0.4 times the feed diameter, with no dependence on wavelength or reflector parameters, although there is a second-order effect from δ_y in (11.1) which becomes significant for $f/D < 1$.

Since we specify an aperture weighting that is small at the reflector edge, the resulting beamwidth is larger than specified by (11.6). Accordingly, the feeds may be greater than specified by (11.8):

$$\begin{aligned}\frac{\delta_x}{D_\mathrm{feed}} &\approx \frac{1.03}{1.22}\\ &\approx 0.8\end{aligned} \tag{11.8}$$

Figure 11.5a represents the feed size required for efficient reflector illumination in (11.1) and (11.3) and feed size required to provide contiguous (3 dB crossover) beams (11.7). Figure 11.5a plots (11.4) and (11.7). The red lines indicate that for an f/D ratio of 0.9, the feed diameter should be approximately 2.25λ, while the element spacing should be about 1.0λ.

While several approximations are necessary to obtain this simple relationship, it is clear that the desired feed spacing is considerably less than the feed size so long as we require reasonable aperture efficiency. The feed diameter size limits our ability to create multiple beams if we used a single feed for each beam. Accordingly, the feed must utilize smaller elements and operate in an overlapped subarray manner

Table 11.3 L-band example

Attribute	Value
Reflector diameter (m)	15.0
Reflector focal length (m)	13.5
Subtended angle at feed (°)	74
Wavelength (m)	0.24
Feed diameter (λ)	2.25
Feed farfield transition (m)	1.7

(see Section 2.7.2). This is the reason that multiple-beam reflector feeds do not use horn feeds although the TecSAR does so successfully as noted in Section 13.3.1.

Using smaller elements is also expedient to correct for phase aberrations in a single beam, as described in Section 11.6. Huber *et al.* [13] provides an excellent description of the design procedure to resolve these objectives.

Equation (11.9) is an exact expression for the coordinate of the central ray intersection with the feed as a function of the beam scan angle (θ) and the distance (L) between the reflector center and the focal point:

$$\delta_x = L \tan(\theta) \tag{11.9}$$

We will evaluate a specific case in Chapter 18 using the reflector and feed characteristics which are enumerated in Table 11.3. Figure 11.5 illustrates the mapping of incident beams to the feed. The desired beam spacing would be approximately 1°, enabling up to about 20 beam positions with a 6-m-long feed for a desired element spacing of 6/20 or 0.3 m, while the feed diameter should be about 0.6 m for efficient reflector illumination.

11.6 Scan limitations

The primary limitation on scan capability is the significant defocussing when the beam is scanned. This limitation may be mitigated to some extent with a larger feed array although the detracts from reflector advantage in T/R module count.

Figure 11.6 illustrates the problem. Figure 11.6a–c shows a cross section of the reflector and feed assembly for scan angles of −8°, 0°, and +8°, respectively.

Figure 11.6d shows the intersection of these rays with the feed for scan angles between ±8°. Only at 0° scan angle do the rays converge to a point, although the proper representation would be an airy disk. At other scan angles, the rays fill a progressively larger region, whose size is indicated in the figure. The overlapping symbols provide a very rough suggestion of field intensity. There are 24 red circles representing overlapping feed elements spaced to provide beam crossover at −3 dB. Azimuth scan requires additional lines of elements but that is unnecessary for this degree of approximation. Huber *et al.* [14, 15] have published a detailed feed analysis for TanDEM-L using a field-based solution.

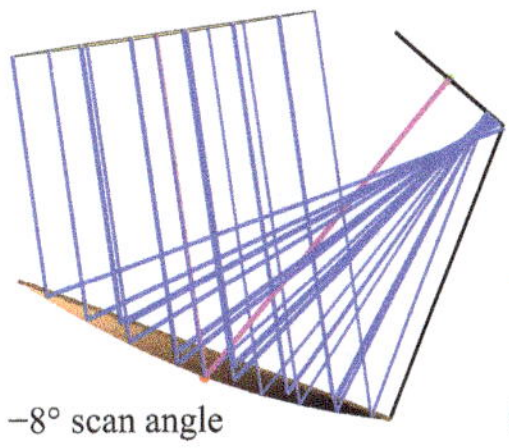

(a) Scan to −8° in elevation

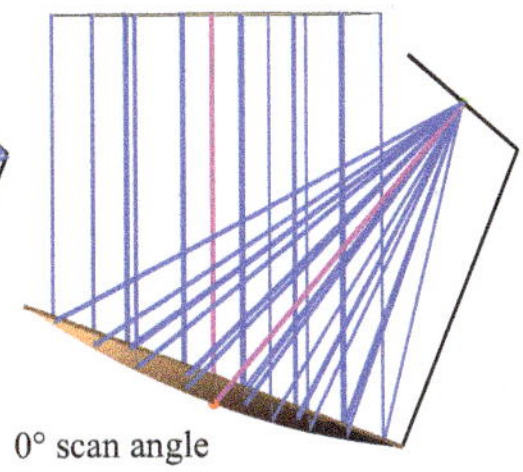

(b) Boresight scan in elevation

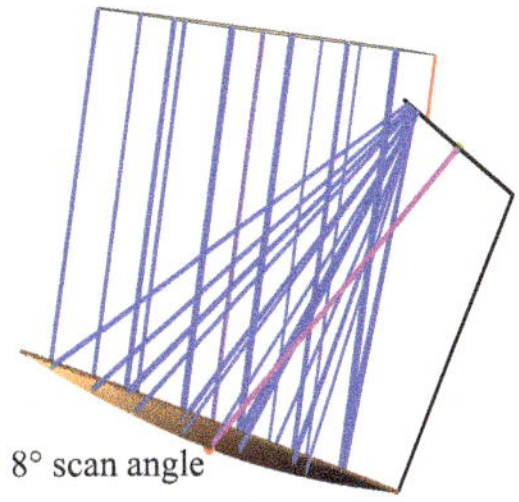

(c) Scan to 8° in elevation

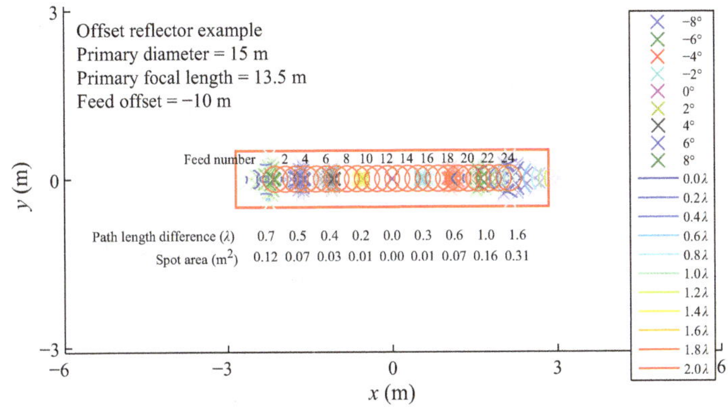

(d) Ray intersection

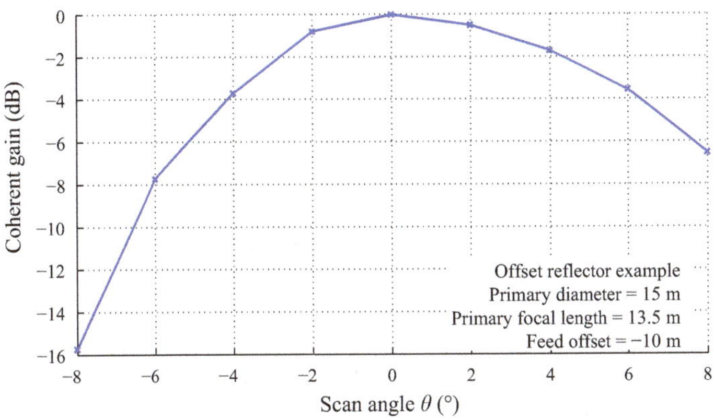

(e) Coherence loss

Figure 11.6 Ray tracing with offset reflector model
ThreeDimensionalRayTracing.m

An additional complication is that the rays arrive at the feed out of phase with each other (except for the boresight case). Figure 11.6d includes contours of constant phase in units of wavelength between 0 λ and 2.0 λ, showing that the phase differences can be significant at maximum scan. The relevant peak-to-peak path length difference is shown in the figure. The path length difference reduces gain because of (partial) destructive interference. Figure 11.6e represents a simple calculation of gain reduction on the assumption that the rays are equally weighted. The gain reduction can be substantially reduced by an ESA feed with proper phase shifts at each element.

Cooley [5] describes some advantages of a defocussed feed. A defocussed feed benefits from a larger number of T/R modules contributing to the signal since perfect focus concentrates the power to a single (or very few) T/R module. He observes:

1. a higher potential total RF transmit power (higher effective isotropic radiated power (EIRP))
2. more technology options for RF transmit power generation (e.g., various solid-state MMIC technologies in lieu of a single traveling wave tube amplifier (TWTA) or microwave power module)
3. simplification of the array thermal/cooling design (due to lower concentrations of RF power), and finally
4. spatial distribution drops the required peak RF power levels in the feed RF electronics, thereby mitigating multipaction design concerns (a notable issue, especially at lower frequencies)

We will revisit this analysis in Chapter 18 with a higher degree of fidelity.

References

[1] Warnick K.F., Maaskant R., Ivashina M.V., *et al. Phased Arrays for Radio Astronomy, Remote Sensing, and Satellite Communications. EuMA High Frequency Technologies Series*. Cambridge, UK: Cambridge University Press; 2018.

[2] Cappellen Wvan., Veidt B., Gough R., *et al.* 'Phased array feeds for the Square Kilometre Array'. *2011 XXXth URSI General Assembly and Scientific Symposium*. August 2011, pp. 1–4.

[3] Cooley M., Davis D. 'Reflector antennas' in Skolnik M. (ed.). *Radar Handbook*. New York, NY, USA: McGraw-Hill Education, Chap. 12; 2008. pp. 12.1–12.43.

[4] Rao S., Shafai L., Sharma S.K. *Handbook of Reflector Antennas and Feed Systems*. 1st edn. Boston, MA, USA: Artech House; 2013.

[5] Cooley M. 'Phased array fed reflector (PAFR) antenna architectures for space-based sensors'. *2015 IEEE Aerospace Conference*. 2015, pp. 1–11.

[6] Mailloux R.J. *Phased Array Antenna Handbook*. 3rd edn. Boston, MA, USA: Artech House; 2018.

[7] Jensen E.B. 'Benefits of AESA-Fed reflectors for spaceborne radar'. *IDGA Military Radar Conference*. 2008.

[8] Croswell W.F., Ball D.E., Taylor R.C. *Phased Array-Fed Antenna Configuration Study. Tech. Rep. NAS 3-23252*. Melbourne, FL, USA: Government Electronics Systems Division, Harris Corporation; 1983.

[9] Ruze J. 'Lateral-feed displacement in a paraboloid'. *IEEE Transactions on Antennas and Propagation*. 1965, vol. 13(5), pp. 660–5.

[10] Mobrem M., Kuehn S., Spier C., *et al*. 'Design and performance of astromesh reflector onboard soil moisture active passive spacecraft'. *2012 IEEE Aerospace Conference*. 2012, pp. 1–10.

[11] Fitzgerald W.D. 'Tech. Rep. TR-484, under electronic systems division contract F19628-70-C-0220. Prepared for the office of the chief of research and development, department of the army'. *Limited Electronic Scanning with a Near-Field Cassegrainian System*. Lexington, MA, USA: Lincoln Laboratory, MIT; September 1971. Available from https://apps.dtic.mil/ dtic/tr/fulltext/u2/735661.pdf.

[12] Fitzgerald W.D. 'Tech. Rep. TR-486, under electronic systems division contract F19628-70-C-0220. Prepared for the office of the chief of research and development, department of the army'. *Limited Electronic Scanning with an Offset-Feed Near-Field Gregorian System*. Lexington, MA, USA: Lincoln Laboratory, MIT; September 1971. Available from https://apps.dtic.mil/dtic/tr/fulltext/u2/736029.pdf.

[13] Huber S., Younis M., Krieger G. 'Tandem-L: SAR system design aspects'. *5th Workshop on Advanced RF Sensors and Remote Sensing Instruments, ARSI'17*. DLR, September 2017. Available from https://elib. dlr.de/113660/1/ARSI 2017 paper.pdf.

[14] Huber S., de Almeida F.Q., Villano M., *et al*. 'Tandem-L: a technical perspective on future spaceborne SAR sensors for earth observation'. *IEEE Transactions on Geoscience and Remote Sensing*. 2018, vol. 56(8), pp. 4792–807.

[15] Huber S., Villano M., Moreira A., *et al*. 'Tandem-L: design concepts for a next-generation spaceborne SAR system'. *Proceedings of EUSAR 2016: 11th European Conference on Synthetic Aperture Radar*. 2016, pp. 1–5.

Chapter 12
Design practice

Designs are never ideal, and compromises are necessary to reconcile requirements with size, weight, power, and cost. ESA antennas provide unique benefits to improve performance which must be weighed against their greater cost and complexity.

The first consideration in the design process is to devise an architecture consistent with system constraints and capable of satisfying system requirements. This requires a functional decomposition and partition, bearing in mind practical considerations such as component cost and availability. It is typically an iterative process, as each design iteration reveals opportunities for improvement and difficulties in realization.

A number of authors have discussed the subject. Tench *et al.* [1] provide an early view of the design process and Moore [2] provides a more contemporary view. Herd and Conway [3] present a recent historical overview of both technology and cost. McLachlan *et al.* [4] describe the development of ESAs for a number of aircraft. Hees *et al.* [5] discuss the design of the SCOUT product line for next generation X-Band SAR instruments for space applications at Airbus Defence and Space GmbH. Renard [6] discusses the design for airborne radar applications.

12.1 Design tradeoffs

The antenna designer has multiple and often conflicting objectives (Table 12.1).

There are a variety of means to improve SAR performance involved in system tradeoffs. Resolution, noise-equivalent sigma zero, area coverage, and area rate are the main objectives. The most fundamental is to increase the integration time to improve resolution and SNR.

12.2 Architecture

Usually, the system requirements will be allocated to subsystems, and the ESA will be described by the attributes in Chapter 2. The ESA requirements must be negotiated with the system engineering function to assure that they are both achievable and affordable.

Figure 12.1 illustrates the elements of an airborne ESA radar described by Moore [2]. This illustration includes the major constituents of an ESA as well as

Table 12.1 Design objectives

Objective	**Issue**
Reduce system losses and noise figure	Hard to do
Decrease data swath	Take coverage hit
Increase transmit power	Limited power budget
Increase pulse duration	May cause pulse timing issues
Decrease pulse bandwidth	For resolved targets
Increase pulse repetition frequency	May cause range ambiguity problems
View target from more favorable angle	Undesirable operational constraint
Increase antenna area	Expensive, may lessen coverage
Decrease slant range	May compromise mission performance

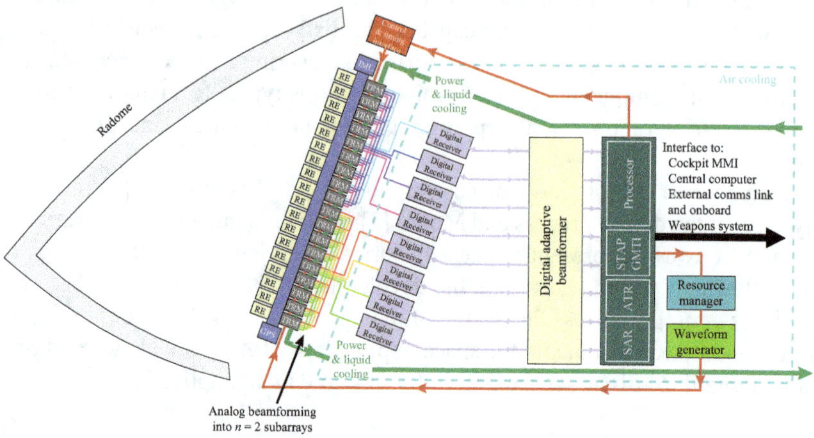

Figure 12.1 Functional and physical partitioning
After Moore [2]

a radome. A similar partitioning is described by Kinghorn *et al.* [7]. Capece and Capuzi [8] describe the antenna partitioning for COSMO-SkyMed.

Array partitioning

An array of elements can be decomposed into an array factor representing the physical location of the elements and the element factor of the individual element (perhaps modified by the effect of mutual coupling). This representation extends to the case where the element is itself a collection of individual elements and the collection is termed a subarray.

With the given array and element selected, the selected element or subarray must control the phase shift or provide a time delay to determine beam pointing and possibly beam shaping. The phase shift and time delay are usually understood to occur in the analog domain, at the RF frequency and sometimes the IF frequency.

Table 12.2 Functional and physical partitioning

Chapter	Function
13	Radiating elements (mutual coupling)
14	T/R modules
15	Assembly, packaging, power, and thermal management
16	Technology base and cost

Alternatively, if the signal is digitized, the time delay may be applied in the memory of the receiver.

Because undersampling is common, we must evaluate the resulting grating lobes. Additionally, as phase and time delays are usually quantized rather than continuously variable, the resulting errors may combine at some angles to provide increased gain, termed quantization lobes. Both grating and quantization lobe magnitude may be reduced by implementing a nonregular (random) spacing of the elements. This does not eliminate the lobe signal power but spreads it out avoiding the appearance of false targets. These errors are discussed in Chapter 6.

Finally, the elements will use both amplitude and phase control for beam pointing and shaping to achieve desired performance such as reduced sidelobes or flat-topped gain.

Each item in Figure 12.1 represents useful building blocks (first-level decomposition) and each will be discussed in subsequent chapters as indicated in Table 12.2. The partitioning may vary according to system requirements, for instance, each T/R module may feed multiple radiating elements, but the overall structure is universal. Table 12.2 summarizes the physical elements of an ESA and includes manufacturing and cost considerations.

12.3 Design verification

The analysis of the preceding chapters has discussed ideal cases. These must be evaluated with respect to errors which may occur concurrently in varying degrees. The nominal design should be evaluated for these error combinations, which increase the analysis time by a significant multiple of the nominal case analysis time. However, the resulting analysis time is generally shorter and less expensive than verification in hardware.

It may be sufficient to evaluate only the extreme values of these errors although that incurs the risk of excessive pessimism and overdesign. Design of experiments (DOE) methodology may be useful for choosing analysis cases.

Ultimately, verification in hardware is necessary before committing to volume production as analysis usually involves approximations, some of which may not be explicit.

References

[1] Tench M.D.R., Claridge P.A., Gates W.G. 'Design of a phased array antenna using solid state transmit/receive modules'. *IEE Colloquium on Mechanical Aspects of Antenna Design*. April 1989, pp. 10/1–10/10.

[2] Moore S. 'UK airborne AESA radar research'. *2009 International Radar Conference "Surveillance for a Safer World" (RADAR 2009)*. October 2009, pp. 1–7.

[3] Herd J.S., Conway M.D. 'The evolution to modern phased array architectures'. *Proceedings of the IEEE*. 2016, vol. 104(3), pp. 519–29.

[4] McLachlan A.D. 'T/R module design and production processes for airborne radar systems'. *2007 IET International Conference on Radar Systems*. 2007, pp. 1–4.

[5] Hees A. 'X-Band phased array antenna frontends for next generation SAR instruments'. *2019 IEEE International Symposium on Phased Array Systems and Technology (PAST)*. 2019.

[6] Renard C. 'Design of active antennas for airborne radar applications'. *Proceedings of the Fourth European Conference on Antennas and Propagation*. 2010, pp. 1–2.

[7] Kinghorn T., Scott I., Totten E., *et al.* 'Recent advances in airborne phased array radar systems'. *2016 IEEE International Symposium on Phased Array Systems and Technology (PAST)*. 2016, pp. 1–7.

[8] Capece P., Capuzi A. 'Active SAR antennas development in Italy'. *2011 3rd International Asia-Pacific Conference on Synthetic Aperture Radar (APSAR)*. 2011, pp. 1–5.

Chapter 13
Radiating elements

Radiating elements are the most visible and in many respects the most important constituent of an ESA. Their function is to create a desired electromagnetic field on the surface of the aperture to generate the desired farfield pattern. To this end, the T/R module provides an RF signal to the antenna port (connector) with the necessary amplitude and phase which the radiating element, as a linear device will impart to its radiated signal.

There are a number of considerations for radiating element design:

- impedance match, power transfer
- radiation resistance, scattering
- mutual coupling and surface waves
- load impedance
- single element
- adjacent element (function of separation)
- scattering in receive
- Q (quality factor) – energy storage.

Hansen [1] provides an analysis of these criteria.

13.1 Requirements

The antenna specification is derived from system requirements. Among the important attributes are

- suitable wide-angle radiation pattern
- compatibility with feed and T/R modules
- readily arrayed
- low cost.

The element function is to radiate all applied power and receive all incident power according to a prescribed function. The element match (return loss Γ or S_{11}) is the critical metric.

Most element types can be arrayed. We will consider the following:

- Dipole elements – Section 13.2
- Waveguides, slots, and horns – Section 13.3
- Notch elements – Section 13.4
- Patch elements – Section 13.5

Section 4.2 discusses efficiency. It is useful to characterize radiating elements by their voltage standing wave ratio or by a scattering matrix. S_{21} is the power transmission from the element port (coaxial cable) to free space which is unity in the ideal case. S_{11} is the ratio of the power reflected back to the input port which is zero in the ideal case. The element itself is assumed nondissipative which is a reasonable approximation. In practice, acceptable performance requires $S_{11} < -10$ dB, i.e., less than 10% of the input power is reflected.

Mutual coupling (Chapter 5) alters element behavior, so the parameters are sometimes measured with the element installed in the array. This is called an embedded element and assumes several elements in all directions from the element under test. Near the edge of the array, the behavior will be different but the number of such elements is usually a small fraction of the total. Pozar [2, 3] discusses this approach.

13.2 Dipole radiating element

Dipoles have been extensively described in the literature. The conventional half-wave dipole is not widely used, perhaps because its construction is difficult or because it is difficult to construct in an array. It is also subject to relatively high mutual coupling, which in fact is a feature of Yagi–Uda antennas which are dipole arrays operating in end-fire.

Figure 13.1a illustrates two examples of radiating elements, dipoles, and loops, which Balanis [4] discusses in detail in Chapters 4 and 5 of his classic work. An infinitesimal dipole has a cosine theta beam pattern but is very inefficient. Quarter-wave dipoles have a more complicated beam pattern (not much different from a cosine theta pattern) and are very efficient at resonance. The dipole radiation pattern is shown in Figure 13.2. Notably, it is omnidirectional in the xy-plane, so when arrayed, it is usually installed over a ground plane in order to restrict radiation to a hemispherical volume. Figure 13.1b represents a dipole array with its associated ground plane shown in green.

Coupled dipole arrays

Coupled dipole arrays have been proposed and demonstrated for efficient, broadband arrays. Figure 13.3a shows a concept for a coupled dipole array. Dipole arrays with enhanced coupling (which could be construed as mutual coupling) improve bandwidth and match. With a slight rearrangement of the sources, as shown in Figure 13.3b, a coupled dipole can be understood as an array of loop antennas. Munk *et al.* [6] describe their benefits.

This array approximates ideal current sheet so it is potentially very broadband and well matched.

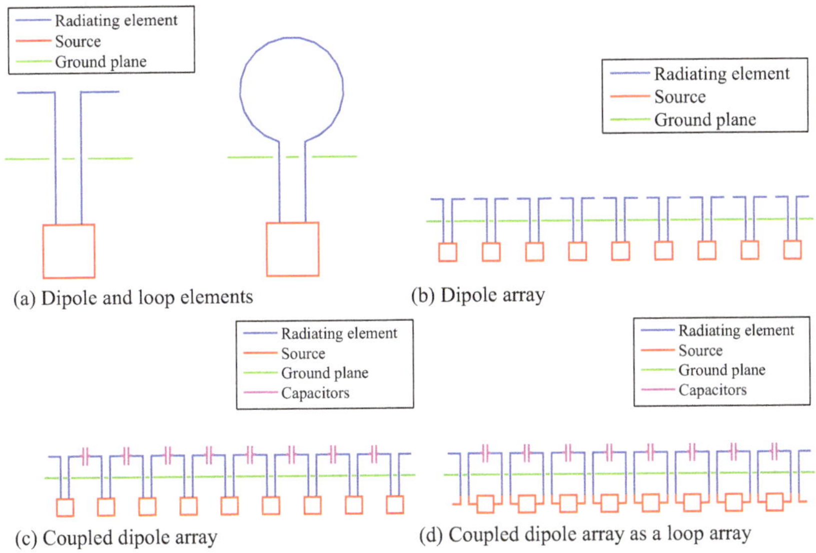

Figure 13.1 Elemental radiators and arrays

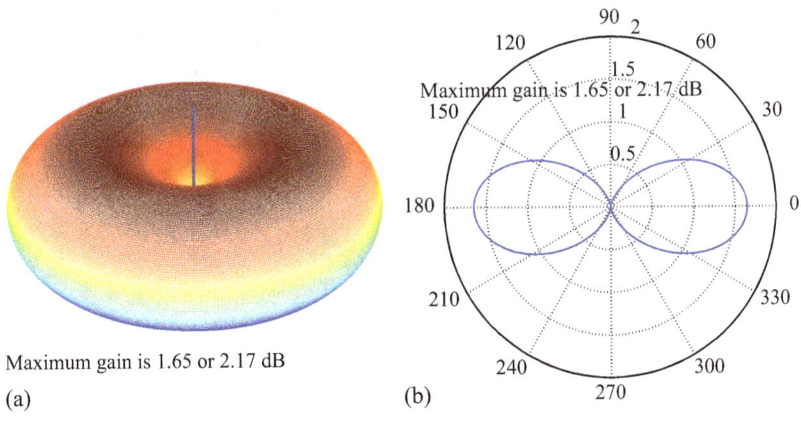

Figure 13.2 Dipole radiation pattern: (a) three-dimensional pattern representation and (b) pattern cut through vertical plane
DipolePattern.m

7–21 GHz dual-polarized array

Figure 13.4 shows a 7–21 GHz dual-polarized planar ultrawideband modular array developed by Holland and Vouvakis [7–10] at the University of Massachusetts Amherst. It implements a current sheet antenna of coupled dipoles using Rogers 5880 circuit material. It does not include active components, so each element is

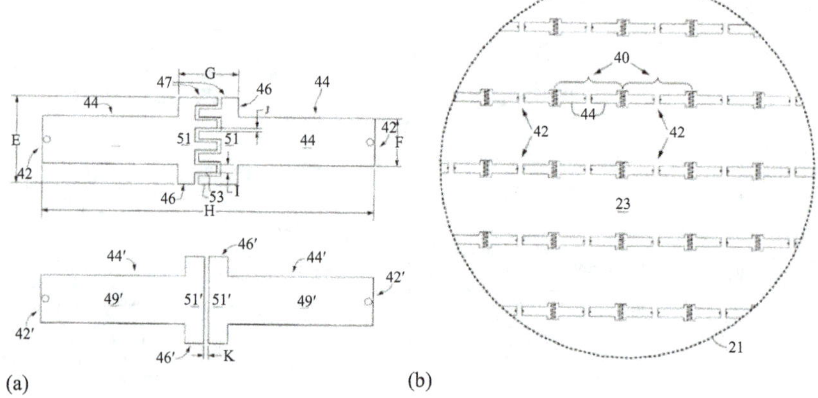

Figure 13.3 *Coupled dipole arrays: (a) coupling detail and (b) dipole array*
US Patent 6,512,498 Taylor [5].

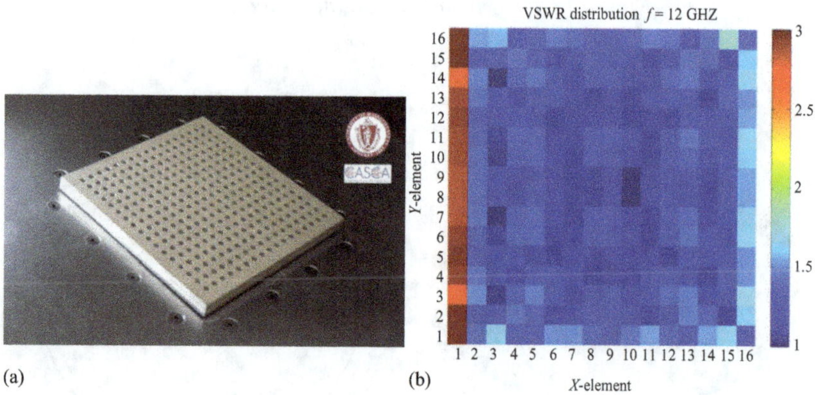

Figure 13.4 *7–21 GHz dual-polarized array: (a) array and (b) VSWR at 12 GHz*
©IEEE 2012. Reprinted with permission from [7], Figures 2 and 12(b)

accessed by a connector. Its construction is compatible with the array design concept described in Section 14.2.3.

Figure 13.5 shows a good element match as a function of frequency. Between 7 and 21 GHz, the VSWR is between 1.5 and 2 and S_{11} is between −10 and −15 dB.

13.3 Horn and waveguide radiating elements

Waveguide is widely used in microwave systems because it is a convenient size and has extremely low loss. It also finds use in antennas where it may radiate from an

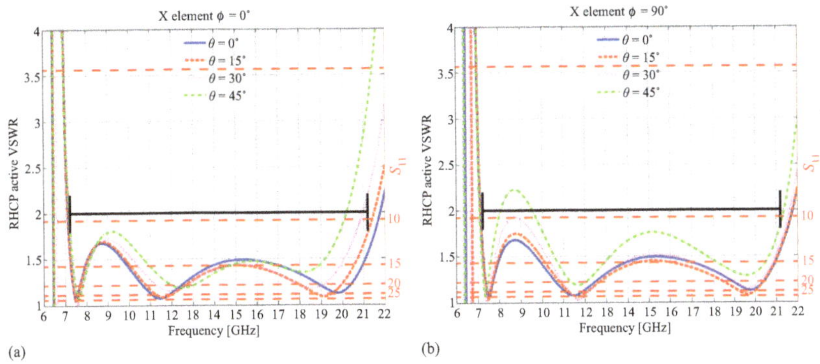

Figure 13.5 VSWR (S_{11}): (a) E-plane and (b) H-plane
Holland [8], Figure 4(a) and (b)

open end, sometimes with a flare for impedance match, or through slots cut into the waveguide which may be configured for a variety of beams and polarizations.

Some of its benefits include

- low loss, light weight, and inexpensive
- easy to machine at high precision
- polarization may be selected by slot orientation
- amplitude taper is readily incorporated by slot design
- feed is integral to waveguide
- frequency scanning is inherent but may be center-fed to avoid frequency scanning – beamwidth increases away from nominal frequency and eventually bifurcates limiting bandwidth.

Open waveguide (sometimes with a tapered horn section) is a good radiator but not often used in arrays because the element size is $\sim 1.5\lambda$, so it is not feasible to use a tight lattice as discussed in Section 11.5. An open-ended waveguide radiates quite well and forming a horn assists with the impedance match between the waveguide and free space. Spacing is less of an issue in thinned arrays where horn directivity helps control grating lobes. Raffoul and Hilburn [11] discuss the radiation efficiency of slotted waveguide arrays.

13.3.1 Horn feeds

Two SAR satellite systems use classic horn feeds: SAR-Lupe with a single feed horn and no electronic scanning and TecSAR with eight feed horns and eight selectable beams. Very little information is published concerning these systems.

Figure 13.6 illustrates two horn antennas. The first is a reference horn and the second is the TecSAR horn array. Not shown is the SAR-Lupe, a single horn-fed reflector described by Braun and Knobloch [13].

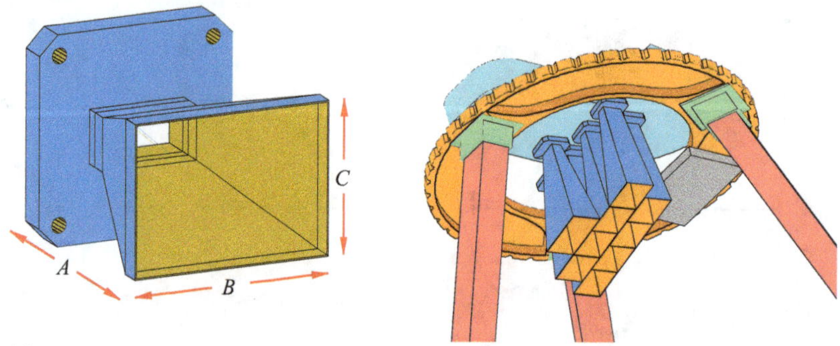

(a) (b)

Figure 13.6 Horn feeds: (a) standard gain horn (10 dB). Source*: Pasternack® PE9856-10 data sheet. (b) TecSAR. After Sharay and Naftaly [12]*

Figure 13.7 shows a commercial product offering a 10 dB gain horn. Its properties are summarized in Table 13.1. Its dimensions exceed a wavelength in both dimensions, so an array of these horns would be subject to grating lobes, which the element pattern would suppress at boresight.

The Israeli TecSAR system uses this approach. The TecSAR satellite feed array of eight horns at the focal point of a reflector is shown in Figure 13.6. The selection of pointing direction is by selection of horn. The two rows each of four horns staggered by one-half their width enable beam cross-over at –3 dB although the beam direction in the orthogonal sense moves slightly.

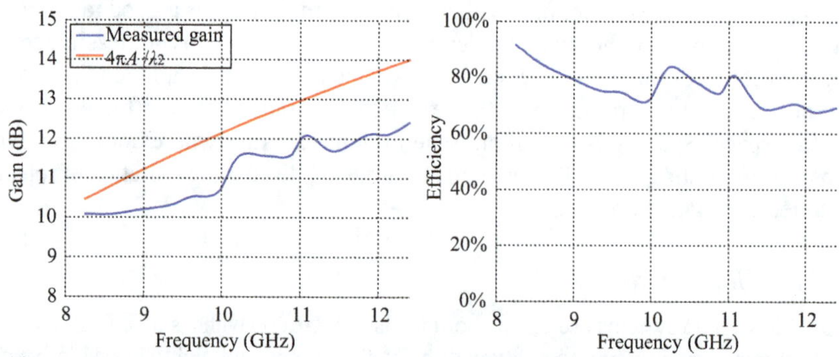

(a) (b)

Figure 13.7 X-band 10 dB standard gain horn: (a) standard gain horn gain and (b) standard gain horn efficiency

Source*: Pasternack® PE9856-10 data sheet [14]*

Table 13.1 Standard gain horn specifications

Attribute	Value
Frequency (GHz)	8.2–12.4
Wavelength (cm)	3.6–2.4
Beamwidth (°)	48
Gain (dB)	10.1–12.4
A (cm)	5.08
B (cm)	4.01
C (cm)	2.92

Source: Pasternack® PE9856-10 data sheet

13.3.2 Slotted waveguide

Waveguides have very low loss and are easily integrated with a radiating element. This type of radiating element is used by RadarSat and TerraSAR-X. Slots in the waveguide couple RF to free space. Size and orientation of slot can be tailored for desired properties. One of the advantages of this approach is that it combines the radiating element with a feed network (see Section 2.6.1).

Some planar arrays have been constructed in this fashion with a corporate feed. The corporate feed has no phase variation with frequency to limit bandwidth. Slotted waveguides are readily combined with waveguide-based corporate feed to provide low-loss RF distribution to hundreds of radiating slots. It is very wide band but not electronically scanned. This approach was used for SeaSAT. It is proposed for the Synspective satellite mentioned in Section 1.4.3 and described by Akbar *et al.* in [15, 16].

The satellite applications appear to use a series feed or possibly a center feed. The series feed from one end introduces frequency scanning of beam limiting its bandwidth. The center feed shown in Figure 10.6 combines two back-to-back series feeds which provide constant boresight pointing but split into two beams as frequency increases.

Josefsson and Rengarajan's [17] *Slotted Wavegide Array Antennas* provide an in-depth discussion of this topic. Römer [18] discusses the theory and design of the subarrays used in TerraSAR-X.

13.3.2.1 TerraSAR-X first generation

Figure 13.8 illustrates the dual-polarized waveguide radiating element used on TerraSAR-X. The performance of both configurations is summarized in Table 13.2.

Derneryd and Lagerstedt [23] describe the antenna design as follows:

> Non-inclined narrow wall slots in one waveguide generate the horizontal polarisation. The slots have to extend into the neighbouring broad walls of the waveguide to be resonant. The edge slots in the narrow wall need to be excited with a pair of wires inside the waveguide and not by slot tilt in order for minimum cross polarisation generation.

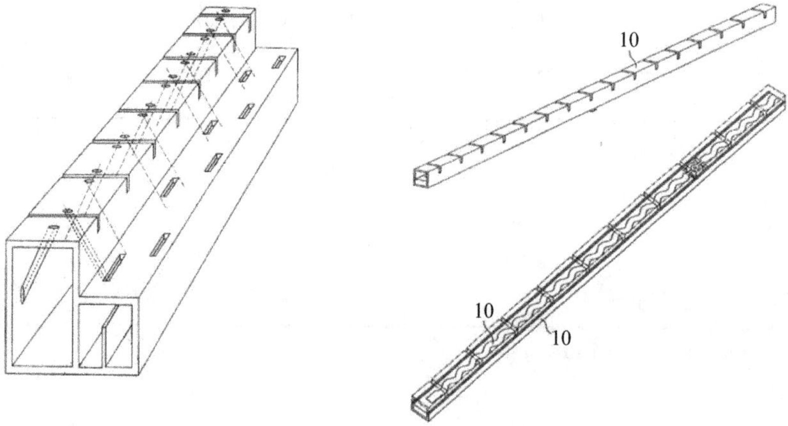

(a) (b)

Figure 13.8 Slotted waveguide radiating element: (a) first generation. Source: *Lagerstedt and Lagerloef [19, 20]. (b) Second generation.* Source: *Römer [21, 22]*

Table 13.2 Slotted waveguide feed array

Attribute	First generation	Second generation
VSWR	<1.5	<1.5
S_{11} (dB)	~−15	~−15
Horizontal polarization bandwidth (MHz)	>120	>650
Vertical polarization bandwidth (MHz)	>400	>700

Offset broad wall slots in the second waveguide generate the vertical polarisation. In order to minimise the waveguide width using longitudinal, broad wall slots, ridge loading is used. Both of the above slot types exhibit pure polarisation generation and high isolation between the ports within a subarray.

Figure 13.9 shows the performance of the radiating element in Figure 13.8a in terms of the return loss of the element.

13.3.2.2 Next-generation TerraSAR-X

A number of improvements are planned for the next-generation TerraSAR-X. One objective is wider bandwidth in accordance with the International Telecommunication Union(ITU) allocation, described in Section 17.2, which requires a new radiating element. A candidate design is described in European Patent EP2100348 [21]. It utilizes

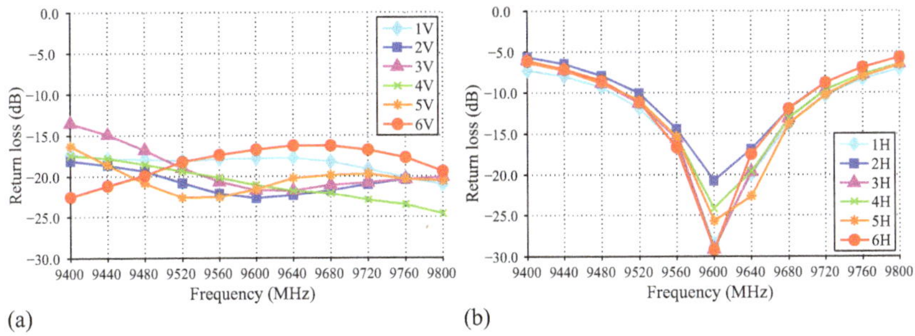

Figure 13.9 Return loss bandwidth (first generation): (a) vertical polarization and (b) horizontal polarization
After Derneryd and Lagerstedt [23]

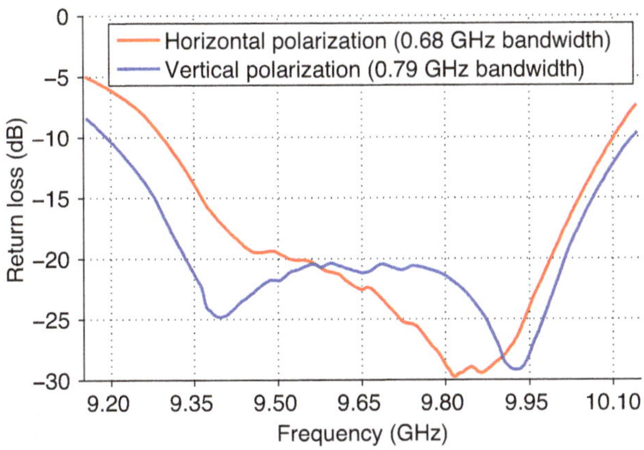

Figure 13.10 Return loss bandwidth – next-generation TerraSAR-X
Römer [21, 22]

a "Serpentine inner conductor alters propagation velocity so that slots are excited in phase. … Propagation modes are not dispersive which broadens bandwidth."

Figure 13.8b illustrates this radiating element and Figure 13.10 shows the performance of the radiating element for next-generation TerraSAR-X. The return loss bandwidth is more than 600 MHz, i.e., more than twice its predecessor.

13.4 Flared notch radiating element

The flared notch offers very wideband and near-perfect aperture match. Mechanically, it is difficult to integrate into arrays because the radiator stands off the array face requiring a right angle interconnect between the T/R module and the radiating element. Accordingly, it is used in applications where its benefits warrant added cost. The flared notch element easily achieves multi-octave bandwidth although it

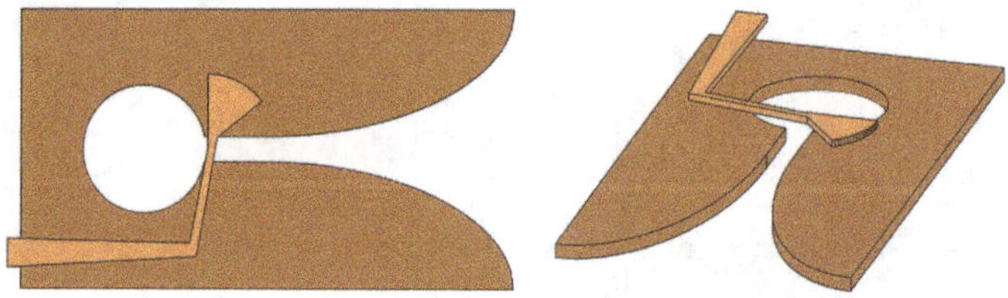

(a) (b)

Figure 13.11 Flared notch radiating element: (a) plan view and (b) isometric view

becomes very deep. It also provides excellent return loss and scans to wide angles. Figure 13.11 shows the major elements for a flared notch radiating element. Flared notch elements have wide application and there are many articles concerning its design including Gjokaj *et al.* [24], Chio *et al.* [25], and Cooley *et al.* [26].

Square Kilometer Array application

The Square Kilometer Array (SKA) is described in Section 1.4.4. Its frequency coverage may be as wide as 50 MHz to 30 GHz. This is a ratio of 600 or approximately 9 octaves. Even with wideband radiating elements, the SKA will require separate antennas for various bands. It represents a good application for flared notch elements since the crossed flared notch elements provide dual polarization for up to 10:1 bandwidth with scan performance of ±45°. Figure 13.12a shows the flared notch element selected by Astron for its Electronic MultiBeam Radio Astronomy ConcEpt (EMBRACE) Its specifications are summarized in Table 13.3.

Figure 13.12b shows the simulated performance for the prototype. The return loss is better than –10 dB across the operating bandwidth at boresight and still good for 45° scan in the H-plane. As it is a receive-only array, return loss is less critical.

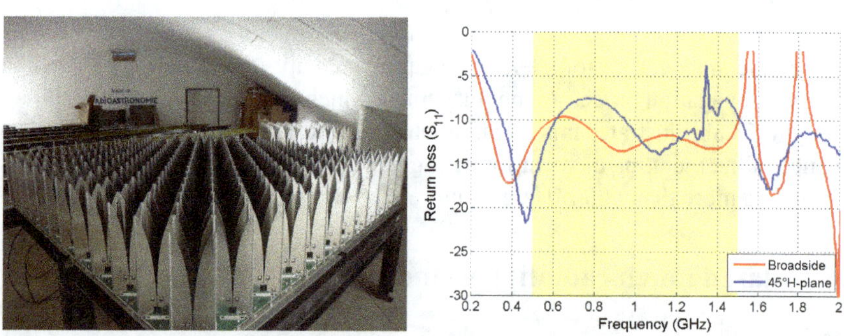

(a) (b)

Figure 13.12 Dual-polarized flared notch proposed for SKA: (a) array (under radome) at Nançay. Courtesy of Astron [27]. (b) Subarray. After de Vaate [28]

Table 13.3 Electronic MultiBeam EMBRACE

Attribute	Specification
Frequency range (MHz)	500–1,500
Polarization	Single (requirement) Dual (capability)
Physical collecting area (m^2)	≈160 – Westerbork Synthesis Radio Telescope (WSRT) ≈90 – Nançay
Aperture efficiency (%)	>80
Electronic scan range (°)	±45
Instantaneous bandwidth (MHz)	40
Number of RF beams	2
Number of digital beams	8+

Source: Kant [29]

13.5 Patch radiating element

The patch antenna is popular because of its reasonable cost and performance and particularly for its ease of integration into an array. It has higher loss than the preceding examples. It resembles a monopole antenna or loop antenna over a ground plane with a metal patch at the top. A second metal patch may be added for additional tuning capability. The planar configuration lends itself to large areas and it is possible to mount electronic components on the back for higher-level integration. Figure 14.6 shows one such example which also takes advantage of the integration capability of SiGe chips.

Patch radiating elements are used by Iridium, COSMO-SkyMed, SEOSAR, and PAZ as well as feeds for L-band reflector systems among the systems discussed here. There are numerous useful articles including Pozar [30], Carver and Mink [31], and Wallace [32].

Figure 13.13 shows the design of a patch radiator from Byström [33]. The vertical scale is exaggerated in Figure 13.13b. This design was analyzed using high-frequency structure simulator (HFSS) by Ansoft®, intended for finite-element modeling of electrically small (few wavelength) devices. The software provides good correlation to measurements.

The design uses a balanced feed with current flowing between the two ports through the lower patch analogous to a current loop. This provides a basic dipole radiation pattern. It is positioned over a ground plane to suppress the back hemisphere radiation. The capacitive patch makes a resonant structure which would normally result in a bandwidth of approximately 20%. A second nonconnected patch resonates at a different frequency and increases the bandwidth to more than 30%.

HFSS calculates voltages and currents at the input port giving a direct representation of $|S_{11}|$. We expect from conservation of energy that the rest of the power is

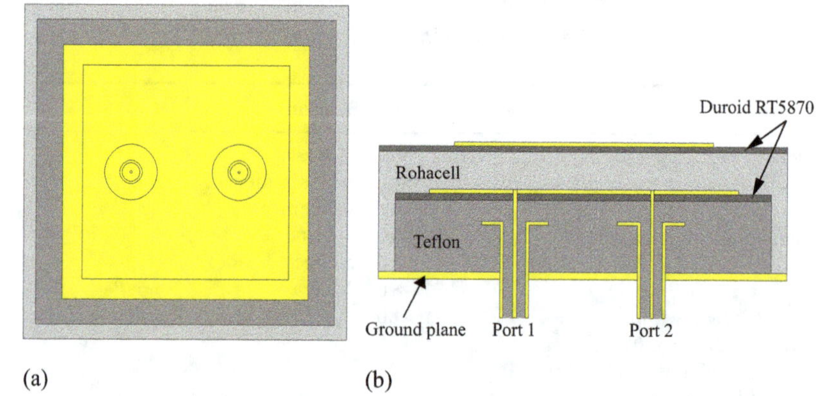

Figure 13.13 Stacked patch balanced feed radiating element: (a) element plan view and (b) element cross section
After Byström [33], courtesy Saab Microwave Systems (formerly Ericsson Microwave Systems)

radiated. These plots present the input return loss at various scan angles and show that up to 40° the return loss is 10 dB or better over an octave bandwidth. The double resonance is clearly visible.

Figure 13.14 shows S_{11} (return loss) as a function of scan angle. This is a measure of power reflected back to the source, i.e., not radiated. The radiating element has relatively low loss, so energy conservation requires that $|S_{11}|^2 + |S_{21}|^2 = 1$. Based on this assumption, Figure 13.15 displays S_{21} as a function of scan angle and frequency. The scan and bandwidth gain is within 1 dB of theoretical.

This radiating element has good predicted performance across an octave bandwidth. The HFSS software enables sensitivity analysis to assess the effects of varying material properties and manufacturing tolerances.

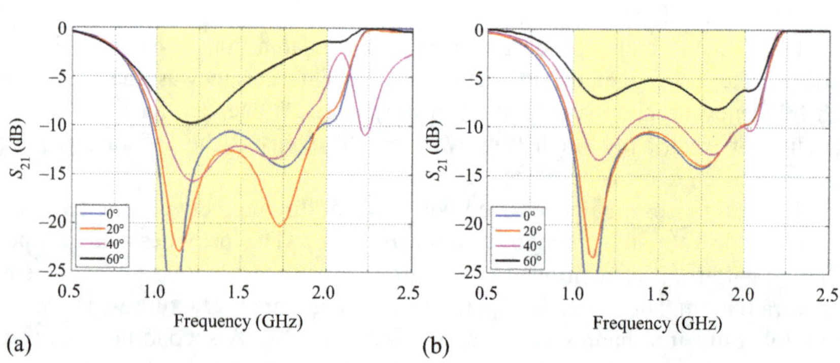

Figure 13.14 Stacked patch return loss (S_{11}): (a) E-plane scan direction and (b) H-plane scan direction
After Byström [33]

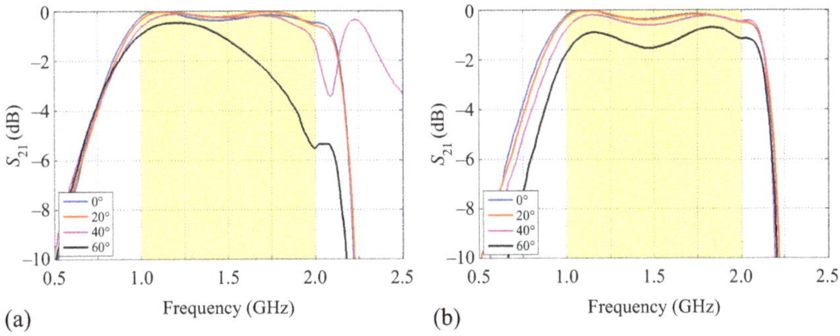

Figure 13.15 Stacked patch gain (S_{21}): (a) element scan direction and (b) H-plane scan direction
After Byström [33]

References

[1] Hansen R.C. 'Use of scan impedance and scan element pattern'. *2001 Antenna Applications Symposium*. 2001, pp. 364–71.

[2] Pozar D.M. 'The active element pattern'. *IEEE Transactions on Antennas and Propagation*. 1994, vol. 42(8), pp. 1176–8.

[3] Pozar D. 'Analysis of finite phased arrays of printed dipoles'. *IEEE Transactions on Antennas and Propagation*. 1985, vol. 33(10), pp. 1045–53.

[4] Balanis C.A. *Antenna Theory: Analysis and Design*. 4th edn. New York, NY, USA: Wiley-Interscience; 2016.

[5] Taylor R.C., Munk B.A., Durham T.E. 'Wideband phased array antenna and associated methods'. United States Patent 6,512,487. January 2003. Available from https://patents.google.com/patent/US6512487B1/en?oq=US6512487.

[6] Munk B., Taylor R., Durharn T., *et al*. 'A low-profile broadband phased array antenna'. *IEEE Antennas and Propagation Society International Symposium. Digest. Held in conjunction with: USNC/CNC/URSI North American Radio Sci. Meeting (Cat. No. 03CH37450)*. 2003, vol. 2, pp. 448–51.

[7] Holland S.S., Schaubert D.H., Vouvakis M.N. 'A 7–21 GHz dual-polarized planar ultrawideband modular antenna (PUMA) array'. *IEEE Transactions on Antennas and Propagation*. 2012, vol. 60(10), pp. 4589–600.

[8] Holland S.S., Vouvakis M.N. 'A 7–21 GHz planar ultrawideband modular array'. *2010 IEEE Antennas and Propagation Society International Symposium*. 2010, pp. 1–4.

[9] Holland S.S., Vouvakis M.N. 'A fully planar ultrawideband array'. *Antenna Applications Symposium*. Vol. I. AFRL-RY-HS-TR-2011-0001. Allerton Park, Monticello, IL, USA, July 2010, pp. 221–32. Available from https://apps.dtic.mil/dtic/tr/fulltext/u2/a538846.pdf.

[10] Holland S.S. 'Low-profile, modular, ultra-wideband phased arrays'. PhD thesis. University of Massachusetts, Amherst, September 2011. Available from https://scholarworks.umass.edu/open access dissertations/450.

[11] Raffoul G., Hilburn J. 'Radiation efficiency of an X band waveguide array'. *IEEE Transactions on Antennas and Propagation*. 1974, vol. 22(2), pp. 355–7.

[12] Sharay Y., Naftaly U. 'TECSAR: design considerations and programme status'. *IEE Proceedings - Radar, Sonar and Navigation*. 2006, vol. 153(2), pp. 117–21.

[13] Braun H.M., Knobloch P.E. 'Sar on small satellites — shown on the SAR-Lupe example'. *Proceedings of the International Radar Symposium 2007*. 2007.

[14] Pasternack Enterprises. WR-90 Waveguide Standard Gain Horn Antenna Operating From 8.2 GHz to 12.4 GHz With a Nominal 10 dBi Gain With UG-135/U Square Cover Flange. 2014. Available from https://www.pasternack.com/images/ProductPDF/PE9856-10.pdf.

[15] Akbar P.R., Saito H., Zang M., *et al*. 'X-bBand parallel-plate slot array antenna for SAR sensor onboard 100 kg small satellite'. *2015 IEEE International Symposium on Antennas and Propagation USNC/URSI National Radio Science Meeting*. 2015, pp. 208–9.

[16] Akbar P.R., Pyne B., Saito H., *et al*. 'Present state of antenna design for X-band SAR sensor onboard 100 kg class satellite'. *2017 11th European Conference on Antennas and Propagation (EUCAP)*. 2017, pp. 2640–1.

[17] Josefsson L., Rengarajan S.R. *Slotted Waveguide Array Antennas*. London, UK: SciTech Publishing, an imprint of the IET; 2018.

[18] RömerC. 'Slotted waveguide structures in phased array antennas, PhD thesis'. 2008, p. 159.

[19] Lagerstedt A., Lagerloef R. 'Waveguide Antenna (see US5831583)'. EP0775373 (A1). May 1997. Available from https://patents.google.com/patent/EP0775373A1/en?oq=EP0775373.

[20] Lagerstedt A., Lagerloef R. 'Waveguide Antenna'. United States Patent 5,831,583 (EP0775373). November 1998. Available from https://patents.google.com/patent/US5831583A/en?oq=US5831583.

[21] Römer C. 'Waveguide radiator, especially for synthetic aperture radar systems, EP2100348'. 2009.

[22] Römer C. 'Waveguide radiator, especially for synthetic aperture radar systems, US8493275'. 2013.

[23] Derneryd A.G., Lagerstedt A. 'Novel slotted waveguide antenna with polarimetric capabilities'. *Geoscience and Remote Sensing Symposium, 1995. IGARSS 95. Quantitative Remote Sensing for Science and Applications, International*. 1995, vol. 3, pp. 2054–6.

[24] Gjokaj V., Chahal P., Papapolymerou J., *et al*. 'A novel 3D printed Vivaldi antenna utilizing a substrate integrated waveguide transition'. *2017 IEEE International Symposium on Antennas and Propagation USNC/URSI National Radio Science Meeting*. 2017, pp. 1253–4.

[25] Chio T.H., Schaubert D., Holter H. 'Experimental radiation and mutual coupling characteristics of dual-polarized tapered slot antenna (TSA) arrays'. *Proceedings of 1995 Antenna Applications Symposium*. 1999, pp. 280–309.

[26] Cooley M., Essman S., Quade S., *et al*. 'Planar-fed folded notch (PFFN) arrays: a novel wideband technology for multi-function active electronically scanning arrays (AESAs)'. *2016 IEEE International Symposium on Phased Array Systems and Technology (PAST)*. 2016, pp. 1–6.

[27] Astron. Prototype array. Available from https://www.skatelescope.org/multimedia/image/ska-pathfinder-embrace-antennas/.

[28] de Vaate J.G.B. 'Antennas design group'. SKA 2010 *Science and Engineering Meeting*. March 2010. Available from https://indico.skatelescope.org/event/2/contributions/1453/attachments/1403/1732/AAVP Vaate 201003. pdf.

[29] Kant D. 'EMBRACE status and results'. Annual Work Package (*WP2) Meeting*. October 2010. Available from https://indico.skatelescope.org/event/3/contributions/2668/attachments/2279/2872/EMBRACE-PrepSKAWP220101028.pdf.

[30] Pozar D.M., Schaubert D.H. *Microstrip Antennas: The Analysis and Design of Microstrip Antennas and Arrays*. IEEE, 1995, pp. i–xi. Available from https://ieeexplore.ieee.org/document/5263386.

[31] Carver K., Mink J. 'Microstrip antenna technology'. *IEEE Transactions on Antennas and Propagation*. 1981, vol. 29(1), pp. 2–24.

[32] Wallace M. 'Mutual elements and substrate effect analysis on patch antenna arrays'. University of South Florida. 2010. Available from http://digital.lib.usf.edu/SFS0028083/00001.

[33] Byström D. 'Broadband patch antenna'. *Delivering Performance Workshop, 29 September -- 13 November 2003*. Ansoft Corporation, November 2003, pp. 1–14. Available from http://www.ansoft.com/markets/HighPerformanceRF/8EricssonBroadbandPatchAntenna.pdf.

Chapter 14
T/R modules

T/R modules are the heart of an ESA and their continued improvement has been the key to ESA development. T/R modules provide distributed gain and phase control, typically at each radiating element. They provide flexibility enabling the attractive performance of the ESA. There are many good examples in systems and many more concepts described in the literature.

The cost of T/R modules has been the most important restriction on their wide use. Since the 1990s, costs have declined precipitously, leading to the vast increase in ESA applications, primarily because of commercial demand for MMICs, application-specific integrated circuits (ASICs), and other components. Cost reduction has been an unceasing endeavor and will be discussed in Chapter 16.

T/R module design and manufacture is largely a matter of experience rather than theory. There are a few books and a number of good articles on the subject. The book by Sturdivant and Harris [1] describes many of the design considerations going into a T/R module including notably packaging and thermal control, based in part on their experience at Hughes Aircraft and Georgia Tech, respectively. Weedon *et al.* [2] present a T/R Module Tutorial at many IEEE symposia, not separately published. A number of designers publish histories of development. Barkdoll *et al.* [3] describe module developments at Northrop-Grumman. Kopp *et al.* [4] and Kolias and Borkowski [5] describe module developments at Raytheon. Parker and Zimmermann [6, 7] describe module developments at the legacy Hughes group at Raytheon. Herd and Conway [8] describe a history of the subject from a Lincoln Laboratory perspective with a discussion of the MPAR module as well as a simple cost model. O'Haver *et al.* [9] present a history of module work at Johns Hopkins University Applied Physics Laboratory. Schuh *et al.* [10] describe the European Aeronautic Defence and Space (EADS) Standardized modular T/R module (SMTR®) whose development started in 2001 and has been used in several programs including the TerraSAR-X. Rathod *et al.* [11] provide an overview of module evolution and trends from an Indian perspective.

14.1 Requirements

The T/R module requirements are derived from the ESA requirements and many are direct and others are derived. Among the important module requirements are size, weight, and power.

Additionally, the module requires appropriate signal, power, and control interfaces as well as thermal management. Cooling is a key element of the design and may be provided through active or passive means. High-power arrays will invariably require liquid cooling with the associated pumps and heat exchangers with the environment. By preference, component temperature will be controlled within a small range for best reliability.

Customarily, the module contains logic and memory, to communicate through a serial interface and provide control lines to the module circuitry. Current semiconductor technology far surpasses the relatively modest computing requirements of the module.

14.2 Module styles

Figure 14.1 shows two fundamental approaches to assembling T/R modules into subarrays and arrays. The brick style is essentially unconstrained in volume, subject to reasonable depth. The tile module is far more constrained in volume but lends itself to smaller volume and lighter weight. Additionally, integration into the array is simpler as there is only one dimension to deal with, although there are many more interconnects required. Recently, the focus has been multiple T/R elements on a planar substrate, sometimes termed a panel. This approach is a refinement of the tile approach. So long as volume is not a major constraint, the brick style is less difficult and less expensive.

The physical configuration of an ESA is largely determined by its T/R module. For high frequencies, e.g., >1 GHz, the lattice dimensions constrain the module volume in two of its dimensions. For this reason, T/R modules are designed with a small cross section in height and width but a large dimension in length in order to accommodate the necessary components and circuitry. The circuitry and signal path are arranged along the length of the module with one end oriented toward the antenna and the other end toward the back-end electronics. Accordingly, this design is characterized by circuit assemblies that are perpendicular to the plane of the antenna. There are examples of one-, two-, and four-channel modules, supporting a like number of radiating elements. The greater the number of elements , the more complexity and difficulty in manufacture.

When assembled into arrays, the ensemble of modules resembles a brick wall, hence the T/R modules are termed brick T/R modules. Individual T/R modules are assembled into linear arrays sometimes referred to as slats or sticks and these assemblies include cooling provisions as well as signal, control, and power distribution. This approach is straightforward to assemble and test but complicates the cooling and manifolding and is not particularly lightweight.

With continuing miniaturization and with multiple substrates, it has been possible to orient the circuit boards parallel to the plane of the aperture. This maintains the small dimensions in height and width but greatly reduces the thickness of the array. This arrangement is termed a tile module by an extended metaphor. There

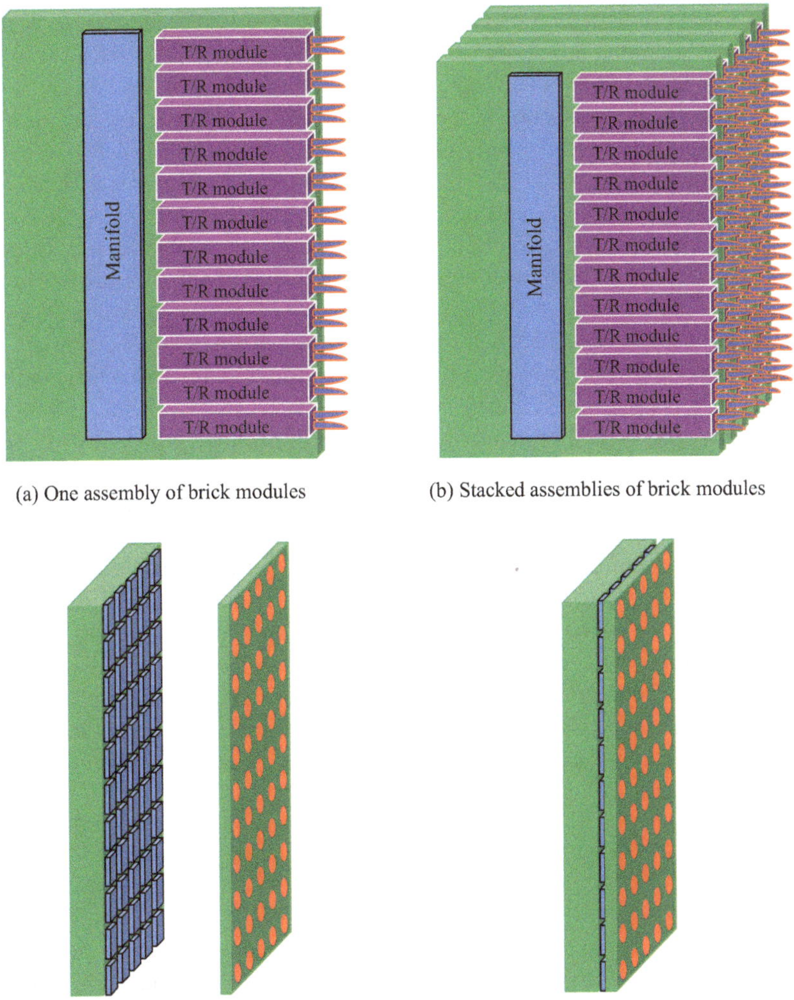

(a) One assembly of brick modules

(b) Stacked assemblies of brick modules

(c) Array of modules and elements

(d) Tile-style array

Figure 14.1 Two-array architectures

are fewer examples of tile modules but the one described in Section 14.2.2 has four channels.

A more extreme example of this approach is to mount all of the components to a single circuit, often flexible, with many T/R channels contained in a single assembly. Examples with 64 and 256 T/R channels have been demonstrated. The complexity of the multiple channels is offset by the simplicity of a single physical assembly. These assemblies offer little or no environmental protection to the components and heat transfer provisions are more difficult.

The electrical configuration of the T/R module is largely standard, constructed with low noise amplifiers, power amplifiers, and gain and phase control. It is also customary to include power and logic distribution and control as well as a means to remove excess heat. Connection of the module to the radiating element and the manifolds is required and is frequently implemented by connectors which also enable access for test and rework. The alternative of a monolithic assembly requires either exceptional manufacturing yields or some means to bypass defective channels.

The electrical configuration of the T/R module is largely standard, constructed with low noise amplifiers, power amplifiers, and gain and phase control. It is also customary to include power and logic distribution and control as well as a means to remove excess heat. Connection of the module to the radiating element and the manifolds providing RF, power, and control is required and is frequently made by means of connectors as these enable access for test and rework. The alternative of a monolithic assembly will require either exceptional manufacturing yields or some means to bypass unusable channels.

The gain and phase control circuits may be configured in one of two ways. In the first instance, the gain and phase controls are bidirectional (reciprocal) for the RF signal. This requires at least one switch to select between transmit and receive functions. In the second instance, the gain and phase controls are unidirectional. This requires three switches to select between transmit and receive functions but it allows for gain stages within the circuit. The choice between the two is determined by allocated requirements, with the second design having the potential for greater dynamic range. Both approaches are illustrated in the examples of brick and tile modules considered. The choice of topology is independent of the choice of physical design.

14.2.1 Brick module configuration

The Raytheon Theater High Altitude Area Defense (THAAD) T/R module represents a noteworthy and successful example of a brick module. Several hundred thousands of these modules have been produced for the THAAD and the SBX described in Section 1.3.3. Intended for ground-based applications, its volume constraints are not as severe as those for airborne systems. Kolias and Borkowski [5] describe the history of the THAAD T/R module from its beginning through 2000. Its design has been improved over the years, notably by the "Best of Breed" module program in 1998–2000 reported in [12] and illustrated in [13]. In 2017, Raytheon received a $10 million contract to retool its AN/TPY-2 production line to accommodate GaN [14] semiconductors into the module design.

Figure 14.2 shows the functional allocation of the circuit. The transmit path is shown in red, the receive path in blue, and shared paths in magenta. The beamformer port is on the left and connects to the attenuator and phase shifter circuits which are reciprocal, i.e., they provide the same function for signals traveling from left to right as for signals traveling from right to left. A switch selects either the transmit path or receive path. Both the transmit and receive paths are connected by a circulator to the antenna port on the right. The circulator is a passive device which requires no

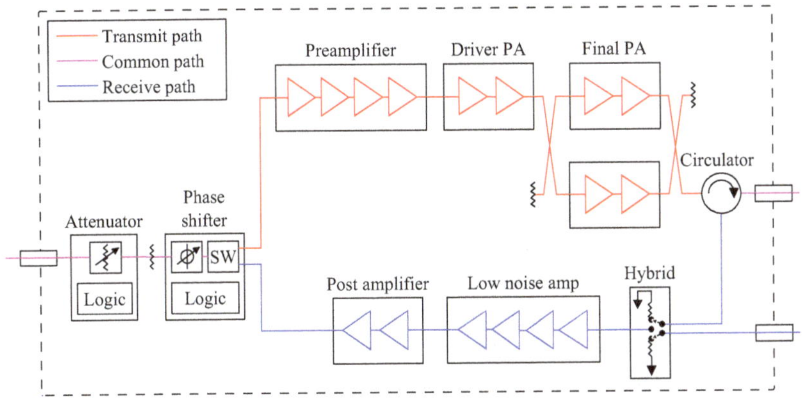

Figure 14.2 Functional diagram
After Kopp et al. [4]

control circuit; the signal is routed depending on its direction such that the transmit path is routed to the antenna port but the antenna port is routed to the receive path. There is a second receive-only port which may be selected by a switch contained in the hybrid circuit. This allows two receive elements which may, for instance, represent different polarizations.

In Figure 14.2, each rectangle represents a MMIC and each triangle within the rectangle represents a gain stage in the MMIC. Each gain stage may contribute 6–9 dB of gain. Accordingly, the transmit path may represent 50–70 dB of gain and the receive path may represent 40–50 dB of gain. This amount of gain requires careful design and circuit layout to avoid oscillation within the package.

This module was designed quite some time ago and has a relatively large number of MMICs. Modern designs may have as few as two or three MMICs representing a control chip, a high power amplifier (HPA), and a low noise amplifier (LNA).

14.2.1.1 European example

The Standardized Modular Transmit Receive (SMTR®) module is good example of a T/R module originally developed for terrestrial use being used in space. Its development by AirBus Defence and Space in Ulm started in 2001. By 2014, they had produced 50,000 for ground-based, airborne, and spaceborne applications. TerraSAR-X is one of its applications.

Adolph *et al.* [15], Opperman *et al.* [1617], and Rieger *et al.* [18] describe the module in good detail. A front-end electronics unit provides a high-power polarization switch as well as a calibration port. Power conditioning and control electronics are also included. These functions and the SMTR module are packaged in a larger package which constitutes the TerraSAR-X T/R module. Its nominal Tx power level of 8 W is more than required for a single element in this application. Accordingly, it feeds a subarray with 12 radiating (slot) elements so element power is about 2/3

W. The use of a non-steered subarray limits the scan capability in azimuth as noted in Table 17.2.

14.2.2 Tile module configuration

The T/R module is a major contributor to the size of the ESA. In order to construct smaller ESAs, the size of the T/R module must be reduced. This requires shortening the length as the other two dimensions are already established by the lattice spacing. For this purpose, an alternate configuration places the circuitry in the plane of the antenna. This arrangement considerably reduces the available circuit area, so multiple, multilayer circuit boards as well as smaller components are required. When assembled into an array, this configuration resembles a tiled surface, so the T/R modules are called tiles.

DARPA sponsored the High Density Microwave Packaging (HDMP) program between 1993 and 1997 to advance this concept. Contracts were awarded to Texas Instruments, Hughes, and Westinghouse. Hughes Aircraft developed the tile module shown in Figure 14.3. The history of the HDMP program is described by Cohen [20] and Lamb *et al.* [21], while the Hughes tile module is described by Sturdivant *et al.* [22], Hauhe and Wooldridge [23], and Wooldridge [24].

This module was a major improvement in packaging density without compromising performance. It required novel techniques to connect its three substrates as well as to connect to the array. Table 14.1 lists some of its characteristics.

Figure 14.4 shows the functional allocation of a single channel of the four. The transmit path is shown in red, the receive path in blue, and shared paths in magenta. The beamformer port is on the left and connects to an attenuator stage. Then three switches route the signal through an additional attenuator stage and a phase shift stage. There are also two gain stages in this path which are nonreciprocal; hence the need for the three switches to configure transmit and receive signal paths. These functions are contained in a single MMIC termed the Common Leg Circuit (CLC). The MACOM chipset shown in Figure 14.5 uses the same configuration. The CLC MMIC provides a drive signal to the transmit path driver MMIC and receives a

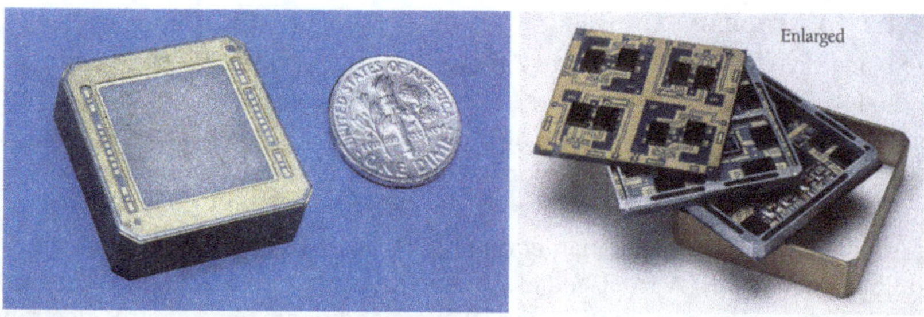

(a) (b)

Figure 14.3 Hughes aircraft tile module: (a) sealed module and (b) exploded view

© SciTech [19], Chapter 38, Figures 15 and 16

Table 14.1 Tile T/R module characteristics

	Transmit	**Receive**
Operating frequency	X-band	X-band
Peak-radiated output power	≈6 W	
Gain		20 dB
Noise figure		≈4 dB
Size	3 cm × 3 cm × 1 cm	

Source: Sturdivant *et al.* [22]

signal from the low-noise amplifier. Notably, there is no output T/R switching. This function must be provided outside the module.

In Figure 14.4, each rectangle represents a MMIC and each triangle within the rectangle represents a gain stage in the MMIC. Each gain stage may contribute up to 10 dB of gain. Accordingly, the transmit path may represent up to 40 dB of gain and the receive path may represent up to 60 dB of gain. The volume of the module cavity is much reduced compared to that of the brick module which reduces the opportunity for resonant modes but increases the likelihood of undesired coupling due to the reduced dimensions.

Figure 14.6 illustrates the collection of four T/R channels and the provision for a four-way power divider/combiner inside the module. The module has three substrates: CLC, driver/LNA, and HPA. Dashed vertical lines show which functions are implemented in which substrates.

A tile array concept building on this module design is shown in Figure 15.2 identifying signal, power, and cooling paths.

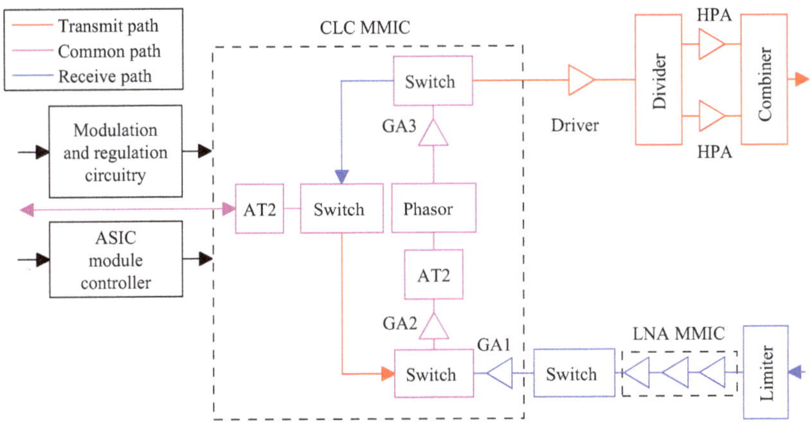

Figure 14.4 Single-channel (one of four) block diagram
After Wooldridge [24]

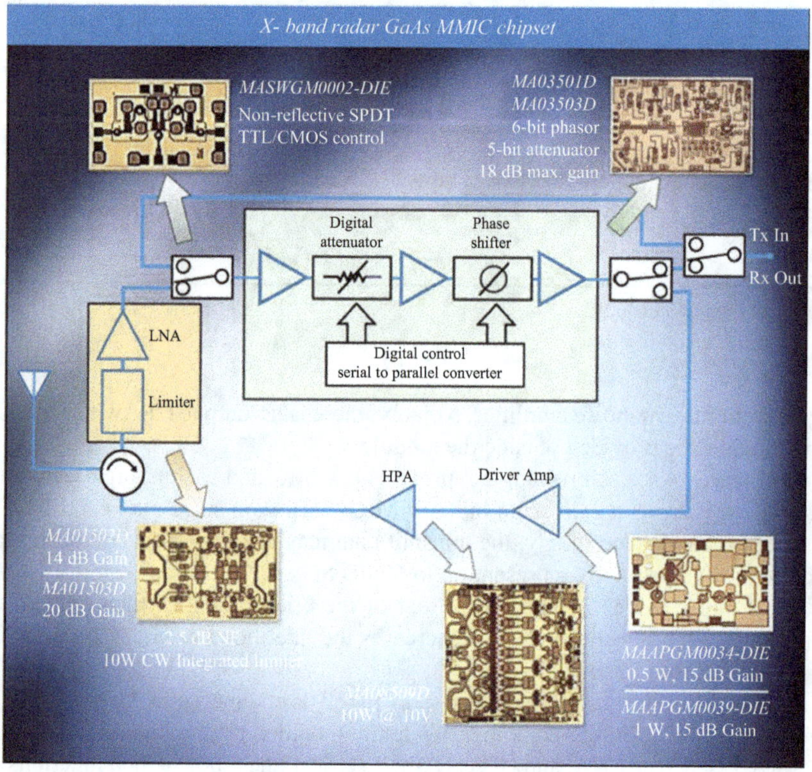

Figure 14.5 T/R module chipset
© MACOM [25]

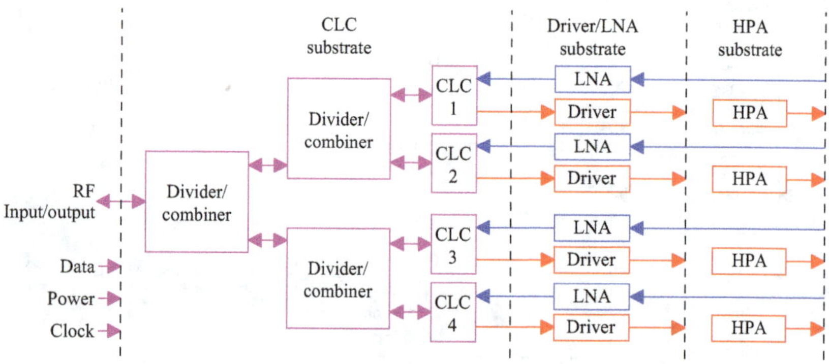

Figure 14.6 Four-channel module block diagram
After Wooldridge [24]

14.2.2.1 UK example

Some UK developments of tile modules are described in Lyon *et al.* [26] and Kinghorn *et al.* [27]. A recent example of this design style is provided by Alessandro Di Carlofelice *et al.* [28] in "Compact and reliable T/R module prototype for advanced space active electronically steerable antenna in 3-D LTCC technology." They provide references to more module articles. The topology of Figure 14.4 is also used by Hensoldt Sensors for their SCOUT design shown in Figure 7 of Hees *et al.* [29].

14.2.2.2 Module packaging

T/R module packaging provides several benefits in ESAs.

- Provides a tested subassembly which ensures that subsequent assembly steps will have "known-good" parts.
- Provides a controlled environment which relaxes some of the environmental constraints of the antenna assembly.
- Protects the T/R module components, e.g., unpackaged MMICs, from damage in handling and use.
- Provides a consistent and effective thermal interface to the next higher assembly.

Its principal drawbacks are size, cost, and weight. Some recent ESAs dispense with packaging although reliability issues may be unacceptable for some applications.

Packaging has been the subject of much research and development. Cohen [30] describes "Trends in the development of MMICs and packages for active electronically scanned arrays (AESAs)." Pilato [31] describes T/R module packages produced by KYOCERA International and discusses interconnection issues. Midford *et al.* [32] describe "The evolution of packages for monolithic microwave and millimeter wave circuits."

14.2.3 Planar configuration

The brick and tile module configurations both rely on high-reliability concepts with very robust hermetic packages. These are still suitable for the defense systems which commissioned them but the expansion of ESAs into new and especially commercial applications cannot sustain the costs, so new packaging approaches are being devised.

A number of companies and organizations are developing PWB-based arrays with the module subassemblies attached using SMT technology or with all components integrated on the PWB.

Figure 14.7 shows a planar array developed by Georgia Tech. It contains both radiating elements and T/R modules and also represents an advanced packaging technique. One of the notable achievements of this program was implementing 64 T/R modules and 64 radiating elements on a single planar substrate. This degree of integration required a highly integrated T/R chip (the T/R integrated circuit or

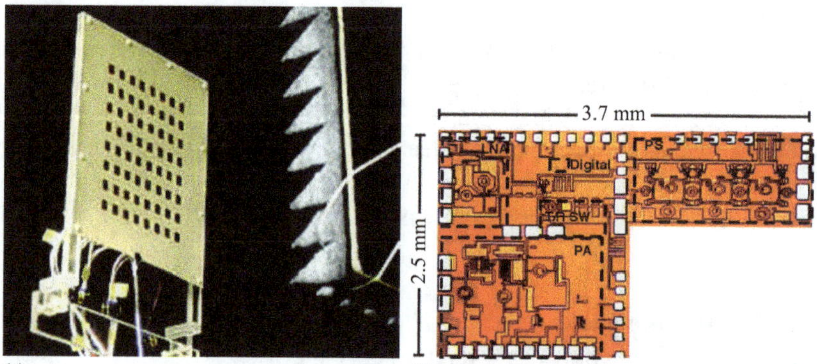

(a) (b)

Figure 14.7 Georgia Tech 64-element antenna: (a) array and (b) chipset
© 2013 IEEE [33], Figures 15 and 24

TRIC). It was implemented in SiGe to provide both the analog and digital functions. SiGe is relatively low power but it is very low cost. The T/R chip has 7 dB gain, a 3-bit phase shifter, 500 MHz bandwidth, and a noise figure of ~2.5 dB. The array uses a flexible circuit (Rogers ULTRALAM® 3850) with the T/R chips attached to one side and patch-radiating elements attached to the other.

The array is described in papers by Papapolymerou [34], Thrivikraman *et al.* [33], and Donado Morcillo [35]. The SiGe chip is described in papers by Thrivikraman *et al.* [36, 37] and Coen [38].

The coupled dipole array by Holland described in Section 13.2 could be substituted for the patch radiators to produce a very wideband array although it uses a different substrate material, Rogers RT/duroid® 5880 instead of Rogers Ultralam® 3850.

Another innovative approach is the MACOM Scalable Planar Array (SPAR™) module used in the MPAR described in Section 1.4.1.2. The modules are separate tested assemblies which are mounted to an antenna panel.

14.3 Monolithic microwave integrated circuits

MMICs are a fundamental enabler of T/R modules and hence ESAs. At microwave frequencies, GaAs has been the semiconductor material of choice. SiGe offers low cost and combined digital and analog functions (mixed signal) enabling integrated devices, even a single chip for a full module functionality.

The first generation SiGe chip is shown in Figure 14.7b and its second-generation version is shown in Figure 14.8. GaN has become popular for its higher voltage and power capability and especially for cost and reliability improvements resulting from commercial production. Niehenke et al. [39] and Bedinger [40] summarize the historical development of the technology

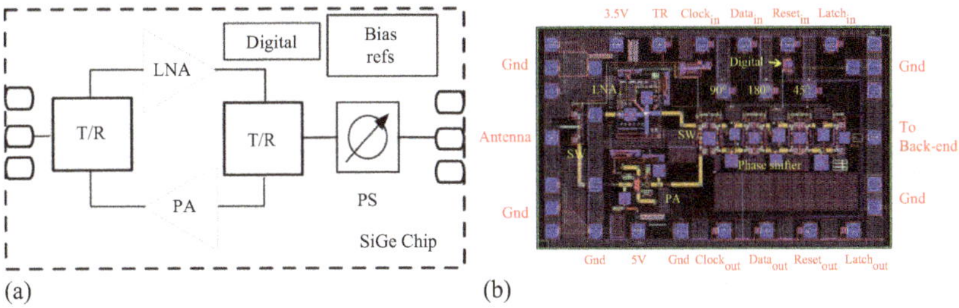

Figure 14.8 Georgia Tech 64-element antenna: (a) TRIC block diagram and (b) TRIC chip: 3.7 mm × 2.13 mm
Coen [38], Figures 4.6 and 4.12

Facility capitalization is very expensive, so the price of these components includes significant amortization, making their price very sensitive to volume. With the advent of cell phones and other consumer devices, production volume picked up nicely. Competitive forces limit the amount of information publicly available, but Cohen [30] provides insight into these trends based on his government experience.

Most T/R modules are made by system houses and most of these utilize in-house foundries. The system houses regard these capabilities as competitive discriminators and highly proprietary; accordingly they do not sell outside. However, some companies including MACOM and Qorvo have developed complete chipsets for T/R modules. Figure 14.5 shows a MACOM T/R module chipset developed some years ago which exhibited a module efficiency in the range of 10%–20%.

No matter the source, the quality and repeatability of the production process has advanced to the point that the devices are very tightly grouped, easing the hardware and systems designers tasks. Table 14.2 summarizes the specifications for MMICs used in a T/R module. Figure 14.2a summarizes industry-wide performance demonstrated in 2002. Table 14.2b lists typical performance for the chipset available in 2015. While the performance claims seem comparable, the former represents the best case and the latter represents the average for a large quantity.

Akilan [41] describes "Packaged chipsets for next-generation X-band phased array radar" and three module configurations based on the chipset listed in Table 14.2b. Note that the core chip includes a silicon chip for T/R module control. Published data suggests a price of several hundred dollars per T/R module for the MMICs.

Table 14.3 shows the power consumption of the chipset listed in Table 14.2b. The calculation shows an efficiency on the order of 30%.

Higher power and efficiency is described in a number of papers. Sardin *et al.* [42, 43] describe 10 W X-band GaAs MMICs with up to 60% PAE. Costrini *et al.* [44, 45] describe 50 W X-band GaN MMICs. Harris *et al.* [46] discuss cooling approaches for 100 W X-band GaN MMICs.

Table 14.2 MMIC performance

Function	Parameter	X-band
(a) Performance ca. 2002		
Power amplifier	Power output	7–12 W
	Gain	6–8 dB/stage
	Efficiency	35%–45%
Low-noise amplifier	Gain	8–9 dB/stage
	Noise figure	0.8–1.5 dB
Phase shifter	RMS phase error	2°–4°
	Loss	3–7 dB
Attenuator	Resolution	0.25–0.50 dB
	Range	32–64 dB, typical
(b) Current commercial performance		
Power amplifier	Power output	15 W
MAAP-015036	Gain	17 dB
	Efficiency	>40%
Driver amplifier	Power output	1 W
MAAM-71100	Gain	18 dB
	Efficiency	30%
Low-noise amplifier	Gain	20 dB
MAAL-010528	Noise figure	1.6 dB
Core chip		
MAMF-011015		
Phase shifter	RMS phase error	5°
MAPS-010166	Loss	6.5 dB
Attenuator	Resolution	0.50 dB
MAAD-011021	Range	31.5 dB
	Loss	7 dB

Source: (a) Kopp *et al.* [4], Table I and (b) MACOM data sheets

Table 14.3 T/R module efficiency calculation

Part type	Part number	I_{DD} (A)	Voltage	Power	Duty factor	Average power
PA	MAAP-015036	4.8	8	38.4	10%	3.8
Driver	MAAM-71100	0.5	8	4	10%	0.4
LNA	MAAL-010528	0.06	4	0.24	90%	0.22
Core	MAMF-011015	0.14	5	0.7	100%	0.7
				Total		5.16
				RF out		1.5
				Efficiency		29%

Source: MACOM data sheets

14.4 Gain control

Gain control is commonly achieved using switched attenuators. There is no inherent frequency dependence of this circuit although the circuit elements may exhibit some frequency dependence. A number of switches are involved in this process. Commercial circuits are available for this purpose.

The gain control circuits, as well as the phase control, are frequently combined with small signal amplifiers and digital interfaces into a single chip. This reduces size, cost, and module complexity. SiGe is a popular choice for this chip and a number of manufacturers offer these chips commercially.

14.5 Phase shifters and time delay units

Section 2.5 describes the role of phase shifters and time delay units in forming and steering beams. This section describes how these functions may be realized in hardware. Romanofsky [47] provides a good summary in "Array phase shifters: theory and technology".

A number of approaches have been assessed and demonstrated. A basic distinction is low or high power. High-power phase shifters are necessary when using a central RF power source. With the advent of solid-state and T/R modules, low-power phase shifters are the norm. There are a variety of low-power circuits but for our purposes the most important are switched line lengths (for time delay) and phase shifters based on filters using inductors and capacitors. Both types require solid-state switches which constrain the power-handling capability of the circuits. There are several topologies for the switched line circuits and the lines may be constructed as a printed circuit or coaxial cable. The filter-based circuits rely on the phase difference between a high-pass filter and a low-pass filter which is reasonably constant over a reasonable bandwidth.

These techniques operate on the RF analog signal. Recently, digital techniques have been developed which digitize the undelayed/unshifted signal and then create a time delay digitally by, for instance, storing it momentarily and retrieving it a short time later. This requires an ADC for each time delay circuit, potentially at the element level, with associated power consumption and some cost. Like other time delay circuits, this is more practical at the subarray level although a few systems have implemented it at the element level.

14.5.1 Time delay units

Time delay units tend to have little bandwidth limitation but requires line lengths the order of the array size, so it tends to be impractical to implement at the element level. Its cost and complexity mean that it is generally applied at the subarray level. Longbrake [48] provides a summary of the benefits of time delay for beam steering.

Traditionally, time delay units have been based on long cables although that is a problem for size and weight. A simple calculation shows the problem. Pasternack RG233-U [49] cable has a loss of 1.8 dB/m and a mass of 0.06 kg/m. If we consider an array which is 1 m^2 and wish to scan to 60°, we must be able to provide $\pm L \cdot \sin\theta$ of line length at the edge of the array which equates to 0.866 m. The average value for all elements is one-half this number or simply 0.433 m. At X-band, a 1 m^2 array contains as many as 1,000 elements, so the total length of cable required is 433 m, representing 26 kg. The volume of the cable and the necessary switches would dwarf the array, so element-level phase shifting with coaxial cable is out of the question.

Large time delays have been realized using lengths of cable by converting the RF signal to light and utilizing fiber optics to provide the delay. The fiber optic is lighter than coaxial cable but the two-way conversion introduces its own issues. It is also possible to convert the RF signal to acoustic waves and utilizing various physical media such as surface acoustic wave devices to produce a time delay. Again conversion issues limit its application. Linearity and dynamic range are primary limitations of both analog approaches.

Solid-state switches combined with stripline or microstrip transmission lines are the basis of current development. One such device was developed by MACOM under contract to the US Air Force Research Laboratories. It is a multiple bit time delay unit which uses on-chip line lengths for short delay and off-chip transmission lines and has flat insertion loss versus frequency and delay state. It incorporates a serial control ASIC. Figure 14.9 shows the device.

Figure 14.10 shows a Hensoldt time delay unit with maximum 500 ps step size, 64 steps (LSB =10 ps), and 656 ps maximum delay representing 20 cm maximum delay. The configuration is unusual because there are four selectable delays at each stage. This has the advantage of halving the number of switches (and associated loss) in the signal path. It has the disadvantage of requiring twice the total time delay. Specifically, the maximum selectable delay is $63\Delta\tau$ but the total line length is $126\Delta\tau$. The difference is the line lengths of $3\Delta\tau$, $12\Delta\tau$, and $48\Delta\tau$ which would be constructed from pairs of time delays in a binary configuration.

The COSMO-SkyMed [51] time delay unit is shown in Figure 14.11. Each TDU has two channels and there are two TDUs for a total of four channels. This supports true time delay for five panels. The time delay unit has a maximum delay of 15 wavelengths.

(a) (b)

Figure 14.9 TELA TDU module: (a) top and (b) bottom
Courtesy US Air Force [50]

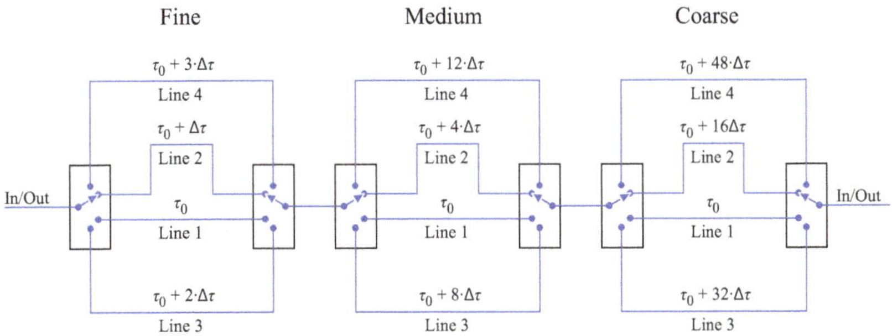

Figure 14.10 Hensoldt time delay unit
After Hees et al. [29]

This represents 45 cm or about 1.5 ns delay at 10 GHz. It uses switched line lengths and provides four bits of resolution with 16 steps and an LSB of 100 ps.

For the COSMO-SkyMed case, we have five azimuth panels, each 1.14 m wide. Using (2.38), this represents $\pm$ arcsin(0.45/4.56) = $\pm 5.6°$ of steering capability. As shown in Table 17.3, the panels are four subarrays wide, so the panel beam may be steered in phase. The subarray width is 0.285 m and the subarray null-to-null beamwidth is 12.0° based on (2.18). Thus, the time delay steering is contained within the subarray mainlobe consistent with the discussion in Section 7.5.

The performance of all three units is summarized in Table 14.4.

14.5.2 Lumped element delay line

A transmission line has no inherent limit on bandwidth. A lumped element circuit, an LC ladder, illustrated in Figure 14.12 approximates a delay line up to a certain frequency termed the cutoff; below the cutoff frequency, the circuit introduces a nearly constant time delay without attenuation while above the cutoff frequency, the circuit attenuates the signal.

The circuit is described in this section following the derivation in Feynman *et al.* [52, Vol. 2, Ch. 22]. This configuration is called a ladder because of its resemblance

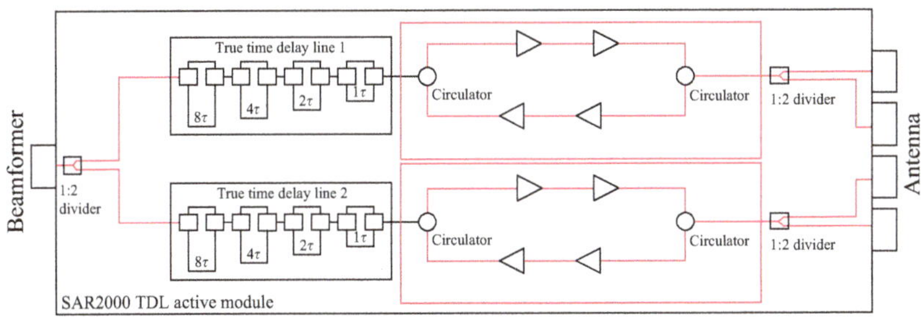

Figure 14.11 Time delay unit
After Torre and Capece [51]

Table 14.4 TDU performance

Attribute	TELA[1]	Hensoldt[2]	Thales Alenia Space[3]
TDU			
Number of bits	8	6	4
Total delay (ps)	764	656	1,500
LSB (ps)	4	10	100
Attenuator			
Number of bits	6	–	–
Total attenuation (dB)	31.5	–	–
LSB (dB)	0.5	–	–
Other			
Gain (dB)	20	–	–
Noise figure (dB)	5	–	–
Input return loss (dB)	18	–	–
Size (mm)	13 × 9	–	–

Source: [1]Dalrymple [50], [2]Hees *et al.* [29], and [3]Torre and Capece [51]

to the apparatus of the same name. The values of the inductance and capacitance represent the corresponding values for a specific length of the transmission line. As the specific length selected becomes smaller, the complexity and bandwidth of the circuit increases. A series inductance and shunt capacitance represent a low-pass filter with no attenuation between DC and a frequency called the cutoff frequency. Beyond the cutoff frequency, the attenuation increases a rate of 6 dB per filter section per octave. Insertion (group) delay is relatively constant up to the cutoff frequency. Accordingly, it may serve as a time delay unit for our purposes. As we will see, however, it requires a large number of components, and a related circuit described in Section 14.5.3 provides a phase shift circuit using many fewer components.

The impedance of this network may be determined recursively, i.e., adding an additional pair of components at the input terminal leaves the circuit impedance unchanged.

To transform a ladder circuit to a tee circuit, we move one-half of the first series component in Figure 14.12 to the end of the ladder and select new reference points as shown in Figure 14.13. Additionally, we equate z_1 to $2z_1{}'$ in the definition of characteristic impedance provided in Section 14.5.2. Consequently, the characteristic impedance z_0 of the ladder circuit defined by (14.1) becomes z_0 for the tee circuit defined by (14.2):

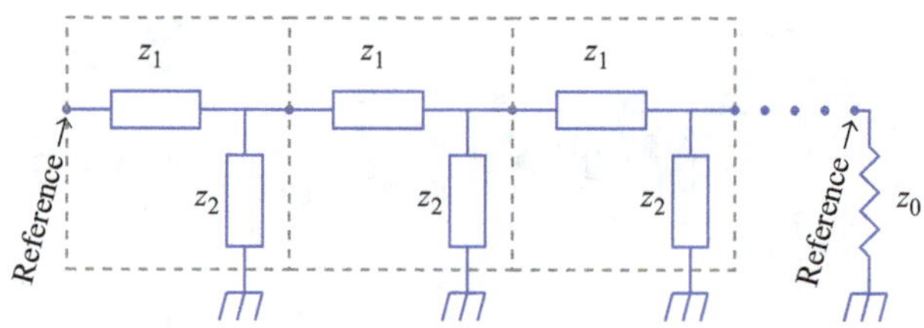

Figure 14.12 General ladder circuit

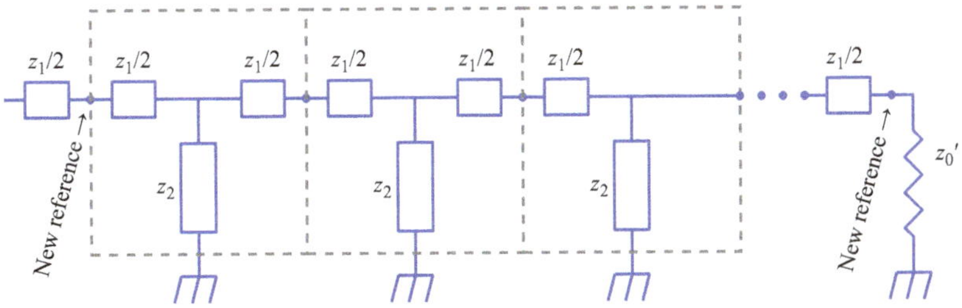

Figure 14.13 Modified ladder circuit
After Feynman et al. [52]

$$z = z_1 + \frac{1}{(1/z_2) + (1/z_0)}$$
$$= z_1 + \frac{z_2\, z_0}{z_2 + z_0}$$

if we assume $z = z_0$ (14.1)

$$\text{then } z_0 = z_1 + \frac{z_2\, z_0}{z_2 + z_0}$$
$$\text{and } z_0 = \frac{z_1}{2} \pm \frac{1}{2}\sqrt{z_1^2 + 4 z_1\, z_2}$$

$$z_0' = \pm\tfrac{1}{2}\sqrt{{z_1}^2 + 4\, z_1\, z_2} \quad (14.2)$$

Our premise is that when the delay line is terminated in a matched load, each section of the ladder has the same effect on the signal. As we will see, however, the characteristic impedance is a function of frequency, so a constant load is matched only at one frequency. We define the gain α of the low-pass filter in (14.3):

$$\alpha = \frac{V_n + 1}{V_n}$$
$$\alpha = \frac{z_0 - z_1}{z_0}$$
$$\alpha = \frac{\sqrt{(z_{12}/4) + z_1\, z_2} - z_1/2}{\sqrt{(z_{12}/4) + z_1\, z_2} + z_1/2}$$
$$\alpha = \frac{\sqrt{(z_{12} + 4\, z_1\, z_2 - z_1)}}{\sqrt{(z_{12} + 4\, z_1\, z_2 + z_1)}} \quad (14.3)$$
$$= \frac{z_1 + 2\, z_2 - \sqrt{z_{12} + 4\, z_1 z_2}}{2\, z_2}$$
$$\alpha = 1 + \frac{1}{2}\frac{z_1}{z_2} - \frac{1}{2}\frac{1}{z_2}\sqrt{z_{12} + 4\, z_1 z_2}$$
$$\alpha = 1 + \frac{1}{2}\frac{z_1}{z_2} - \frac{1}{2}\sqrt{\frac{z_{12}}{z_{22}} + 4\frac{z_1}{z_2}}$$

For a low-pass filter, $z_1 = \mathrm{j}\omega L$ and $z_2 = 1/\mathrm{j}\omega C$ resulting in (14.4):

$$\begin{aligned}
&\textit{low} - \textit{pass}\\
&\alpha = 1 - \frac{\omega^2 LC}{2} + \frac{1}{2}\sqrt{\omega^4 L^2 C^2 - 4\,\omega^2 LC}\\
&\alpha = 1 - \frac{2\,\omega^2}{\omega_{02}} + 2\sqrt{\frac{\omega^4}{\omega_{02}} - \frac{\omega^2}{\omega_{02}}}\\
&\textit{high} - \textit{pass}\\
&\alpha = 1 - \frac{1}{2\,\omega^2 LC} + \frac{1}{2}\sqrt{\frac{1}{\omega^4 L^2 C^2} - 4\frac{1}{\omega^2 LC}}\\
&\alpha = 1 - \frac{2\omega_{02}}{\omega^2} + 2\sqrt{\frac{\omega_{04}}{\omega^4} - \frac{\omega_{02}}{\omega^2}}
\end{aligned} \tag{14.4}$$

So long as the expression under the radical (i.e., z_0) is negative or zero, $|\alpha|$ is unity. This is the case for

$$\begin{aligned}
&4\,\omega^2 LC \geq 1\\
&\omega \leq \sqrt{4/LC}
\end{aligned} \tag{14.5}$$

We define the "cutoff frequency" ω_0 corresponding to the equality in (14.5) and rewrite the expression for z_0 in terms of ω_0 in (14.6). Cutoff has a physical meaning. For a low-pass filter, it is the highest frequency that passes without attenuation:

$$\omega_0 = \sqrt{\frac{4}{LC}} \tag{14.6}$$

So we may simplify (14.4) substituting $4/\omega_0{}^2$ for the product of L and C giving (14.7):

$$\alpha = (1 - 2\,\omega^2/\omega_{02}) - \mathrm{j}\,2\,\omega/\omega_0\,\sqrt{1 - \omega^2/\omega_{02}} \tag{14.7}$$

The phase of α will be of interest and is defined in (14.8):

$$\begin{aligned}
\phi &= \arctan\left(\frac{\mathrm{Im}(\alpha)}{\mathrm{Re}(\alpha)}\right)\\
&= \arctan\left(\frac{-2\,\omega}{\omega_0}\,\frac{\sqrt{1 - \omega^2/\omega_{02}}}{1 - 2\,\omega^2/\omega_{02}}\right)
\end{aligned} \tag{14.8}$$

The quadratic terms of ω/ω_0 may be approximated by a Maclaurin expansion and synthetic division as shown in (14.9):

$$\frac{\sqrt{1 - \omega^2/\omega_{02}}}{1 - 2\,\omega^2/\omega_{02}} = 1 + \frac{3}{2}\frac{\omega^2}{\omega_0^2} + \frac{23}{8}\frac{\omega^4}{\omega_0^4} + \cdots \tag{14.9}$$

Accordingly, for sufficiently small frequencies $\omega \ll \omega_0$, phase shift is proportional to frequency as indicated in

$$\phi \approx \arctan\,(-2\,\omega/\omega_0) \tag{14.10}$$

Our objective with this circuit is to delay our signal of interest. This delay is the "group delay," defined as the derivative of the phase with respect to frequency in (14.11). The derivative creates another quadratic term in ω which we again disregard:

$$\begin{aligned}\tau(\omega) &= -\frac{d\phi(\omega)}{d\omega} \\ &\approx \left(\frac{1}{1+4\,\omega^2/\omega_0^2}\right)\left(\frac{-2}{\omega_0}\right) \\ &\approx -\frac{2}{\omega_0}\end{aligned} \tag{14.11}$$

For frequencies as a sufficient distance from cutoff, consistent with our approximation in (14.10), the group delay and characteristic impedance are both essentially constant and their values are given by (14.6) and (14.11).

The distance δ corresponding to the time delay τ is $c\tau$, so we see that the basic delay distance is $\delta = \lambda_0/\pi$, e.g., less than one wavelength of the cutoff frequency (valid for frequencies much less than ω_0). We are generally interested in delays representing many wavelengths. In order to construct delay lines with longer delays, multiple sections are required. This may be warranted in cases where wideband is important. For bandwidths less than one octave, the approach described in Section 14.5.3 is probably more interesting.

For a high-pass filter, we exchange the roles of inductor and capacitor, e.g., $z_1 = 1/j\omega C$ and $z_2 = j\omega L$. A similar analysis provides the characteristics of a high-pass ladder circuit which are summarized in (14.12) corresponding to (14.3) through (14.11). Note that phase is no longer proportional to ω and is instead proportional to ω_0/ω, so that both $\phi, \tau \to 0$ as $\omega \to \infty$.

Table 14.5 represents a specific example. The performance of this circuit is shown in Figures 14.14 and 14.15. The frequency scale is logarithmic, so the low-pass and high-pass curves are not symmetric. A geometric frequency scale would show symmetry. Of particular note, the low-pass time delay has a positive asymptotic value as previously noted while the high-pass time delay is asymptotic to zero.

Table 14.5 Lumped element example with ω_0 = 10 GHz

Parameter	**Units**	**Low pass**	**High pass**
ω_0	GHz	10	10
Z_0	Ohms	50	50
L	NanoHenrys	1.59	0.40
C	PicoFarads	0.64	0.16
τ_0	Picoseconds	31.8	
δ	Centimeters	0.95	

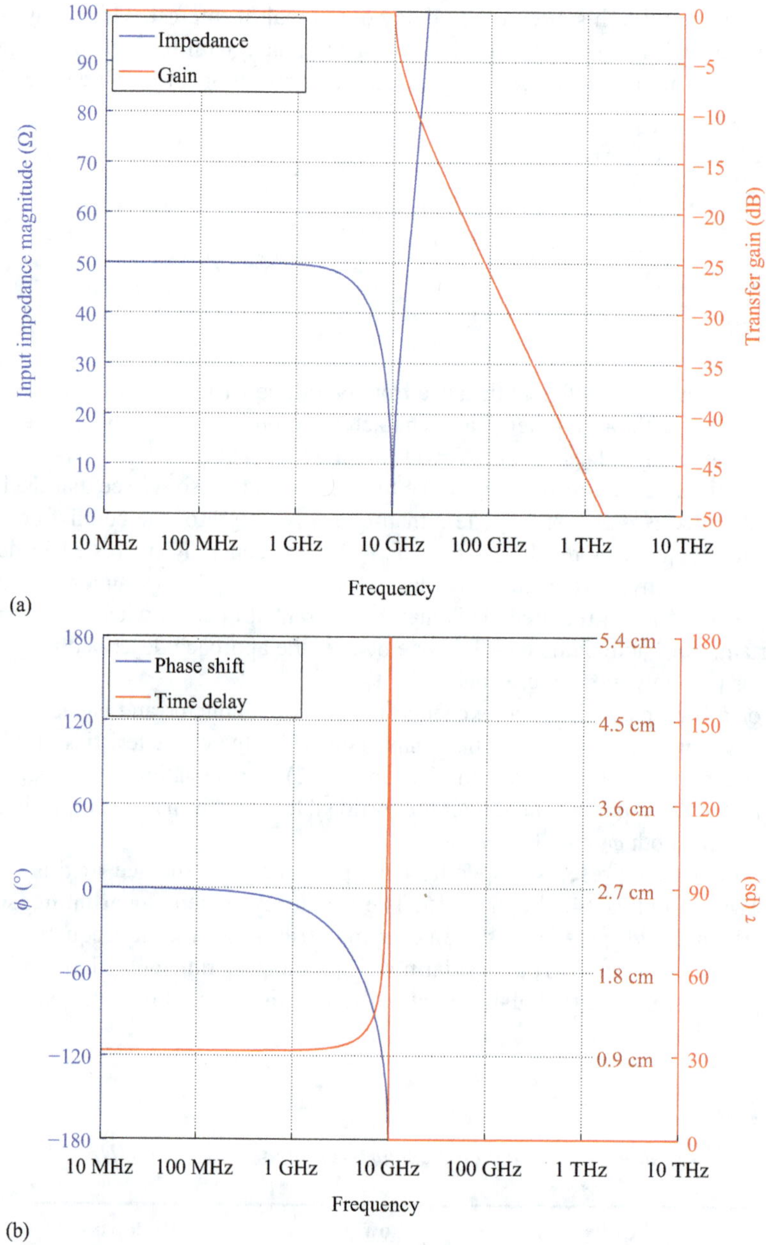

Figure 14.14 Low-pass filter response (parameters in Table 14.5): (a) impedanceand gain and (b) phase shift and time delay

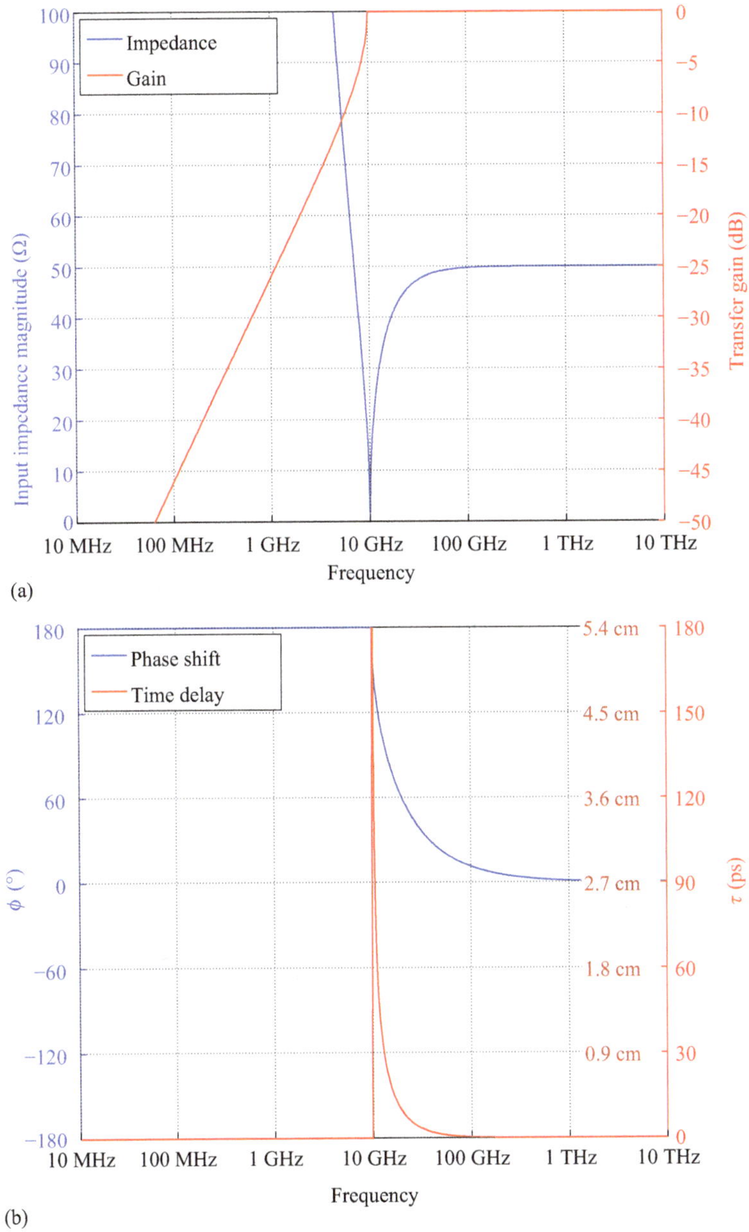

Figure 14.15 *High-pass filter response (parameters in Table 14.5): (a) impedance and gain and (b) phase shift and time delay*
DelayLine.m

$$\alpha = 1 - \frac{1}{2}\frac{1}{\omega^2 LC} - j\frac{1}{\omega L}\sqrt{\frac{L}{C} - \frac{1}{4}\frac{1}{\omega^2 C^2}}$$

$$\omega_0 = \sqrt{\frac{1}{4LC}}$$

$$\alpha = (1 - 2\,\omega_{02}/\omega^2) + j\,2\,\omega_0/\omega\,\sqrt{1 - \omega_{02}/\omega^2} \tag{14.12}$$

$$\phi \approx \arctan\,(2\,\omega_0/\omega)$$

$$\tau(\omega) \approx -\frac{2\omega_0}{\omega^2}$$

14.5.3 High-pass/low-pass phase shifter

The lumped element circuit described in Section 14.5.2 may be used to introduce a phase delay $\Delta\phi$ over a relatively narrow frequency band. The appropriate circuit is described in Section 14.5.3. The associated phase shift provides the desired delay at only one frequency, so its bandwidth is limited. The circuit is described in this section following the derivation in Feynman *et al.* [52, Vol. 2, Ch. 22].

The ladder circuit is not an obvious candidate for a phase shifter because characteristic impedance varies considerably over the frequency range where the phase shift varies as shown in Figures 14.14 and 14.15 Garver [53] was the first to propose a lumped element phase shifter based on his observation that high-pass and low-pass filters had phase shifts that maintained a nearly constant difference for an appreciable bandwidth. Accordingly, switching between the two circuits would produce a consistent phase change.

This design is a popular choice for the phase shifter for T/R modules. It is readily implemented in MMIC technology, easily integrates into the next higher assembly, is small and lightweight, and has good performance. Its disadvantages are (1) it is band-limited due to errors and mismatch away from the design frequency and (2) it is a phase shifter, not a time delay unit which imposes a bandwidth limitation on the system. At microwave frequencies, the lumped element values are both realizable and small and compatible with MMIC devices and processing

Both for analysis and for implementation purposes, it is convenient to reconfigure the ladder circuit into its "tee" filter or its closely related "pi" counterpart. Performance of the two configurations is identical and their analysis is similar. The choice between them is usually made on implementation considerations, especially treatment of parasitic impedances, but for this discussion, we will use the tee configurations shown in Figure 14.16.

These configurations are also reciprocal, i.e., the inputs and outputs are interchangeable. This attribute is utilized in the T/R module shown in Figure 14.2 in which the transmitted signal and the received signal pass through the phase shifter in opposite directions:

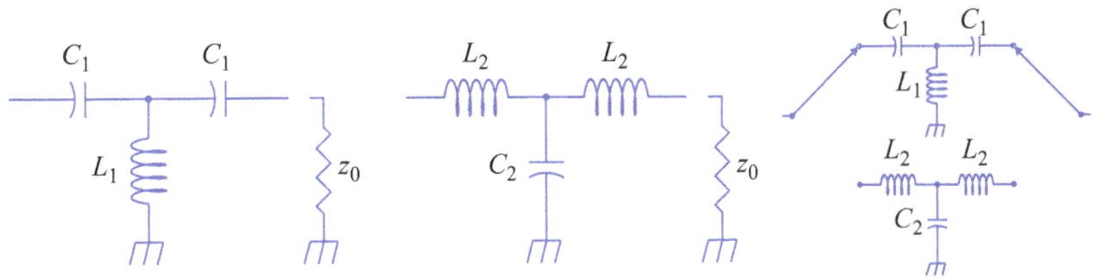

(a) High-passtee filter (b) Low-passtee filter (c) Switched tee filter

Figure 14.16 High-pass/low-pass filters
After Garver [53]

$$z_{0'} = \sqrt{z_{1'2} + 2\, z_{1'}\, z_2}$$

$$z_{0'} = \sqrt{-\omega^2 L^2 + 2\frac{L}{C}} \quad \text{Low} - \text{pass tee filter} \tag{14.13}$$

$$z_{0'} = \sqrt{-\frac{1}{\omega^2\, C^2} + 2\frac{L}{C}} \quad \text{High} - \text{pass tee filter}$$

Similarly, the cutoff frequency loses a factor of 2 so that (14.6) and 14.12 become (14.14):

$$\omega_0' = \sqrt{\frac{1}{2\, LC}} \quad \text{Low}-\text{pass tee filter}$$

$$\omega_0' = \sqrt{\frac{2}{LC}} \quad \text{High}-\text{pass tee filter} \tag{14.14}$$

14.5.3.1 Tee filter analysis

The design challenge is to select component values based on the nominal frequency and nominal impedance recognizing that this design will be band-limited. A different analysis is required because we require a specific phase shift ϕ at the nominal frequency ω. We further require that VSWR ρ (or return loss $|S_{11}|$) be minimized.

The analysis in this section follows that of Garver [7]. Our (14.16), (14.17), (14.19), and (14.20) correspond to (52–54) of his paper. Garver used the ABCD formulation for his analysis. This is a convenient formulation to represent cascaded lumped elements as the equations otherwise become unwieldy. This formulation relates input voltage and current V_1 and I_1 to output voltage and current V_2 and I_2 as shown in (14.15):

$$\begin{bmatrix} V_1 \\ I_1 \end{bmatrix} = \begin{bmatrix} A & B \\ C & D \end{bmatrix} \begin{bmatrix} V_2 \\ I_2 \end{bmatrix} \tag{14.15}$$

Equation (14.16) represents a tee filter with one series, one shunt, and one series component corresponding to Figure 14.16b:

$$\begin{bmatrix} A & B \\ C & D \end{bmatrix} = \overbrace{\begin{bmatrix} 1 & jB \\ 0 & 1 \end{bmatrix}}^{\text{series}} \times \overbrace{\begin{bmatrix} 1 & 0 \\ jX & 1 \end{bmatrix}}^{\text{shunt}} \times \overbrace{\begin{bmatrix} 1 & jB \\ 0 & 1 \end{bmatrix}}^{\text{series}} = \begin{bmatrix} 1 - BX & j(2X - BX^2) \\ jB & 1 - BX \end{bmatrix} \tag{14.16}$$

where

$$\begin{aligned} B &= \text{shunt impedance} = \frac{1}{\omega C} \\ X &= \text{series impedance} = \omega L \end{aligned} \tag{14.17}$$

Up to this point, the tee filter may be high pass or low pass. The only difference is the choice of capacitor and inductor for the shunt and series impedances, B and X:

$$\begin{array}{lcc} & \text{Low pass} & \text{High pass} \\ B & \dfrac{1}{\omega C} & \omega L \\ X & \omega L & \dfrac{1}{\omega C} \end{array} \tag{14.18}$$

Converting from an ABCD representation to a scattering matrix representation, we obtain (14.19) and (14.20). The transmission term (S_{21}) is

$$\begin{aligned} S_{21} &= \frac{2}{A + B + C + D} \\ &= \frac{2}{2(1 - BX) + j(B + 2X - BX^2)} \end{aligned} \tag{14.19}$$

The associated transmission phase characteristic is

$$\begin{aligned} \Delta\phi = \tan^{-1} &\left[-\frac{B_{\text{high}} + 2X_{\text{high}} - B_{\text{high}} X_{\text{high}}^2}{2(1 - B_{\text{high}} X_{\text{high}})} \right]_{\text{high}} \\ - \tan^{-1} &\left[-\frac{B_{\text{low}} + 2X_{\text{low}} - B_{\text{low}} X_{\text{low}}^2}{2(1 - B_{\text{low}} X_{\text{low}})} \right]_{\text{low}} \end{aligned} \tag{14.20}$$

There are four variables, two inductances and two capacitances. We specify a nominal frequency, nominal phase shift, characteristic impedance for the high-pass and low-pass circuits, and solve the resulting equations.

Solving the equations for S_{21} allows evaluation of the amplitude and phase response of each of the two circuits: low pass and high pass. The difference in phase between the two paths tends to be fairly constant at the crossover frequency and forms the basis for a switched phase shift.

14.5.3.2 High-pass filter analysis

The high-pass filter exchanges the series and shunt circuit elements and provides an equal phase shift with the opposite sign. The net phase shift difference between the two circuit paths is

$$\Delta\phi = \tan^{-1}\left[-\frac{B_{\text{high}} + 2X_{\text{high}} - B_{\text{high}}X_{\text{high}}^2}{2(1 - B_{\text{high}}X_{\text{high}})}\right]_{\text{high}} - \tan^{-1}\left[-\frac{B_{\text{low}} + 2X_{\text{low}} - B_{\text{low}}X_{\text{low}}^2}{2(1 - B_{\text{low}}X_{\text{low}})}\right]_{\text{low}} \tag{14.21}$$

14.5.3.3 Input and output matching

Each circuit is matched if

$$|S_{21}| = 1, \qquad |S_{11}| = \sqrt{1 - |S_{21}|^2} \tag{14.22}$$

Under these conditions

$$B = \frac{2 > X}{X^2 + 1}, \qquad X = \tan\left(\frac{\Delta\phi}{4}\right) \tag{14.23}$$

It is also important to match the circuit to its source and load impedance. Although an exact match is possible at only one frequency, performance is satisfactory over octave bandwidths with reasonable tolerances.

Figure 14.17 represents circuit performance using (14.19) and (14.20); impedance, frequency, and phase shift are independent variables. These are normalized to a frequency of 1 Hz ($\omega = 2\pi$) and characteristic impedance of 1Ω.

Figure 14.17a plots phase shift and impedance match as a function of series and shunt impedance. Return loss $|S_{11}|$ is shown in dashed lines for values of $-10, -20, -30$ dB. The central line indicates no reflection or $|S_{11}| = -\infty$. Phase shift for a number of values is shown in solid lines. As phase shift increases, the distance between return loss limits decreases signifying reduced bandwidth and/or component tolerance. Figure 14.17b shows that the phase shift accuracy is about 2° over an octave.

Figure 14.17c shows the normalized insertion loss (S_{12}) as a function of frequency where ω_0 represents the nominal frequency. The loss is smaller than 0.3 dB for all phase shifts over an octave. Input return loss is directly related to insertion loss and Figure 14.17b demonstrates it is better than 10 dB over an octave.

Component values for the inductors and capacitors are calculated according to (14.25) which derives from (14.19) and (14.20).

One approach equates the ABCD matrix for the tee filter shown in (14.16) with the ABCD matrix for a transmission line shown in (14.24) and solves the family of multiple equations in multiple unknowns:

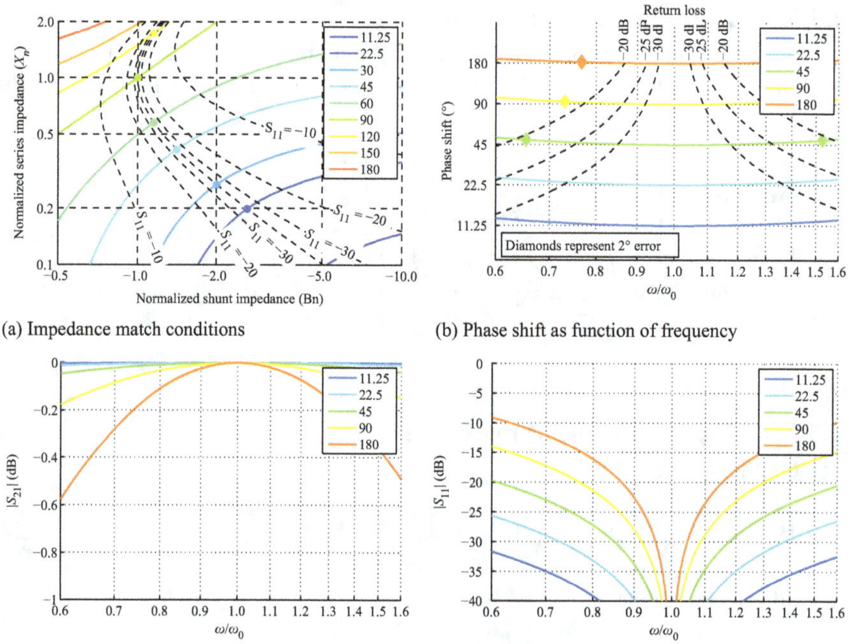

(a) Impedance match conditions

(b) Phase shift as function of frequency

(c) Insertion loss

(d) Input return loss

Figure 14.17 Phase shifter characteristics
HighPassLowPassFilter.m

$$\begin{bmatrix} \cos\phi & j\, z_0 \sin\phi \\ j\, \sin\phi/z_0 & \cos\phi \end{bmatrix} \tag{14.24}$$

$$\begin{array}{ll} \text{High-pass tee} & \text{Low-pass tee} \\ L = \dfrac{Z}{\omega \sin(\phi)} & L = Z\dfrac{1-\cos(\phi)}{\omega \sin(\phi)} \\ C = \dfrac{\sin(\phi)}{\omega\, Z(1-\cos(\phi))} & C = \dfrac{\sin(\phi)}{\omega\, Z} \end{array} \tag{14.25}$$

Choosing an operating frequency of 10 GHz and impedance of 50 Ω, the required component values are calculated using (14.25) and listed in Table 14.6.

14.5.4 Phase shifter benefits and limitations

High-pass/low-pass phase shifters are widely used because of their combination of performance and simplicity. They are small (for gigahertz applications) and easily fabricated and integrated in MMIC process. As noted in Section 2.5.2, phase shifters cause beam squint creating a bandwidth limit in their use. They have an additional bandwidth limit due to the variation of phase shifter with frequency. The latter limitation is acceptable for many applications and may be somewhat mitigated by using

Table 14.6 Lumped element example with $\omega_{nominal}$ = 10 *GHz*

Phase shift (°)	Inductance (nH)	Capacitance (pF)	Cutoff (GHz)
High-pass tee			
22.5	4.079	3.232	0.98
45	2.079	1.6	1.95
90	1.125	0.768	3.83
180	0.796	0.318	7.07
Low-pass tee			
22.5	0.078	0.062	102
45	0.158	0.122	51.3
90	0.33	0.225	26.1
180	0.796	0.318	14.1

Component values calculated to produce one-half of the stated phase shift value

a different circuit topology for the larger bits (e.g., 90° and 180°) which have the narrowest bandwidths. The circuit is linear over a wide dynamic range but requires some make-up gain to compensate for circuit loss.

Component yield will be a function of how close the system requirement is to the theoretical performance and how well the manufacturing process is controlled. With the large production volume of MMICs, foundries have good control of their process, so yields are high.

14.6 Switches and isolators

T/R modules require some means of selecting transmit or receive mode. Switches and/or isolators are used for this purpose. There are several switch types available in MMIC processes, notably FET switches and diode switches, and the switches are commonly integrated with the adjacent circuit chip. It is sometimes desirable to switch the power going to the HPA to improve efficiency. A switch in the power supply line is one option. It may be possible to achieve satisfactory performance by applying a DC bias to the input of the HPA although this procedure still requires a switch.

Switches introduce some restrictions due to voltage/power-handling capability. At low signal levels, FET switches are usually satisfactory. At high power levels, e.g., the output of the T/R module, a circulator may be preferred.

References

[1] Sturdivant R., Harris M. *Transmit Receive Modules for Radar and Communication Systems*. 1st edn. Norwood, MA, USA: Artech House; 2016.

[2] Winslow T., Gross F.B., Weedon W.H., *et al.* 'Advances in GaN technology and design for active arrays'. *2013 IEEE International Symposium on Phased Array Systems and Technology*. Boston, MA, USA; October 2013. pp. 1–231.

[3] Barkdoll T.L., Wartenberg S.A., Mumper R.W., *et al.* 'Wideband transmit modules designed for production'. *1998 IEEE MTT-S International Microwave Symposium Digest (Cat No. 98CH36192)*. 1998, vol. 3, pp. 1707–10.

[4] Kopp B.A., Borkowski M., Jerinic G. 'Transmit/receive modules'. *IEEE Transactions on Microwave Theory and Techniques*. 2002, vol. 50(3), pp. 827–34.

[5] Kolias N.J., Borkowski M.T. 'The development of T/R modules for radar applications'. *2012 IEEE/MTT-S International Microwave Symposium Digest*. 2012, pp. 1–3.

[6] Parker D., Zimmermann D.C. 'Phased arrays–Part 1: Theory and architectures'. *IEEE Transactions on Microwave Theory and Techniques*. 2002, vol. 50(3), pp. 678–87.

[7] Parker D., Zimmermann D.C. 'Phased arrays–Part II: Implementations, applications, and future trends'. *IEEE Transactions on Microwave Theory and Techniques*. 2002, vol. 50(3), pp. 688–98.

[8] Herd J.S., Conway M.D. 'The evolution to modern phased array architectures'. *Proceedings of the IEEE*. 2016, vol. 104(3), pp. 519–29.

[9] O'Haver K.W., Barker C.K., Daniel Dockery G., Huffaker J.D. 'Radar development for air and missile defense'. *Johns Hopkins APL technical digest*. 2018, vol. 34(2), pp. 140–53.

[10] Schuh P., Sledzik H., Reber R., *et al.* 'T/R-module technologies today and future trends'. *The 40th European Microwave Conference*. 2010, pp. 1540–3.

[11] Rathod S., Sreenivasulu K., Beenamole K.S., Ray K.P. 'Evolutionary trends in Transmit/receive module for active phased array radars'. *Defence Science Journal*. 2018, vol. 68(6), pp. 553–9.

[12] Raytheon Corporation. *Raytheon Delivers* 100,000*th Best of Breed Module in Support of Missile Defense Systems*. October 2004. Available from http://investor.raytheon.com/news-releases/news-release-details/raytheon-delivers-100000th-best-breed-module-support-missile.

[13] Raytheon Corporation. *Sea-Based X-Band Radar (SBX) for Missile Defense*. 2016. Available from http://www.raytheon.com/capabilities/products/stellent/groups/public/documents/content/cms04018670.pdf.

[14] Raytheon Corporation. *World's Most Capable X-Band Ballistic Missile Defense Radar Upgrades to Gallium Nitride Technology*. 2017. Available from https://raytheon.mediaroom.com/2017-04-06-Worlds-most-capable-Xband-ballistic-missile-defense-radar-upgrades-to-Gallium-Nitride-Technology.

[15] Adolph M., Hackenberg U., Reber R., *et al.* 'High-precision temperature drift compensated T/R-module for satellite based SAR applications'. *2005 European Microwave Conference*. 2005, vol. 2, pp. 1–4.

[16] Oppermann M., Schroth J., Thurow F. 'Transmit/receive (T/R) modules –key elements for phased array antennas'. *Proceedings of the 9th IMAPS/ACerS International Conference and Exhibition on Ceramic Interconnect*

and Ceramic Microsystems Technologies (CICMT). 2013. CICMT. American Ceramic Society; 2013. pp. 54–8. Available from https://doi.org/10.4071/CICMT-2013-TP24.

[17] Oppermann M., Schroth J., Thurow F. 'Transmit/receive (T/R) modules—key elements for phased array antennas'. *Journal of Microelectronics and Electronic Packaging*. 2013, vol. 10(3), pp. 116–9.

[18] Rieger R., Schuh P., Oppermann M. 'SMTR® module–evolution towards airborne applications'. *2014 International Radar Conference*. 2014, pp. 1–6.

[19] Stimson G.W. *Introduction to Airborne Radar*. 2nd edn. Raleigh, NC, USA: SciTech Publishing; 1998.

[20] Cohen E.D. 'High density microwave packaging program'. *Proceedings of 1995 IEEE MTT-S International Microwave Symposium*. 1995, vol. 1, pp. 169–72.

[21] Lamb F., Paul B., Lesniak C. 'High density microwave packaging technology development for department of defense applications' in Das N.K., Bertoni H.L. (eds.). *In Chapter 5 Directions for the Next Generation of MMIC Devices and Systems–1997 (page 46)*. Boston, MA, USA: Springer; 1997. pp. 37–44. Available from https://apps.dtic.mil/dtic/tr/fulltext/u2/a365100.pdf#page=46.

[22] Sturdivant R., Ly C., Benson J., *et al.* 'Design and performance of a high density 3D microwave module'. *1997 IEEE MTT-S International Microwave Symposium Digest*. 1997, vol. 2, pp. 501–4.

[23] Hauhe M.S., Wooldridge J.J. 'High density packaging of X-band active array modules'. *IEEE Transactions on Components, Packaging, and Manufacturing Technology: Part B*. 1997, vol. 20(3), pp. 279–91.

[24] Wooldridge J.J. 'High density interconnect (HDI) packaging for microwave and millimeter wave circuits'. *1998 IEEE Aerospace Conference Proceedings (Cat. No. 98TH8339)*. 1998, vol. 1, pp. 369–76.

[25] M/A-Com. *Microwave and Millimeter Wave IC Products for Military and Commercial Radar*. Tech. Rep. MACOM. 2003. Available from http://www.macom.com/Markets/Aerospacedefence/pdf/TRBlockDiagramWEB.pdf.

[26] Lyon R.W., Kinghorn A.M., Morrison G.D., *et al.* 'Active electronically scanned tiled array antenna'. *2013 IEEE International Symposium on Phased Array Systems and Technology*. 2013, pp. 160–4.

[27] Kinghorn T., Scott I., Totten E. 'Recent advances in airborne phased array radar systems'. *2016 IEEE International Symposium on Phased Array Systems and Technology*. 2016, pp. 1–7.

[28] Di Carlofelice A., de Paulis F., Fina A., *et al.* 'Compact and reliable T/R module prototype for advanced space active electronically Steerable antenna in 3-D LTCC technology'. *IEEE Transactions on Microwave Theory and Techniques*. 2018, vol. 66(6), pp. 2746–56.

[29] Hees A., Stangl M., Schwämmlein T., *et al.* 'X-Band phased array antenna Frontends for next generation SAR instruments'. *2019 IEEE International Symposium on Phased Array Systems and Technology*. 2019.

[30] Cohen E.D. 'Trends in the development of MMICs and packages for active electronically scanned arrays (AESAs)'. *Proceedings of International Symposium on Phased Array Systems and Technology*. 1996, pp. 1–4.

[31] Piloto A.J., Yamada R., Burgess J., *et al*. 'Advancement in T/R module interconnects'. *11th GaAs Symposium*. 2003, pp. 232–5.

[32] Midford T.A., Wooldridge J.J., Sturdivant R.L. 'The evolution of packages for monolithic microwave and millimeter-wave circuits'. *IEEE Transactions on Antennas and Propagation*. 1995, vol. 43(9), pp. 983–91.

[33] Thrivikraman T., Patterson C.E., Coen C., *et al*. 'Integrated, digitally controlled, 64-element sigE on multilayer organic X-band phased-array receiver antenna for snow measurements'. *IEEE Aerospace and Electronic Systems Magazine*. 2013, vol. 28(3), pp. 26–39.

[34] Morcillo C.A.D., Patterson C., Thrivikraman T.K. 'A lightweight, 64-element, organic phased array with integrated transmit-receive SiGe circuitry in the X Band'. *Earth Science Technology Forum*; 2011. Available from https://esto.nasa.gov/conferences/estf2011/papers/PapapolymerouESTF2011.pdf.

[35] Donado Morcillo C.A. 'Development of lightweight and low-cost microwave components for remote-sensing applications'. PhD thesis. Georgia Institute of Technology, May 2013. Available from https://smartech.gatech.edu/handle/1853/51733.

[36] Thrivikraman T.K., Grens C.M., Kuo W.-M.L., *et al*. 'A high-linearity, X-band, sigE low-noise amplifier for improved dynamic range in next-generation radar and wireless systems'. *2009 IEEE Topical Meeting on Silicon Monolithic Integrated Circuits in RF Systems*. 2009, pp. 1–4.

[37] Thrivikraman T.K., Kuo W.M.L., Cressler J.D. 'A two-channel, ultralow-power, sigE BiCMOS receiver front-end for X-band phased array radars'. *2009 IEEE Bipolar/BiCMOS Circuits and Technology Meeting*. 2009, pp. 43–6.

[38] Coen C.T. 'Development and integration of silicon-germanium front-end electronics for active phased-array antennas'. PhD thesis. Georgia Institute of Technology. 2012. Available from https://smartech.gatech.edu/handle/1853/48990.

[39] Niehenke E.C., Pucel R.A., Bahl I.J. 'Microwave and millimeter-wave integrated circuits'. *IEEE Transactions on Microwave Theory and Techniques*. 2002, vol. 50(3), pp. 846–57.

[40] Bedinger J.M. 'Microwave flip chip and BGA technology'. *2000 IEEE MTT-S International Microwave Symposium Digest (Cat No. 00CH37017)*. 2000, vol. 2, pp. 713–6.

[41] Akilan M.K. 'Packaged chipsets for next generation X band phased array radar'. *9th International Radar Symposium India*. 2013. Available from http://www.radarindia.com/irsi13papers/13-FP-226.pdf.

[42] Sardin D., Reveyrand T., Popovi´c Z. 'X-band 10W MMIC high-gain power amplifier with up to 60% PAE'. *2014 44th European Microwave Conference*. 2014, pp. 1337–40.

[43] Sardin D., Reveyrand T., Popovi´c Z. 'X-band 10 W MMIC high-gain power amplifier with up to 60% PAE'. *2014 9th European Microwave Integrated Circuit Conference*. 2014, pp. 393–6.

[44] Costrini C., Cetronio A., Romanini P., *et al.* '50 W X-band GAN MMIC HPA: effective power capability and transient thermal analysis'. *The 5th European Microwave Integrated Circuits Conference*. 2010, pp. 408–11.

[45] Costrini C., Cetronio A., Romanini P., *et al.* '50 W X-band GAN MMIC HPA: effective power capability and transient thermal analysis'. *The 40th European Microwave Conference*. 2010, pp. 1650–3.

[46] Harris M., Wagner B., Tong W., *et al.* 'Thermal analysis and validation of packaging for 100W and higher GAN MMIC power amplifiers'. *WAMICON 2011 Conference Proceedings*. 2011, pp. 1–5.

[47] Romanofsky R.F. *Array Phase Shifters: Theory and Technology*. Tech. Rep. TM-2007-214906. Cleveland, OH, USA: NASA Glenn Research Center. 2007. Available from https://ntrs.nasa.gov/archive/nasa/casi.ntrs.nasa.gov/20080001449.pdf.

[48] Longbrake M. 'True time-delay beamsteering for radar'. *2012 IEEE National Aerospace and Electronics Conference (NAECON)*. 2012, pp. 246–9.

[49] Enterprises P. *Flexible RG223 Coax Cable Double Shielded with Black PVC Jacket*. 2014. Available from https://www.pasternack.com/images/ProductPDF/RG223-U.pdf.

[50] Dalrymple T.W. 'TELA wideband array demonstration'. *2010 RF Alliance Conference at Purdue University*. April 2010. Available from http://www.microwaves101.com/downloads/03%20Thomas.Dalrymple.pdf.

[51] Torre A., Capece P. 'COSMO-SkyMed: the advanced SAR instrument'. *Proceedings of 5th International Conference on Recent Advances in Space Technologies–RAST2011*. 2011, pp. 865–8.

[52] Feynman R.P., Leighton R.B., Sands M. 'Chapter 22 AC circuits'. In *The Feynman Lectures on Physics*. Vol. II. Addison-Wesley. 1964. Available from http://www.feynmanlectures.caltech.edu/II22.html.

[53] Garver R.V. 'Broad-band diode phase shifters'. *IEEE Transactions on Microwave Theory and Techniques*. 1972, vol. 20(5), pp. 314–23.

Chapter 15
Assembly, packaging, power, and thermal management

The previous chapters have dealt mainly with performance in terms of antenna beam characteristics and quality. This chapter deals with the necessary provisions and infrastructure for the ESA to perform to its potential at the lowest cost. There are a few papers covering the general subject including Kemkemian *et al.* [1, 2] and Renard *et al.* [3].

15.1 Assembly

Assembly is a significant cost element in the overall ESA cost. Automated procedures are preferred wherever feasible. Costs are inversely proportional to manufacturing yield. Automated assembly is useful in this context since it reduces variability due to inconsistent parts placement and interconnection. Complexity compromises yields, so some amount of rework is usually employed. This in turn constrains design.

Rework is expensive, so every effort must be made to use known good parts rather than detect and remove them after assembly.

Significant numbers of T/R modules in each ESA facilitates development of automated manufacture. This is also the case for mobile telephones and Wi-Fi units, so many manufacturing methods can be adapted from commercial practice. ESA quantities are small in comparison to commercial electronics so the economics of automation are not as favorable for T/R modules.

The US government invested in manufacturing technology for T/R modules beginning in the 1980s.

15.2 Test

There are a variety of test types, engineering or development, qualification, and acceptance, requiring varying degrees of fidelity and thoroughness.

T/R module test accounts for much of the cost of the T/R module. This resulted, in part, from the low level of automation but test costs also included considerable engineering labor (rework) to correct deficiencies observed in test. Yields at the first test opportunity (first pass yield) might be only 50%. Since at this point in production, there was already a considerable investment in the device it would be reworked, potentially at a labor cost equivalent to its initial assembly. After retest, it might fail again. Studies of this process showed that some parts never exited the rework loop. Hirsch *et al.* [4] describe early progress in automated testing used for the L- and C-band T/R modules for SIR-C.

Test cost has diminished because the tests have been substantially automated and because first pass yields are much improved due to the higher quality of parts and processes.

Component and subassembly variation compromise system performance. Test equipment compiles phase and amplitude data for the T/R module if required at the unit or system (ESA) level to achieve required antenna performance. Test equipment companies such as Keysight (former HP) and Rhode & Schwarz have adapted their microwave test equipment to this purpose.

Antenna ranges are utilized to characterize the system performance. Near-field scanners have particular utility to develop individual calibration tables for each T/R module.

ESA system parametric performance is relatively stable, and spaceborne systems, for instance, do not typically require recalibration. This is certainly partly a result of their benign environment (once launched). Airborne and other mobile systems are subject to more vibration and impact which potentially affect their stability.

The complexity and degrees of freedom in an ESA mean that testing is similarly complex. Near-field scanners, see Section 3.2, are the preferred approach, in effect testing an ESA with another ESA. There are several objectives of test, primarily to verify compliance with specification. In this process, defective subassemblies or components will be identified. Additionally, it may be necessary to calibrate the unit because of variability in components and assembly. Calibration generates a table of errors which may be incorporated in gain and phase commands to achieve the best possible antenna pattern. Hassett [5] describes some of these concepts in her paper "Phased array antenna calibration measurement techniques and methods." Mulcahey and Sarcione [6] describe this process for the THAAD UOES array described in section 1.3.3 and analyzed in section 6.2.

Antenna measurement and characterization is necessary for a variety of reasons. Development testing verifies that actual performance matches predictions and identifies any assembly problems. Production testing verifies that the antenna meets its requirements. Testing may occur with sensors in the farfield of the antenna although the required separation may be unwieldy for large antennas. Compact ranges support farfield volume in a smaller volume by use of a collimating reflector but may not accommodate large antennas. Near-field measurements are preferred requiring less real-esatate and offering better environmental control, avoiding electromagnetic interference (EMI) and weather.

15.3 Installation

The ESA system is part of a system with numerous interfaces. The interfaces are defined during the requirements allocation process and must be verified as part of a qualification program. Interactions with the radome, power, and cooling subsystems may cause problems. The radome will distort the beam pattern and will reflect a small part of the signal back to the array.

15.4 Packaging

In devising a physical architecture for an ESA, most designers have used T/R modules which are individually packaged and separately testable. This provides an intermediate building block, comprising much of the value of the ESA, which may be assembled into larger assemblies with the assurance that each module is free of defects. The package also provides mechanical protection and a controlled electromagnetic environment for its components.

Two of the three package dimensions (height and width) are constrained by the antenna lattice dimension which is typically less than one wavelength (3 cm at X-band). The third dimension (depth) has no similar fundamental limit. Modules in which the third dimension is significantly larger than the other two are termed brick modules by their appearance when combined into arrays. The depth largely determines the thickness of the ESA. The circuit substrate of the brick module is aligned with the long axis of the module and its area may be as large as required. Figure 15.1 illustrates the assembly of a brick module into a stick or slat. A number of sticks must be stacked to form the array. The illustration shows 18 T/R modules in a single stick, so 18 sticks would be required to form an 18×18 array. The thickness of the heat sink must be accommodated within the lattice spacing.

Alternatively, it is possible to align the circuit substrate with the plane of the aperture. This considerably constrains the area available for the circuitry. If the area of a single substrate is insufficient, substrate(s) may be added, parallel to the first. This module type is termed a tile, as the depth is typically the smallest dimension and multiple modules combined into an array resemble a tiled surface. Tile modules offer some advantages in assembly as their thermal and electrical interfaces are in a single plane, permitting large planar signal, power and control manifolds and potentially a single heat sink, which does not have to conform to the lattice spacing.

Figure 15.2 illustrates the assembly of a tile module into an array. There is no additional assembly step as is the case with brick modules. The tile design in the illustration was initially developed at Hughes Aircraft Company under the DARPA High Density Microwave Packaging Program. Numerous journal articles discuss the innovative techniques to package a complex circuit into a very small package.

The most common package type is metal/ceramic enclosure with signal/power feed-throughs and a soldered or welded lid. Yamada [9] provides a good overview of material selection criteria from the viewpoint of one of the largest suppliers, who

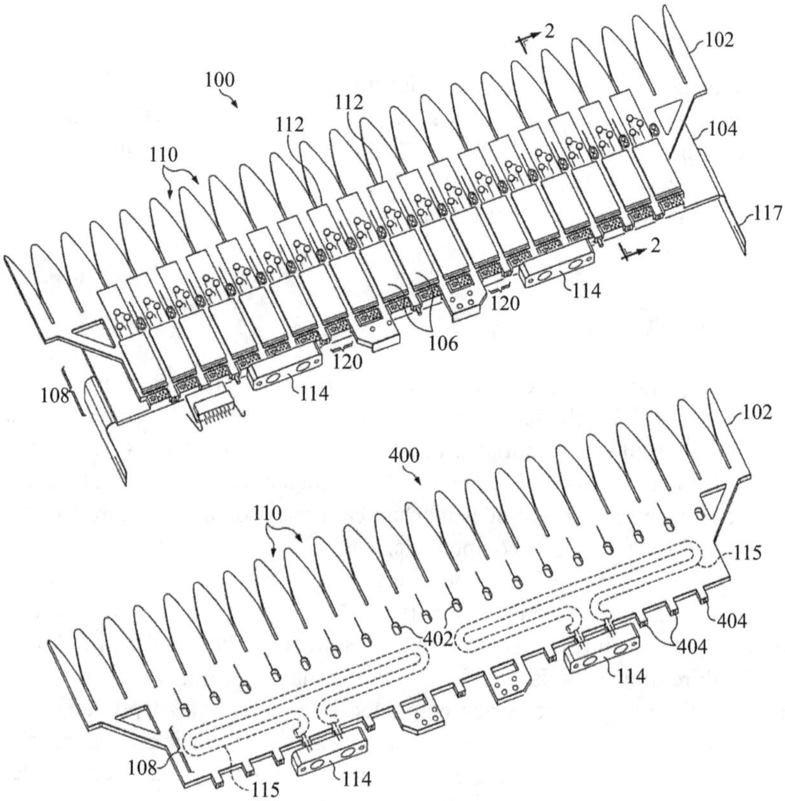

Figure 15.1 Brick module array
Source: US Patent 7,511,664, Mason et al. [7]

has delivered millions of T/R module packages. Jerinic and Borkowski [10] provide an overview of these.

These are well understood but they represent a good part of the weight and cost allocation of the module. Packaged modules, of either type, were essential when components were of inconsistent quality necessitating performance and rework as part of the manufacturing process.

Planar arrays eliminating discrete module packages represent an opportunity for significant cost reduction although their lesser mass complicates heat dissipation. There are numerous examples of ESAs which omit individual module packages and instead assemble the components on large circuit boards with many (commonly 64) T/R channels. An example developed by Georgia Tech is shown in Figure 14.6. This approach is feasible because of the improved quality of components and very repeatable, automated assembly techniques, both a result of the huge increase in commercial electronic and microwave products. The packageless approach eliminates the expensive package and associated connectors which are difficult to assemble

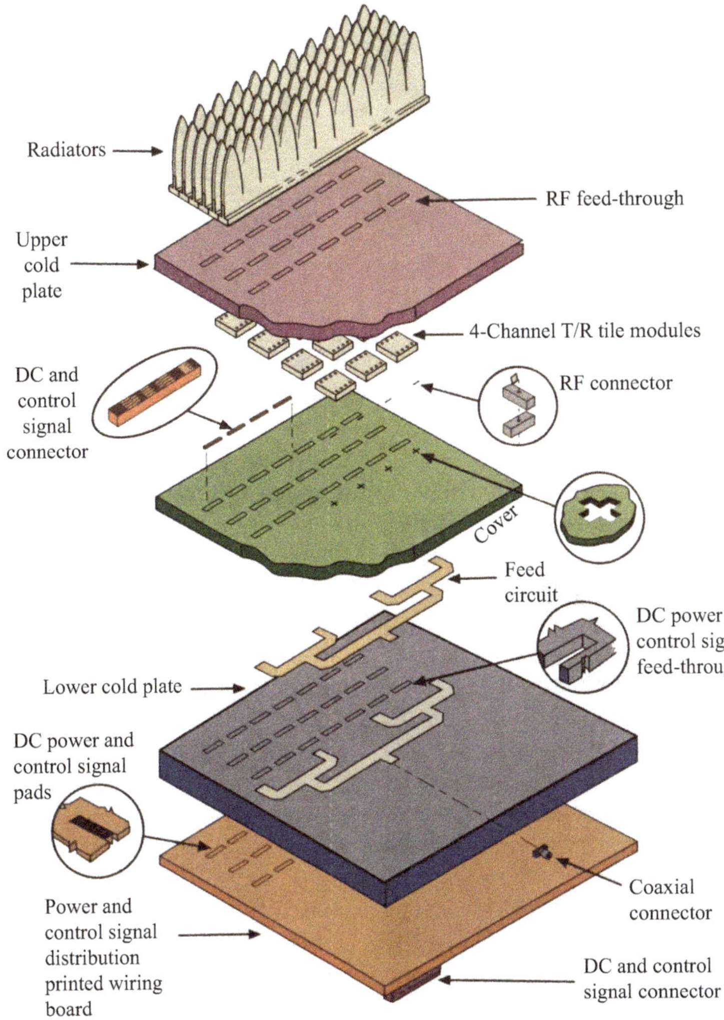

Figure 15.2 Tile module array
@SciTech [8], Chapter 38, Figure 17

automatically. However, the challenge of heat removal limits this technique to relatively low-power systems to date. Also, these assemblies with their exposed components are not sufficiently reliable for some applications.

15.5 Power

An ESA, operating in the transmit mode, consumes considerable power. In communications applications, the RF output is generally constant or slowly varying

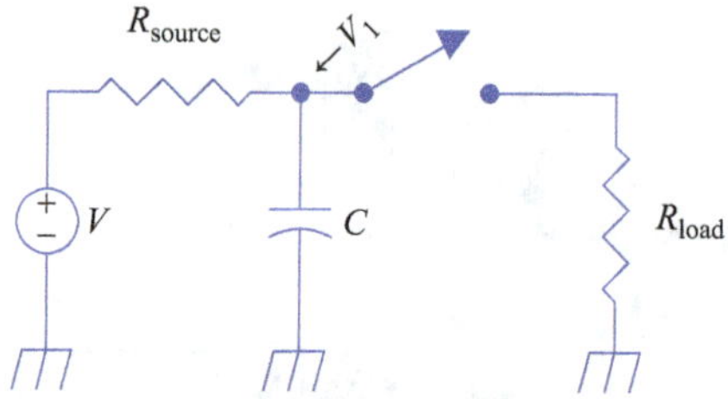

Figure 15.3 Equivalent circuit

representing a constant load to the power supply. In a radar, however, the ESA represents a pulsed load which causes several problems. This is not uniquely an ESA issue. However, the solid-state transmitters operate at much lower voltage than vacuum devices which complicates the design.

It is not generally economic to size the power supply for the peak power which is up to 10 times the average requirement, so capacitors are commonly used to reduce the peak current requirements. Sizing the power supply for the average load requires energy storage in the form of capacitors to supply the peak load. The amount of storage depends on the tolerance for voltage variation at the T/R module. For example, a 10% drop in supply voltage will generally result in a 20% drop in RF power. If the received signal is to be compressed in time, the pulse droop will introduce an amplitude variation which will distort the compressed pulse.

Figure 15.3 shows a greatly simplified power circuit where the power supply is characterized by an open-circuit voltage V and internal impedance R_{source} . C represents a large energy storage capacitor near the T/R module, and the T/R module is represented by a switch and a resistive load R_{load}. Table 15.1 lists parameters which are generally consistent with the cases we will consider in Chapters 17 and 18. Figure 15.4 shows the voltage measured at point V_1 as well as the currents flowing through R_{source} and R_{load}. Note that we have imposed an arbitrary limit on the maximum current available from the power supply, equal in this case to twice the average current. The power supply limits at this maximum current during the transmit pulse as well as for some time after for capacitor recharge. In the case shown, the capacitor is completely recharged between transmit pulses. The current through R_{source} is continuous and varying; the current through R_{load} occurs only during the transmit pulse.

The pulse repetition frequency (PRF) and transmit duty factor are the principal parameters. As the pulse width gets longer, it is necessary to select a power supply

Table 15.1 Nominal parameters

V (V)	10
R_{source} (Ω)	1
R_{load} (Ω)	10
C (μF)	20–100
PRF (kHz)	1

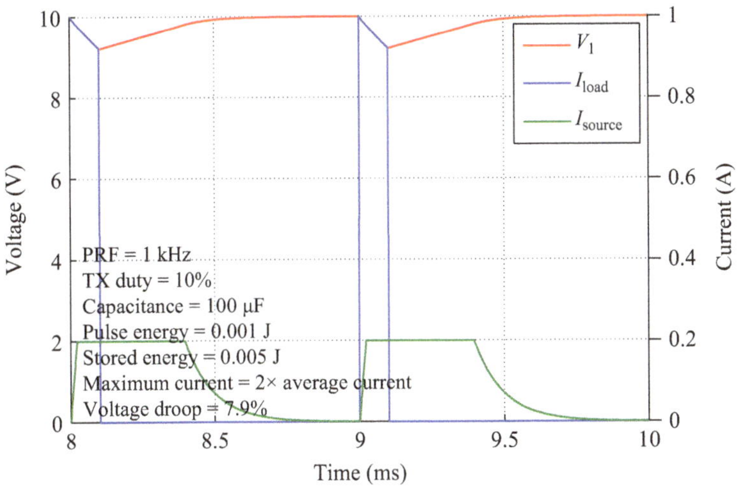

Figure 15.4 Power circuit simulation

PulseDroop.m

closer to the peak current requirement as the size of the capacitors is increasingly unmanageable. In this case, it is assumed that the power supply can supply twice the average current requirement in contrast to the peak current requirement which is 10× the average current. The difference is supplied by the energy storage capacitor. The analysis sizes the capacitor to store 5× the energy contained in the pulse. This excess is required to limit the voltage droop during the transmit pulse which, in this case, is about 8%. Since the transmit power is roughly proportional to the supply voltage, this difference represents about −0.7 dB peak-to-peak variation. This variation must be incorporated into the system error budget. Figure 15.5 shows how the voltage droop varies with the capacitance. Clearly, it is a serious problem for some of the cases considered.

As a test of reasonableness, a Vishay® 591 [11] series dry tantalum chip capacitor might be suitable. A 100-μF, 16-V part is 7.5 × 4.3 × 2 mm (LWH) and offered at a price of $1.60 in quantities of 1,000. Both the size and price are a significant part of the system budgets.

Other concerns are loss and stability. Pulsed loads introduce higher ohmic loss into power distribution which may be mitigated with lower resistance (heavier) cabling or higher voltage distribution. The pulsed load may interact with whatever control circuitry exists to stabilize the voltage. If the PRF is similar to the bandwidth of the feedback circuit, there may be serious instabilities in the power system. A detailed analysis is required to select the best power distribution topology.

The power analysis involves a differential equation of first order in time, in contrast to the wave equation which is second order. Thermal modeling also involves a first-order equation in time so both have solutions which are exponential with time. In the thermal case, we have a pulsed heat source which is attached to a heat sink

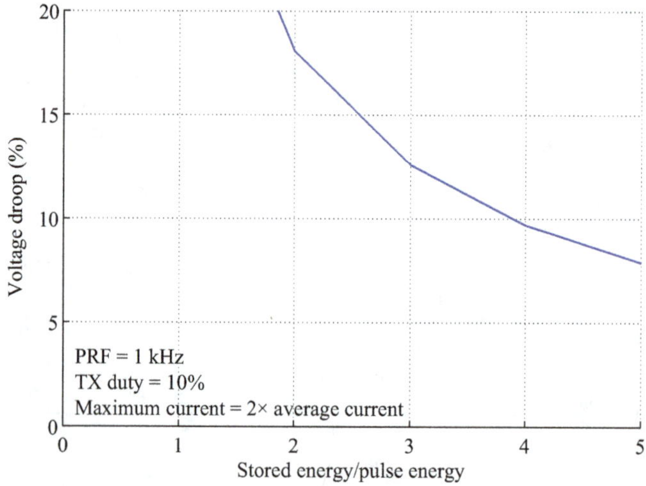

Figure 15.5 Pulse droop

which is roughly constant temperature. The thermal resistance between the source and the sink is typically significant, so it is often necessary to include some local thermal capacity adjacent to the pulsed heat source analogous to the capacitor in the power supply case. Local heating in the power amplifier will reduce the gain introducing an additional source of droop in the transmit signal.

15.6 Thermal management

Thermal management has a major effect on the ESA physical design, so it is important to establish power and dissipation budgets early. If volume is at a premium, more complex thermal arrangements will be required. Air cooling requires substantially more volume but is simpler. Liquid cooling handles higher power at increased weight and cost. Conductive or radiative cooling is feasible only for low-power systems.

Thermal management is a major ESA design constraint due to performance and reliability issues. The power amplifiers generate considerable heat which must be removed and dissipated in some fashion, in order not to damage or destroy the electronics. High operating temperatures shorten component lifetime, reduce amplifier gain, and increase noise figure. Additionally, the performance varies with temperature, so it is desirable to maintain a uniform temperature to minimize parametric errors and reduce or eliminate the necessity of compensation. The transmit function normally generates most of the heat but for low-power systems the receive function may dominate.

Because transmit efficiency DC to RF ranges from 10% to 50%, the resulting heat generation ranges from 10× to 2× the average radiated power. This is an important additional motivation to increase module efficiency whose importance matches its role in reducing prime power requirements.

There are three means of heat transport: conduction, convection, and radiation and various combinations are utilized. An important figure of merit is the power density of the ESA as convection and radiation are effective only for low-to-moderate power densities. At high power levels, pumped cooling liquids are required.

The ESA designer requires an interface specification to guide his design. This includes the ambient thermal conditions and volume and temperature of coolant available.

Fiala [12] and Pluymers and Reese [13] discuss ESA thermal design issues.

Thermal design

The role of thermal design is to transfer the heat from its source in the ESA to a heat sink where it is dissipated into the environment. Table 15.2 summarizes the issues.

For low-power devices, this may require no special measures. For many of the arrays described in this book, however, thermal design is critical to achieving best performance and high reliability.

Low-power arrays may be defined as those whose power per element is less than 1 W and high-power arrays are those that exceed that threshold. Low-power arrays may not need any special provision for cooling because convection or radiation may suffice; high-power arrays may require pumped liquid or forced air cooling.

The heat is predominantly generated in the amplifiers. While obviously important to maximize the efficiency of the amplifier, the attainable efficiency is limited by the amplifier class (A, AB, B, etc.) and the intrinsic characteristic of the device. For purposes of this discussion, we consider semiconductor devices, which may be GaAs, GaN, SiGe, or some more exotic material.

The high power amplifier is the primary source of heat and the primary device requiring cooling. The device manufacturer will typically specify performance at a given baseplate temperature. The peak temperature within the device is difficult or impossible to measure so it is assessed by thermal models. For all devices, the heat is generated in the gate region which is quite small. The manufacturer of the device will offer advice on how to install it so as to achieve best performance usually specified as a maximum baseplate temperature.

Table 15.2 Thermal environments

Environment	Characteristics
Ground based	Ultimately transfer heat into air – problem in the desert
Airborne	Liquid cooled to heat exchanger – lots of airflow
Ship based	Float in an infinite heat sink
Spaceborne	Radiation only option – directly or transferred to dedicated thermal radiators

Heat generation of active devices is concentrated between the drain and the source of the semiconductor device. The drain and the source are typically separated by a fraction of a micron (μ or 10^{-6} m) which is a function of the operating frequency; for X-band, it is typically 0.5–0.25 μ. This distance is termed gate length. The maximum current capability depends on the other dimension, termed gate width; current density is on the order of 1 A/mm. Accordingly, power densities are on the order of 10^3 W/mm^2 at the point of heat generation. If the device is 5 mm^2, the power density at the mounting surface is only on the order of 1 W/mm^2 assuming a 10-W device. For best thermal performance, the initial heat spreading is very important and depends on the thermal properties of the device material.

Semiconductor materials have relatively poor thermal conductivity and are difficult to handle due to their fragility. They are frequently mounted to carriers to provide for heat spreading as well as to make a more robust assembly. The heat source in the semiconductor is on the order of micrometers which causes very high thermal gradients within the semiconductor. The function of the heat spreader is to provide a larger area for heat transfer, on the order of millimeters, so that subsequent transfer of heat is more efficient.

There are two techniques for attachment: adhesive (epoxy) and solder. Epoxy is not a good heat conductor, so it is not used for high-power devices. High dissipation devices are mounted to a heat spreader with solder which provides a good heat transfer.

The semiconductor device is attached to a substrate which provides for mechanical stability, good thermal transfer, and interconnection for signal, power, and control lines. The substrate must be a good thermal conductor, a poor electrical conductor, and low loss tangent at the operating frequency of the module. The substrate is typically integrated into the module package to provide a thermal path to the exterior surface of the module for direct contact with a heat sink.

Material properties are enumerated in *Materials for High-Density Electronic Packaging and Interconnection* (1990) Office [14]. See also *American Institute of Physics Handbook* (McGraw-Hill) by Gray [15].

Thermal measurement

There are a number of techniques to measure temperature. Infrared thermography has limited use because the hot spot is smaller than the infrared wavelength, so the hot spot is not well resolved. Additionally, the hot spot is partially covered by metal whose low emissivity complicates the measurement. Temperature-sensitive liquid crystals allow observation in the visible spectrum improving spatial resolution considerably. It is possible to use the temperature-dependent characteristic of a diode to measure the temperature nearby. Ultimately, a combined electrical and thermal model may be the best approximation of temperature. Device temperature is of interest primarily as it affects device lifetime so device manufactures specify and masure device lifetime at various baseplate temperatures to assess reliability directly.

Liquid cooling has a long history and is widely used in internal combustion engines. Water is the most common liquid due to its availability and good thermal properties. Experience has shown that anticorrosion additives are desirable and antifreeze is used in low-temperature applications.

Heat pipes are an interesting hybrid technique, which rely on liquid transfer and the latent heat of vaporization for greater heat transfer. The basic heat pipe does not utilize a pump to move the coolant from the cold to the hot end, so in a gravity field, it must be horizontal. This restriction does not exist in space although ground testing must take it into account.

Array cooling

We will consider the case of a satellite ESA. The ESA is relatively low power density but it is restricted to radiative cooling. The ESA is typically the device radiating the heat, both from the RF face which is oriented towards Earth and the non-RF side which is facing space. Heat sources include internally generated heat, Earth blackbody radiation (IR) impinging on spacecraft, earth reflected sunlight (albedo) reflected toward spacecraft, and solar radiation shining on spacecraft. There are several considerations. The Earth and sun add heat to the ESA in a deterministic but varying fashion. The ESA may be operating, dissipating heat, or more often not operating. The design challenge is to maintain the ESA within survival temperature limits and preferably within normal operating temperature limits. Designer tools are restricted to surfaces with specified emissivity, which may be a function of (optical) wavelength and heaters.

Radiative equilibrium

The ESA will achieve radiative equilibrium when its radiated heat is equal to its dissipation plus the external heat load. When the ESA is nonoperational and in shadow, a heater may be required to protect the circuitry against low temperature.

Thermal radiation is based upon the Stefan–Boltzmann law shown in (15.1) which represents the thermal radiation intensity $j*$:

$$j* = \varepsilon \cdot \sigma \cdot T^4 \tag{15.1}$$

where

ε = emissivity
$\sigma = 5.6704 \times 10^{-8}$ J/s m^{-2} K^{-4}
T = temperature

ESA temperature as a function of areal RF density

We can evaluate the equilibrium temperature of an ESA which utilizes radiative cooling. Figure 15.6 shows the results for an ESA which incurs only self-heating due to electronics waste heat and the result for self-heating plus the thermal input from Earth radiation.

We consider two cases with the following assumptions:

- Cases are self-heating only and self-heating + Earth IR.
- Low earth orbital altitude, array points 45° off nadir.

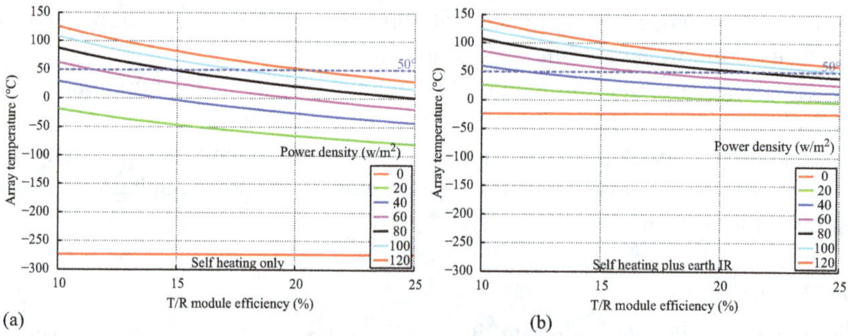

Figure 15.6 Array equilibrium temperature: (a) self-heating only and (b) self-heating + earth IR
Thermal.m

- Maximum temperature of ESA is 75°C supports.
- ESA RF density is 0–120 W/m^2.
- ESA efficiency ranges from 10% to 25%.

The results are shown in Figure 15.6 and indicate that RF power densities up to 120 W/m^2 are consistent with achievable T/R module efficiency of 20%–25%. Table 17.4 summarizes RF power densities for several radar satellites. The largest is TerraSAR-X at 129 W/m^2.

References

[1] Kemkemian S., Larroque A., Enderli C. 'The industrial challenges of airborne AESA radars'. *IET International Radar Conference 2013*. 2013, pp. 1–5.

[2] Kemkemian S., Enderli C., Larroque A. 'Advances in production testing and operational support of new AESA radars'. *IEEE International Symposium on Phased Array Systems and Technology*. 2013, pp. 145–51.

[3] Renard C., Barbier T., Hérault J., *et al.* 'Design of active antennas for airborne radar applications'. *Proceedings of the Fourth European Conference on Antennas and Propagation*. 2010, pp. 1–2.

[4] Hirsch V.A., Miers T.H., Coffin J.R. 'Automated T/R module testing for the spaceborne imaging radar C (SIR-C) antenna subsystem'. *33rd ARFTG Conference Digest*. 1989, vol. 15, pp. 46–55.

[5] Hassett K. *'Phased array antenna calibration measurement'*. 2016 10*th European Conference on Antennas and Propagation (EuCAP)*. 2016. Available from https://www.nsi-mi.com/images/Technical2016/EUCAP2016 KH Phased Array Antenna Calibration Measurement. pdf.

[6] Mulcahey J.K., Sarcione M.G. 'Calibration and diagnostics of the THAAD solid state phased array in a planar nearfield facility'. *Proceedings of*

International Symposium on Phased Array Systems and Technology. 1996, pp. 322–6.

[7] Mason J.S., Fletcher M.C., Brown M.D., *et al.* '*Subassembly for an active electronically scanned array*'. US 7,511,664 B1. 2009. Available from https://patents.google.com/patent/US7511664B1/en.

[8] Stimson G.W. *Introduction to Airborne Radar*. 2nd edn. Raleigh, NC, USA: SciTech Publishing; 1998.

[9] Yamada R. 'Material selection for ceramic T/R module packages'. *Proceedings of the 1st IMAPS/ACerS International Conference and Exhibition on Ceramic Interconnect and Ceramic Microsystems Technologies (CICMT)*. American Ceramic Society; 2005. pp. 239–44. Available from https://americas.kyocera.com/assetskii/001/5261.pdf.

[10] Jerinic G., Borkowski M. 'Microwave module packaging. *1992*'. *IEEE MTT-S Microwave Symposium Digest*. 1992, vol. 3, pp. 1503–6.

[11] Vishay. *Solid tantalum chip capacitors Tantamount*™. 2020. Available from http://www.vishay.com/docs/40012/591d.pdf.

[12] Fiala J., Overholt M.R., Motakef S., Carlson D. *Thermal Considerations in Low Cost T/R Module Design. 2010 GomacTech*. 2010. Available from www.capesym.com/docs/gomactech2010.pdf.

[13] Pluymers B.A., Reese R.M. 'Thermal management of active electronically scanned array transmit/receive LRU line replaceable unit'. *IEEE Radar Conference*. 2007, pp. 150–5.

[14] Office P. *Materials for High-Density Electronic Packaging and Interconnection*. Tech. Rep. NMAB-449. National Materials Advisory Board, national Research Council. 1990. Available from https://apps.dtic.mil/dtic/tr/fulltext/u2/a261852.pdf.

[15] Gray D.E. *American Institute of Physics Handbook*. 1957. Available from https://archive.org/details/AmericanInstituteOfPhysicsHandbook,https://patents.google.com/patent/McGraw-Hill.

Chapter 16

Technology base and cost

Cost is a dominant consideration with ESAs and was historically the principal reason that ESAs were not more prevalent. Cost represents a significant competitive discriminator, so this information is rarely published. Accordingly, the cost discussion in this chapter is necessarily inferential.

The challenges described in Chapters 13–15 represent the cost drivers. Large investments in R&D have mitigated most of these issues to the point that ESAs are cost competitive with other antenna types, particularly in view of their additional capabilities difficult or impossible to achieve with traditional designs.

Historically, T/R modules constituted about 50% of the cost of an ESA and components were about 50% of the cost of the T/R module. Accordingly, considerable effort was expended on both component and module cost. These estimates were predicated on high reliability (military) requirements and standards. Commercial systems relaxe these constraints with commensurate cost savings.

Cost is a strong function of both production rates and quantities. In 2001, a US Defense Science Board Study titled *Future DoD Airborne High-Frequency Radar Needs/Resources* [1] forecast a requirement for a total of 6 million T/R modules for current and planned US military programs through 2021. In 2015, Raytheon [2] stated that they had "delivered more than 2 million T/R modules and more than half a million module-less T/R channels for TRIMMs and panels" and had a capacity of "150,000–200,000 RF modules per year." Although other manufacturers have not published their quantities, these numbers appear to substantiate the DSB prediction. Quantity alone is responsible for significant cost reduction.

Grim *et al.* [3] provide an overview of classical costing methodology. This methodology relies on a database of past programs to extrapolate costs according to similarity of radar parameters. With a historical perspective, the model does not incorporate possible future improvements in design and fabrication technology except by postulating an improvement or learning curve.

Herd and Conway [4] suggest that "Phased array costs can be minimized by utilizing high-volume commercial microwave manufacturing and packaging techniques, coupled with digital array architectures." They describe T/R module evolution to modern architecture and provide a cost model. They cite the multifunction phased array radar module, developed by Lincoln Laboratory and MACom and described in Section 1.4.1.2, using commercial best practice, as a model.

A Rand study *Cheaper, Faster, Better? Commercial Approaches to Weapons Acquisition* [5] published in 2000 suggests that "Civil-military integration (CMI) lies at the core of current DoD efforts to reduce the costs of procuring and maintaining modern weapon systems." Chapter 3, "Commercial Technology Trends Relevant to Military Radars," asks "Is the commercial market in military-relevant microwave electronics large enough to encompass an adequate range of technologies, parts, and components required to support a comprehensive CMI strategy...." As we know now, the answer is resoundingly affirmative.

Ultimately, the demand for ESAs was sufficient to support dedicated production lines, using similar components, processes, and equipment as their commercial counterparts and achieving commensurate cost savings.

16.1 Government investment laid foundation

Technology base advances are the key enabler for T/R modules. The US government invested more than a billion dollars in these programs during the 1980s providing seed money for what followed. Table 16.1 identifies a few of the more prominent examples. Marchese [15] provides details about the U.S. Air Force (USAF) programs. A current DARPA program, "Arrays at Commercial Timescales" (ACT), will incorporate commercial design and manufacturing concepts into ESA development.

Subsequent proliferation of commercial microwave devices, based on these technologies, provided the impetus and more importantly the funding for large-scale investment (many billions) to realize its full potential. Personal computers, mobile phones, and wireless networking dwarfed government investment starting in the 1990s.

Table 16.1 US government R&D

Sponsor	Program	Dates	Value
DARPA	Very High Speed Integrated Circuit (VHSIC) [6, 7]	1979–89	$1 billion
USAF	Solid State Phased Array [8]	1982–86	$20 million
USAF	Ultra Reliable Radar [9]	1984–89	
DARPA	Microwave and Millimeter Wave Integrated Circuit (MIMIC) [10]	1987–95	$600 million
USAF	T/R Module Manufacturing Technology	1989–92	$10 million
DARPA	High Density Microwave Packaging [11]	1993–97	$32 million
USN	T/R Module Manufacturing Technology	1997–2000	$8 million
DARPA	GaN [12] Wide Band Gap Semiconductor-RF (WBGS-RF) and Nitride Electronic NeXt Generation Technology (NEXT)	2002–2008 and 2010–2013	~$100 million
DARPA	Arrays at Commercial Time Scale [13, 14]	2013 to present	$100 million

Source: as noted

16.2 Industrial consolidation for military products

During the 1990s, a wave of consolidation swept the aerospace and defense industry segment. Consolidation was motivated by reduced defense budgets which followed the dissolution of the USSR. Industry consolidation, combined with competitive forces, concentrated AESA capability in a relative handful of companies.

Figure 16.1 from Watts [16] summarizes some of the major mergers in the United States. Companies with major radar expertise and large volume production are shown in red. The resultant aerospace companies all possess the requisite skills and technology to develop ESAs; however, Raytheon and Northrop Grumman are dominant in deploying military systems and probably in aggregate cost.

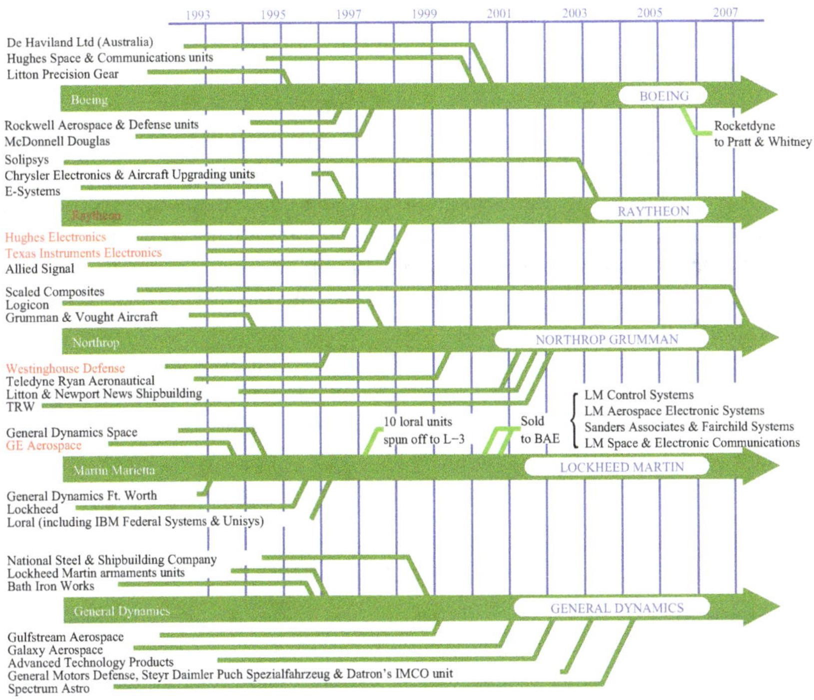

Figure 16.1 The consolidation of US defense manufacturing, 1993–2007[a]

[a]*Acquisitions of professional services and information technology (IT) firms have been mostly excluded. Exceptions include Logicon, E-Systems, and Solipsys. For consolidation diagrams including professional services and IT, see Pierre Chao* et al., *Structure and Dynamics of the U.S. Federal Professional Services Industrial Base: 1995–2005 (Washington, DC: CSIS, May 2007), pp. 75–86. BAE, a British-Italian defense prime with a US subsidiary, is not shown except for units it acquired from Lockheed Martin in 2000. The heavier lines indicate acquisitions of larger defense firms*

Source: Watts [16]

Some companies, such as Ball Aerospace, possess radar expertise but did not participate in industry consolidation. Universities and FFRDCs (federally funded research and development centers) including the Jet Propulsion Laboratory, Georgia Tech Research Institute, Lincoln Laboratory, and Sandia National Laboratories were likewise exempt.

Semiconductor products entail both design and manufacture. Many companies retain the design capability but contract out the fabrication because of the huge capital cost of the foundries. Most of the foundries are in Asia. A vast majority of semiconductor devices are used in consumer devices which are likewise produced in Asia. ESA developers retain an in-house capability to fabricate specialty parts albeit at higher cost.

16.3 Commercial investment for consumer products

As discussed in Chapter 14, T/R module design has been shaped by the available component and manufacturing technology.

Application of commercial components has greatly reduced the cost of T/R modules. A certain convergence of function, both analog and digital, has been a big factor.

Another key cost enabler is reducing the labor content of each module. Early T/R modules, using inconsistent components and manual assembly, required considerable skilled technician labor to meet specification. Apart from the cost of the labor, each manual intervention represented an opportunity to introduce additional defects into the module being reworked. Additionally, the repairs reduced the reliability of the device.

Automated assembly, using some of the same techniques of commercial electronics, has reduced variability and cut the hands on labor content. Manual tuning is gone. Automated test equipment with single touch connection has also contributed and the full characterization delivers cal tables if required. Production volume enables the required capital investment.

16.4 T/R module cost reduction since 1980

There is no consensus on the cost of T/R modules in part because costs depend critically on quantity but also because there is no actual market for T/R modules. The major producers are all focused on selling complete systems based on proprietary modules and show little interest in outsourcing this critical capability. The cost of AESAs is similarly unclear. Costs are treated as proprietary and competition sensitive. The most reliable numbers are government contracts but it is rare to find examples whose scope relates to an ESA; usually the scope is for a system in which the ESA is only a part of the contract value. A few examples are discussed in this section.

It seems clear in any event that the cost of T/R modules has decreased, to the point that they no longer dominate AESA and system costs.

Section 14.2.1 describes the THAAD T/R module. It has evolved considerably since its original design in 1992. In 1998, the government funded a "Best of Breed" study to incorporate technology from the former Hughes Aircraft and

Texas Instruments. Lewis [17] describes this program and reports that it achieved an almost 50% cost reduction.

16.4.1 Congressional Budget Office report

At a time when the USAF was considering the development of a space radar constellation at a cost of many billions of dollars, the Congressional Budget Office (CBO) was directed to analyze and report to congress their assessment of the probable costs.

In 2007, the CBO issued their report. It considered among other things the costs of MMICs and T/R modules. In this instance, however, they seem to have relied on outdated technical and cost data. Their module cost estimate of $5,800 is compared to historic prices instead of contemporary and forward prices which were less than $1,000 at the time of the report.

Alternatives for Military Space Radar – A CBO Study [18, p. 51]

> *Transmit/Receive Modules.* An industry study in the late 1990s concluded that a T/R module costs about $10 to $20 per square millimeter. In the radar designs that CBO considered, each T/R module would be 17.5 millimeters wide and 22 millimeters high, or 385 square millimeters. Assuming an average cost of $15 per square millimeter would put the cost of those T/R modules at about $5,800 apiece. By comparison, T/R modules built for various programs over the past 10 years have cost between $4,000 and $11,000 each, CBO estimated.

Ultimately, the military space radar system was terminated, in part because of its large and uncertain cost.

16.4.2 Manufacturing technology

The United States Department of Defense (USDOD) was an early advocate for the use of ESAs and its research laboratories played an active development and risk reduction role. A few of their many reports are cited in this book.

Several programs were conducted under the aegis of "manufacturing technology" (ManTech), a category used by the USDOD to improve the productivity and cost in areas known to be problematic.

The USAF sponsored a ManTech program for T/R modules from 1989 to 1993. Hughes Aircraft Company and a team of Texas Instruments and Westinghouse each received contracts. The contract objective was described as follows:

Manufacturing Technology For Radar Transmit / Receive Modules [19]

> Active element phased array systems utilizing transmit/receive (T/R) modules are considered to be one of the most promising technologies for future

> ground-based. airborne. and space based radar applications. Benefits include performance improvements and reductions in size and weight. Feasibility and validation T/R microwave modules for many new systems have been built in small prototype quantities or very limited production quantities. Costs are extraordinarily high as a result of complex designs, the need for precision fabrication, the cost of parts and materials, and the general lack of adequate assembly, test and automation equipment. This program is needed to reduce T/R module costs and demonstrate that the technology is producible. The objective of this project is to establish and demonstrate a low cost manufacturing capability for large quantities of complex microwave T/R modules for inclusion in active element-phased array radar systems. Contractors will be required to demonstrate high volume producibility at a projected production rate requirement of 1,000 modules per day and a target cost not to exceed $400 per module. Additionally, efforts are currently underway to transfer this technology to a third contractor, ITT, in support of Navy requirements.

Subsequently, the US Naval Air Warfare Center sponsored a similar program between 1996 and 1998. Table 16.2 summarizes the contracts received by Hughes Aircraft Company and Raytheon. Raytheon's acquisition of Hughes was announced on January 17, 1997, but the contracts continued separately albeit with enhanced coordination between the two. The contract goal was a $300 T/R module described as follows:

Manufacturing Research and Development of X-Band Active Electronically Scanned Array Transmit/Receive Modules [20]

> AESA radars demonstrate excellent potential for expanding the capability, improving the reliability and lowering the life cycle costs of ship based, aircraft and missile radars. However, current production costs for T/R modules make the initial acquisition of AESA radars cost prohibitive for most of these applications. To make AESA radars affordable, a major technology paradigm shift must be effected to bring module production costs down to an average of $300 or less after the first Engineering Model Development (EMD) next generation shipborne radar is completed. The purpose of this program is to develop T/R module designs and manufacturing methods, meeting the performance and interface requirements for future Navy combatants, while demonstrating that these designs are capable of meeting the $300 cost goal. This effort must complete within 18 months so that low cost T/R module technology can be inserted into near-term shipboard and airborne radar system development schedules.

No public reports were issued for these activities, but it is clear that they served their purpose of inculcating cost awareness into the design process. The relatively low contract cost was recovered many times over in the production contracts that followed.

Table 16.2 Manufacturing research and development of X-band AESA T/R modules

Contractor	Value	Date
Request for proposal		July 15, 1996
Hughes Aircraft Company	$3,554,246	November 22, 1996
Raytheon Electronic Systems	$4,498,223	December 17, 1996

16.4.3 ESA prices

While no manufacturer advertises production costs, some information on ESA radar systems prices has been published pursuant to government contract awards. Four relevant contracts are summarized in this section.

The USAF is upgrading most, if not all, of their aircraft radars to ESAs. This is necessitated by the very long in-service life of these aircraft. The contract value includes installation and test, so dividing cost by quantity will overstate the ESA price. With that caveat, the system unit prices are $8 million and $3.5 million for the F-15 and F-16, respectively. MDA continues to purchase THAAD AN/TPY-2 ESA radars at a cost on the order of $200 million. This cost includes considerable ancillary equipment, so it is also overstated.

Based on this data, the T/R module cost remains conjectural, but the unit price is certainly well below $1,000 each.

Boeing Company, Space & Security awarded $558,462,269 (November 30, 2016) [21]

> The Boeing Co., Space & Security, St. Louis, Missouri, has been awarded a $558,462,269 modification (P00015) to previously awarded contract FA8634-16-C-2653 for F-15 Combined APG-63 Version 3 Radar Improvement Program Version 3 and APG-82 Version 1 Radar Modernization Program (RMP) radar upgrades. Contractor will provide procurement, installation, initial spares and support for 42 Version 3, and 29 RMP radars. Work will be performed at St. Louis, Missouri, and is expected to be complete by Jan. 15, 2019. This award is the result of a sole-source acquisition. Fiscal 2015 and 2016 procurement funds in the amount of $397,907,824 are being obligated at the time of award. Air Force Life Cycle Management Center, Wright-Patterson Air Force Base, Ohio, is the contracting activity (FA8634-16-C-2653).

Northrop Grumman Systems Corporation awarded $243,873,277 (May 31,2017) [22]

> Northrop Grumman Systems Corp., Linthicum Heights, Maryland, has been awarded a $243,873,277 predominantly fixed-price contract

> with cost-plus-fixed fee and fixed-price incentive portions for 72 Active Electronically Scanned Array radars, spares and support services. Work will be performed at Linthicum Heights, Maryland, and is expected to be complete by Jan. 31, 2019. This award is the result of a competitive acquisition with one offer received. Fiscal 2016 research, development, test and evaluation funds in the amount of $30,714,025 are being obligated at the time of award. Air Force Life Cycle Management Center, Wright-Patterson Air Force Base, Ohio, is the contracting activity (FA8615-17-C-6047). Northrop Grumman will upgrade 72 U.S. Air National Guard F-16s to meet a U.S. Northern Command Joint Emergent Operational Need for homeland defense.

Raytheon Awarded $212 Million Contract for Terminal High Altitude Area Defense Radar [23]

> TEWKSBURY, Mass., Feb. 15, 2007 /PRNewswire/ – Raytheon Company (NYSE: RTN) has been awarded a $212 million contract by the Missile Defense Agency for the manufacture, delivery and integration support of one Terminal High Altitude Area Defense radar, also called the AN/TPY-2 radar.
>
> > 'The AN/TPY-2 radar plays a vital role in the Ballistic Missile Defense System, protecting the U.S., deployed forces and allies from ballistic missile threats,' said Pete Franklin, vice president of Raytheon Integrated Defense Systems' (IDS) Missile defense business. 'This award underscores the Missile Defense Agency's confidence in Raytheon's ability to deliver affordable, advanced sensor capabilities to meet the Ballistic Missile Defense System mission.'
>
> Under the terms of the contract, Raytheon Integrated Defense Systems will manufacture one AN/TPY-2 radar. Work will be performed at Raytheon's Missile Defense Center in Woburn, Mass., and at the company's Integrated Air Defense Center in Andover, Mass. The AN/TPY-2 is a high-power, transportable X-Band radar designed to detect, track and discriminate ballistic missile threats. Raytheon IDS designed and built the AN/TPY-2 radar for the Ballistic Missile Defense System and the Terminal High Altitude Area Defense weapon system drawing on extensive sensor knowledge from its X-Band 'Family of Radars.'

Recently, Raytheon [24] has utilized a modular approach to building large arrays for the Navy's SPY-6. They describe this approach as follows:

> SPY-6 radars are built from individual "building blocks" called Radar Modular Assemblies, or self-contained radars that come in $2' \times 2' \times 2'$ [60 cm × 60 cm × 60 cm] boxes. These boxes stack together to fit the mission requirements of any ship – a feature that makes the SPY-6 family the Navy's first truly scalable radars.

The Medium Extended Air Defense System (MEADS) was a joint US/German/Italian development utilizing an X-band ESA. Pursuant to that development, a separate contract was issued for T/R modules with an apparent price of €3,000 each.

EADS Delivers 5,000 T/R Modules for MEADS Radar, February 3, 2009 [25]

> The Defence Electronics business unit of EADS Defence & Security (DS) has delivered a first batch of 5,000 transmit/receive (T/R) modules and related electronics for the joint US-Italian-German Medium Extended Air Defence System (MEADS) Multifunction Fire Control Radar (MFCR). EADS DS is thus the first European company to mass produce the core components for the new Active Electronically Scanned Array (AESA) radar technology on an industrial scale. Within the framework of the MEADS design and development program, Defence Electronics had been awarded a contract worth about €120 M for the production of approximately 40,000 T/R modules and associated electronic components, so-called manifolds, which will make the MEADS fire control radar the most powerful radar of its type in the world. These modules are key elements in the new generation of radar systems which are based on AESA technology.

16.4.4 Commercial examples

Commercial applications of ESAs and ESA-like technology are numerous. The mobile telephone is a close relative to the T/R module. Both contain T/R functions. The telephone contains in addition a receiver, frequency synthesizer, digital processing, user interface, and power supply which are not part of the T/R module. Despite the additional functionality, simple mobile telephones are sold at prices much less than $100 for quantities of millions. Accordingly, this price represents a lower bound on achievable T/R module cost.

The 5G networks rely on aggressive frequency reuse and require higher gain to support the bandwidths. Wi-Fi access points are subject to the same constraints of frequency reuse and wide bandwidth. Accordingly, both contain some of the functionality of T/R modules and ESAs.

Commercial electronics are relatively inexpensive because they are produced at extremely high volume in highly automated factories. Notably their design lifetime is only a few years as they will be obsolete before they are nonfunctional. The manufacturing technology developed by these manufacturers and their equipment suppliers has been adapted by the high-reliability suppliers.

References

[1] Briggs D.L., Everett R.R. *Future DoD Airborne High-Frequency Radar Needs/Resources*. Tech. Rep. A391893. Defense Science Board, Under

Secretary of Defense for Acquisition and Technology. 2001. Available from https://www.acq.osd.mil/dsb/reports/2000s/ADA391893.pdf.

[2] Raytheon. *Corporation*; 2015.

[3] Grim P., Stamm D., Bell G. 'A method for estimating the development and production cost of advanced radar systems at the subsystem and component level'. *ICEAA/SPA/SCEA Professional Development & Training Workshop*. 2009. Available from http://www.iceaaonline.com/ready/wp-content/uploads/2017/09/MD012-Stamm.pdf.

[4] Herd J.S., Conway M.D. 'The evolution to modern phased array architectures'. *Proceedings of the IEEE*. 2016, vol. 104(3), pp. 519–29.

[5] Lorell M.A. *Cheaper, Faster, Better? Commercial Approaches to Weapons Acquisition.* Tech. Rep. MR1147.Santa Monica, CA, USA: Rand Corporation. 2000. Available from https://www.rand.org/pubs/monograph reports/MR1147.html.

[6] GAO. *GAO Assessment of DOD's Very High Speed Integrated Circuits (VHSIC) Technology Program.* Tech. Rep. NSIAD-85-37. Government Accountability Office. 1985. Available from https://www.gao.gov/products/NSIAD-85-37.

[7] VHSIC Program Office. *Very High Speed Integrated Circuits – VHSIC – Final Program Report*. Tech. Rep. Office of the USD for Acquisition, DARPA. 1990. Available from https://archive.org/details/DTICADA230012/page/n11.

[8] McQuiddy D.N., Gassner R.L., Hull P., Mason J.S., Bedinger J.M. 'Transmit/receive module technology for X-band active array radar'. *Proceedings of the IEEE*. 1991, vol. 79(3), pp. 308–41.

[9] Lingle D.E., Mikszan D.P., Mukai D. 'Advanced technology Ultra-Reliable Radar (URR)'. *Proceedings of the IEEE National Radar Conference*. 1989, pp. 1–6.

[10] Cohen E. 'The MIMIC program—A retrospective'. *IEEE Microwave Magazine*. 2012, vol. 13(4), pp. 77–88.

[11] Lamb F., Paul B., Lesniak C. 'High Density Microwave Packaging Technology Development for Department of Defense Applications' in Das N.K., Bertoni H.L. (eds.). *Directions for the Next Generation of MMIC Devices and Systems – 1997*. Boston, MA, USA: Springer; 1997. pp. 37–44.

[12] Rosker M.J., Albrecht J.D., Cohen E., Hodiak J., Chang T. 'DARPA's GaN technology thrust'. *2010 IEEE MTT-S International Microwave Symposium*. 2010, pp. 1214–7.

[13] Epstein B., Olsson R.H., Bunch K. 'Arrays at commercial timescale. Addressing development and upgrade costs of phased arrays. *2018*'. *IEEE Radar Conference*. 2018, pp. 327–32.

[14] Keller J. 'DARPA hires seven companies, spends more than $100 million, to reinvent the RF phased array antenna'. *Military & Aerospace Electronics*. 2014, vol. 25(2), pp. 6–11.

[15] Marchese J. *"Breakthrough" Technologies Developed by the Air Force Research Laboratory and its Predecessors*. Tech. Rep. ADA528970. US Air

Force Research Laboratory. 2005. Available from https://apps.dtic.mil/dtic/tr/fulltext/u2/a528970.pdf.

[16] Watts B.D. *The US Defense Industrial Base Past, Present and Future*. Tech. Rep. Center for Strategic and Budgetary Assessments. 2008. Available from https://csbaonline.org/research/publications/the-us-defense-industrial-base-past-present-and-future.

[17] Lewis J.W. 'THAAD rada examination of a cost saving initiative'. MA thesis. Naval Postgraduate School. 1999. Available from https://calhoun.nps.edu/handle/10945/9066.

[18] CBO. *Alternatives for Military Space Radar*. Tech. Rep. 1609. Congressional Budget Office. 2007. Available from https://www.cbo.gov/ sites/default/files/cbofiles/ftpdocs/76xx/doc7691/01-03-spaceradar.pdf.

[19] Project Office. *ManTech Project Book 1992*. Tech. Rep. ManTech Technology Transfer Center. 1992. Available from https://apps.dtic.mil/dtic/tr/fulltext/u2/a261881.pdf.

[20] CBD. 'Manufacturing research and development of X-band active electronically Scanned array Transmit/Receive modules'. *Commerce Business Daily*. 1998.

[21] USDOD. Boeing Company, Space *& Security awarded* $558,462,*269*. 2016. Available from https://www.defense.gov/Newsroom/Contracts/Contract/Article/1016911/.

[22] USDOD. *Northrop Grumman Systems Corporation awarded* $243,873,*277*. 2017. Available from https://www.defense.gov/Newsroom/Contracts/Contract/Article/1198370/.

[23] Raytheon. *Raytheon Awarded $212 Million Contract for Terminal High Altitude Area Defense Radar*. 2007. Available from http://investor.raytheon.com/news-releases/news-release-details/raytheon-awarded-212-million-contract-terminal-high-altitude.

[24] Raytheon. *U.S. Navy's SPY-6 Family of Radars*. Available from https://www.raytheon.com/capabilities/products/spy6-radars.

[25] EADS. *EADS Delivers 5*,000 T/R Modules for MEADS Radar. 2009. Available from http://www.eads.net/1024/es/pressdb/pressdb/20090126_eads_ds_de_tr_module_meads.html.

Chapter 17
ESAs in space

ESAs are widely used in space applications, and numerous satellite SAR use ESAs. These examples provide an interesting opportunity to compare and contrast alternative design concepts. ESAs are a good match to the application because their costs do not seem as daunting compared to the cost of satellites. Their reliability and graceful degradation are attractive for missions expected to last many years.

Communications satellites, discussed in Section 1.4.2, were quickly realized after the first satellite launch in 1958. Radar satellites, discussed in Section 1.4.3, were developed some years later.

17.1 Iridium communications satellite

The Iridium satellite constellation was launched between 1997 and 2002. The system development was managed by Motorola, while the antenna is a Raytheon product. The original concept planned for 77 satellites. Figure 17.1 shows a prototype of the first-generation Iridium satellite. The Iridium constellation offered mobile telephone service anywhere in the world. The antenna is described in "The Iridium main mission antenna concept" by Schuss *et al.* [2]. The original service provider experienced financial difficulties, entered bankruptcy, and was refounded as a viable commercial enterprise.

Commercial technology was applied to reduce cost, accepting shorter design life and allocating additional satellites to achieve required availability. The satellites have been quite long-lived: 98 were constructed, 95 launched, and 72 were operational after 15 years. Iridium's experience demonstrates that design life may be overspecified and/or underestimated. The second-generation "Iridium Next" was developed by Thales Alenia Space and replaced the first generation in orbit between 2017 and 2019.

Iridium represents an early application of arrays to a large constellation of satellites. Each satellite has 3 arrays and each array is 8 elements wide and 16 elements tall with 22 corner elements omitted. Figure 17.2 shows the element distribution. The lattice is equilateral triangular.

The antenna array provides multiple beams with a fixed beamformer. It utilizes a two-dimensional Butler matrix (see Section 2.6.2.1) to create a number of switched

(a)

Attribute	Value
Constellation	66 satellites (72 with spares)
Launch dates	5 May 1997 to 7 May 1998
Altitude	781 km
Orbit Inclination	86.4°
Frequency	1.62 GHz
Array boresight	50° from nadir
Antenna size	0.86 m × 1.88 m
Radiators	106 patches
Array feed	8 × 16 Butler Matrix

(b)

Figure 17.1 Iridium satellite: (a) first-generation test article and (b) system data

Photo by Eric Long, Smithsonian National Air and Space Museum (NASM2006-934) [1]

beams within its field of view. Pattan [3] describes Butler matrix beamformers and their application to Iridium. There are 8 azimuth beam positions and 16 elevation beam positions for a total of 128 beams. Since the beam space is rectangular and the target space is not, many available beam positions are not required, and only 10 elevation beam positions are utilized for a total of 80 beams.

In addition, at near range, elemental beams are combined to form "cells" on the ground with roughly equal area. The constellation of satellites provides continuous global coverage. There is approximately 35% overlap in coverage which facilitates hand-off of the conversation from one satellite to the next.

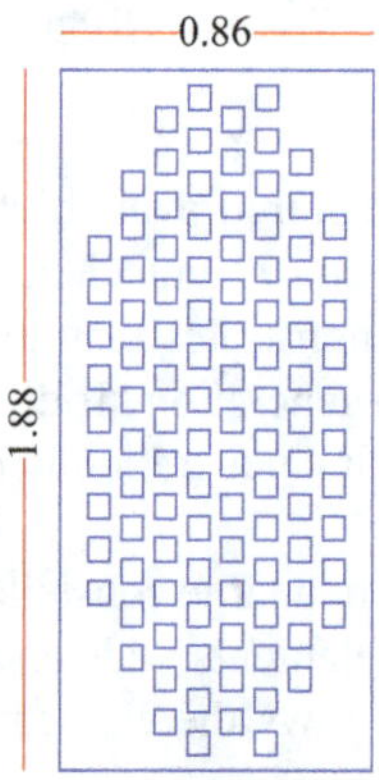

(a)

Attribute	Value
Wavelength	18.5 cm
Width	86 cm
Patch width	6.0 cm
DeltaX	9.3 cm
Height	188 cm
Patch height	6.4 cm
DeltaY	11.6 cm
Number of elements	106

(b)

Figure 17.2 Iridium array (one of three): (a) drawing and (b) parameters

After Schuss et al. [2]

Beams in (*u*, *v*) *space*

Figure 17.3 illustrates Iridium beam positions for a single north-facing array located over the equator. Figure 17.3a uses antenna coordinates ($u - v$) represented by Figure C.1; Figure 17.3b and c uses ground coordinates (latitude–longitude) and Figure 17.3d represents the beams in a rectangular frame of reference (nautical miles). This format will also be used in Figures 17.4 and 18.12.

Figure 17.3a shows the 80 beams generated by one of the three antenna panels in antenna coordinates. Because the axes of Figure 17.3a are u and v, each beam is the same shape, irrespective of its pointing direction. The beamwidth at boresight is approximately 14° in azimuth and 6° in elevation. Away from boresight, beams increase in solid angle with a corresponding decrease in gain. The black outline shows the system requirement for coverage, which requires all or part of 46 beams.

The red dot shows the physical boresight of the antenna corresponding to $u = 0$ and $v = 0$. The circles centered on boresight represent the beam directions of constant scan angle θ. Antenna scan restrictions set the left and right limits at approximately

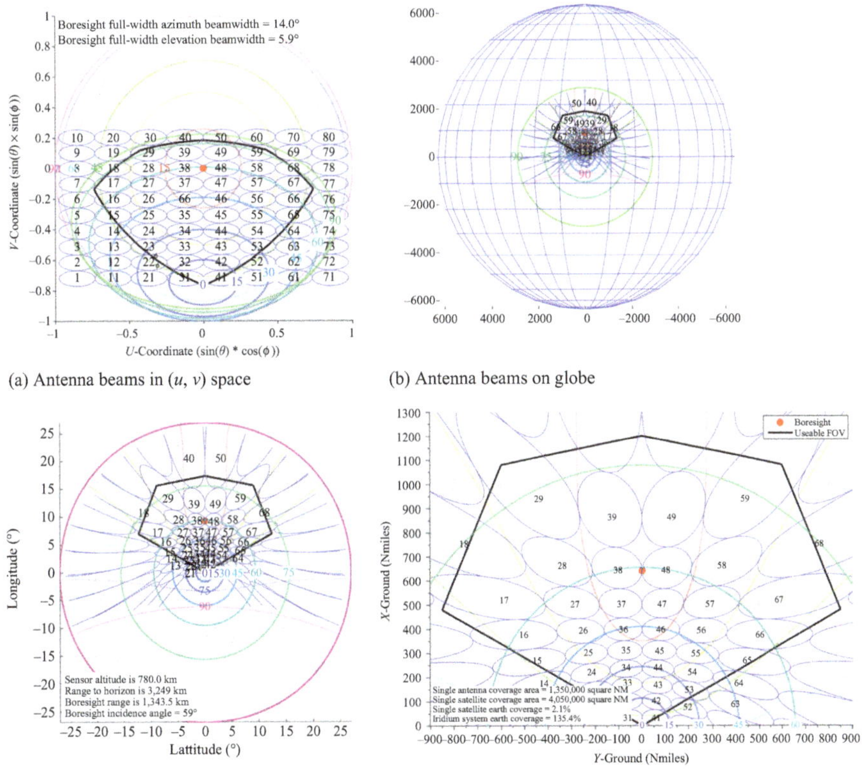

(a) Antenna beams in (*u*, *v*) space

(b) Antenna beams on globe

(c) Antenna beams on globe (detail)

(d) Antenna beams projected to ground

Figure 17.3 Iridium beam projection

BeamProjection.m

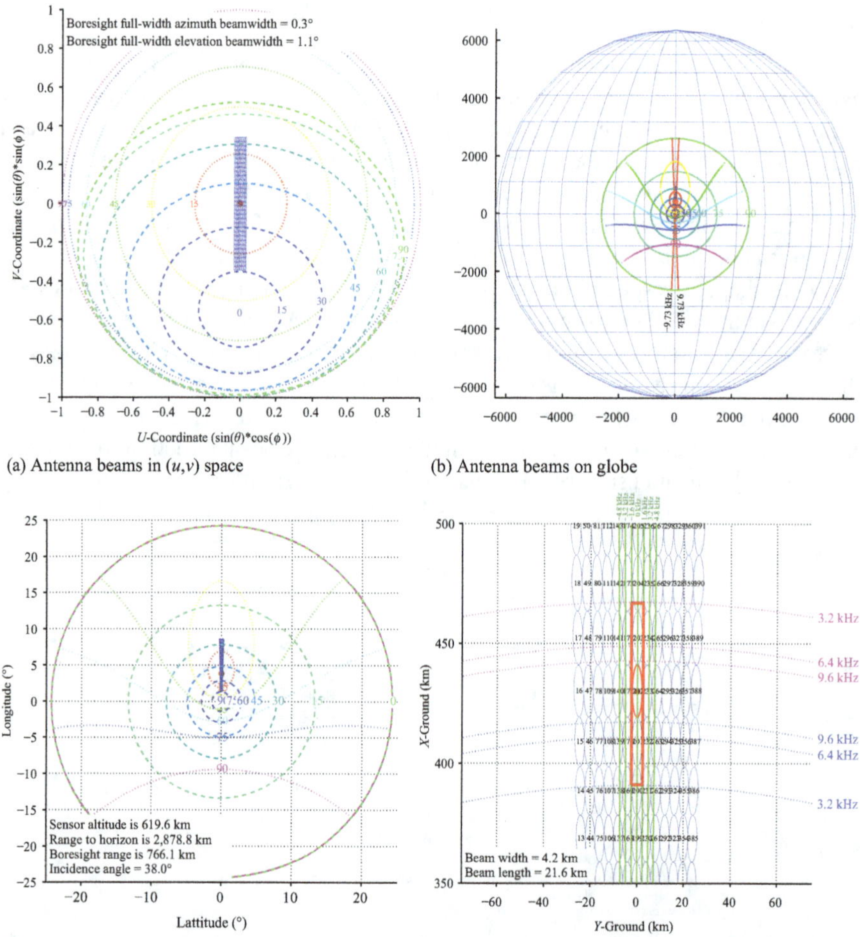

(a) Antenna beams in (u,v) space

(b) Antenna beams on globe

(c) Antenna beams on globe (detail)

(d) Antenna beams projected to ground

Figure 17.4 COSMO-SkyMed beam projection
BeamProjection.m

45°. The lower scan limit is nadir, labeled 0° incidence angle or equivalently 90° grazing angle. Ovals of equal incidence angle are shown centered on nadir. The upper scan limit is shown by the black line which is approximately 75° incidence angle (15° grazing angle). Beam directions outside the circle of incidence angle = 90° do not intersect the earth.

Beams on globe

Figure 17.3b replots the data in Figure 17.3a into latitude–longitude onto a sphere representing the earth. The satellite location is above Earth at position 0° latitude and 0° longitude. The ovals of equal incidence angle in the previous figure appear as circles in this figure. The green circle labeled 90 represents the horizon seen from the satellite, while the smaller circles represent smaller incidence angles as indicated.

The magenta arc labeled 90 represents the antenna scan angle in antenna coordinates, e.g., the plane of the array. Lesser scan angles are also indicated but more visible in the subsequent figures.

The beams within the black coverage area match those of the previous figure.

Beams on globe (detail)

Figure 17.3c enlarges the area of the beams that intersect the earth. Incidence angle circles (centered on nadir) and scan angles (from antenna boresight) are shown as previously. It is clear that the beams illuminate different size patches on the earth. For the mobile telephone application, it is desirable to have similar sized cells (with similar gain), so some of the smaller beams are combined before processing.

Beams projected to ground

Figure 17.3d shows the beams in ground coordinates centered on nadir. The black six-sided outline represents the coverage area of a single array. The included angle of coverage area at nadir is 120°. The three arrays provide full 360° azimuthal coverage from nadir to approximately 75° incidence angle.

The numbers at the bottom left show the coverage area computation. The earth coverage averages 135% providing the overlap to hand-off between satellites. The original system had 77 satellites (11 in each of 7 orbital planes). This was reduced to 11 in each of 6 orbital planes after determining that the smaller number provided sufficient overlapping coverage.

Figure 8 of Schuss *et al.* [2] illustrates how approximately 50 elemental beams are combined to form 16 beams.

17.2 Radar satellites

Earth observation has emerged as an important application for satellites. Section 1.4.3 illustrated some active satellite radars with ESAs. These systems have been comprehensively described and analyzed in the literature. van 't Klooster [4] provides an overview as of 2008. Ouchi [5] provides a history of this development as of 2013. Most of these systems are imaging radars (SAR) but altimeters Rudenko [6], atmospheric profilers Tanelli [7], scatterometers, and precipitation measurement radars have been flown. Cubesats have been proposed for some of these missions (see Peral [8]). Reflector antennas are sufficient for many of these missions and even preferable due to size, weight and power considerations.

For our purposes, the large SAR satellites provide the most interesting ESA applications. The SAR systems operate in a variety of radar bands, primarily L, S, C, and X. Corresponding frequency ranges are shown in Table 17.1. The demand for bandwidth has motivated the International Telecommunication Union to approve wider bandwidths for earth observation. At X-band, the World Radiocommunication

Table 17.1 ITU spectrum allocation for radar

Band	Frequency (MHz)
L	1,215–1,300
S	3,100–3,300
C	5,250–5,570
X	9,300–9,900

Source: National Academies of Sciences, Engineering, and Medicine [9], Table 3.3

Conference (WRC) 2007 approved 600 MHz bandwidth and the WRC 2016 approved 1.2 GHz bandwidth between 9.2 and 10.4 GHz [10].

17.2.1 SAR from space

The coverage area from space is much larger and the distances are considerably greater in comparison to airborne radar systems. The satellite motion is very predictable, so platform motion compensation is not as difficult as it is for airborne systems. Earth's rotation introduces a systematic bias in the Doppler spectrum of the returned signal not experienced by airborne systems.

The design choice starts with system requirements: orbit height and incidence angle; frequency band and radar performance requirements: swath width, area rate, NESZ, and resolution. The orbit height determines the orbital period which in turn determines the revisit frequency of the sensor. Incidence angle is a function of desired ground resolution and the target characteristics. Both ESAs and reflector systems provide useful performance.

Important parameters include orbital altitude, resolution, swath width, and area coverage rate. Figure 17.5 illustrates the sensor geometry with the satellite

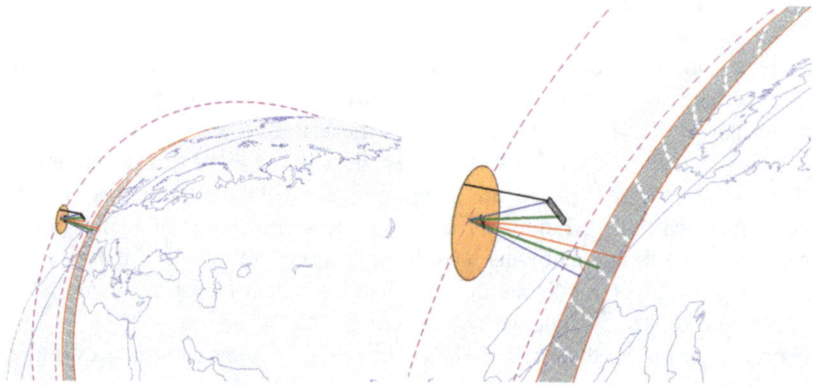

(a) (b)

Figure 17.5 Satellite SAR ground swath: (a) large area and (b) detail
Swath.m

orbit and nadir path on the Earth shown as dashed lines and the SAR swath shown as a shaded band offset from the nadir path. Range resolution is constrained by the ITU bandwidth allocation (see Table 17.1) and azimuth resolution is determined by the synthetic aperture length (coherent integration time). It is common to select range and azimuth resolutions that are similar because the orientation of the observation point to the ground area is effectively random. Range and azimuth sidelobes will create spurious images overlaying the image of interest, so it is necessary to reduce sidelobes, both in the antenna pattern and in the signal waveform. All SAR textbooks and numerous papers, notably Freeman *et al.* [11], provide additional insight into these criteria.

The swath width is primarily determined by the elevation beamwidth. The pulse repetition frequency (PRF) must be sufficiently low to avoid ambiguities in range, which introduce spurious images, but sufficiently high to fully sample the Doppler-shifted return signal. These two constraints limit the area coverage rate of the system. As the nadir return is often stronger than the signal of interest, it is preferable to select a PRF which causes the received nadir return to coincide with the transmit time window. The corresponding area rate is the product of swath width and velocity of the ground point, $v = v_{sat}\, r_e/(r_e + h)$.

17.2.1.1 Resolution

For SAR in stripmap mode, the best along-track resolution is one-half of the length of the physical array so long as the PRF is sufficiently large to sample the full Doppler spectrum of the azimuth beamwidth. PRF and azimuth resolution are related by (17.1), while PRF and ground swath are related by (17.2); these two equations are (3) and (5) in McCandless [12].

In Section 18.3, we will describe an L-band ESA which is 4.0 m high and 12.8 m wide. Based on this design, Figure 17.6 shows the relationship of azimuth resolution, swath width, and PRF (Doppler bandwidth) using (17.1) and (17.2). The figures represent these equations with a common abscissa in the form of a nomogram.

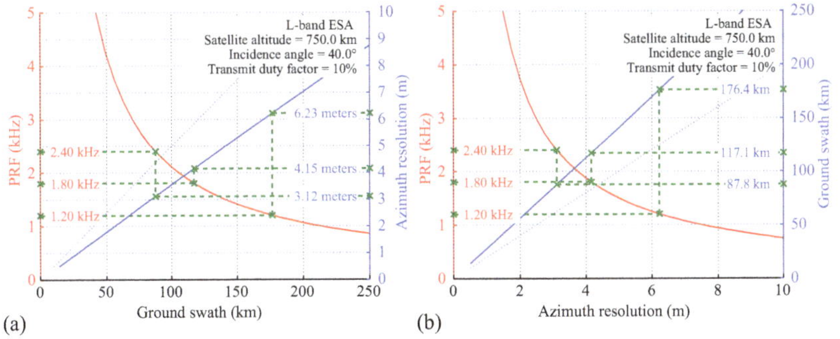

Figure 17.6 Swath width and resolution constrained by PRF: (a) azimuth resolution vs swath width and (b) swath width vs azimuth resolution *StripMap.m*

The dashed green line connects parameter pairs. Note that these numbers must be reduced by some amount to mitigate the effects of range and azimuth sidelobes as well as receive blanking during transmit:

$$\frac{\text{Velocity of radar}}{\text{Azimuth resolution}} = \frac{V}{\delta_{\text{AT}}} < \text{PRF} \tag{17.1}$$

$$\text{PRF} < \frac{c}{2 \times \text{swath width}} \tag{17.2}$$

The cross-track (range) resolution δ_r is related to the bandwidth BW of the system by (17.3), (1) in McCandless [12]:

$$\delta_{\text{r}} \gtrapprox \frac{c}{2\,BW} \tag{17.3}$$

This resolution represents the slant-range resolution. For the equivalent ground-referenced resolution, the slant-range resolution must be divided by the sine of the incidence angle, worsening the resolution by a factor ranging from 2.0 to 1.3 for incidence angles ranging from 30° to 50°.

Figure 17.7 represents the antenna pattern of the same L-band ESA incident on the Earth's surface. The pattern is based on uniform illumination which would probably be unsatisfactory because of the sidelobes. The transmit will be broadened (spoiled) in elevation in the SweepSAR mode described in Figures 17.8 and 17.9. Figure 17.7a shows the pattern in latitude and longitude coordinates. The antenna is on a satellite at an altitude of 750 km and the antenna is pointed such that the angle between the boresight and the local vertical is 40°. This is termed the grazing angle.

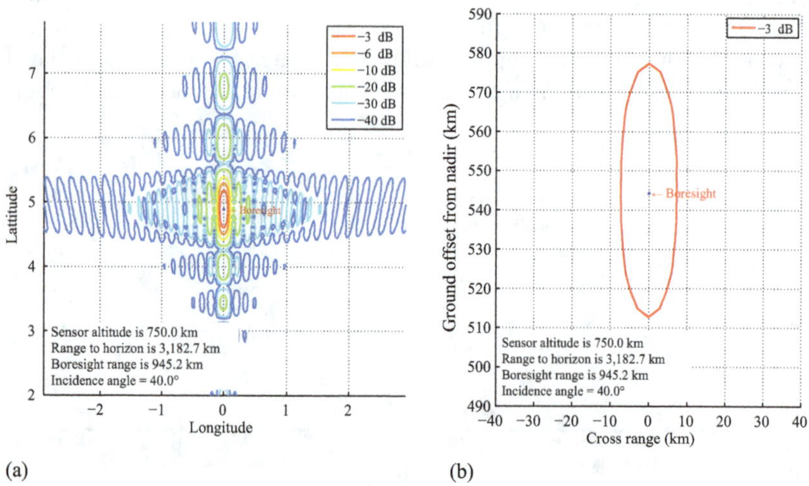

Figure 17.7 Beam projection on earth surface: (a) beam pattern contours and (b) 3 dB beam contour

MultipleBeam.m

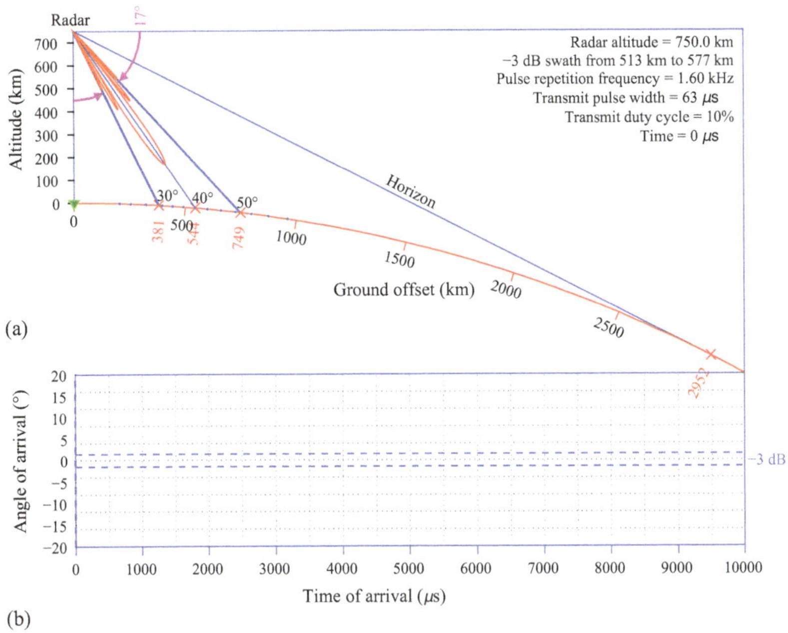

Figure 17.8 SweepSAR at time = 0: (a) radar geometry and (b) received signal

Figure 17.7b shows the mainbeam only from Figure 17.7a and represents range and cross-range on the ground in units of kilometers. The beamwidth is approximately 3° in elevation and 1° in azimuth.

17.2.2 Radar satellite geometry and timing

Multiple beams may be used to increase the ground coverage of a SAR system as noted in Sections 1.2.1 and 2.6.2. Younis [13] proposes this concept in his thesis. Multiple elevation beams increase the ground swath and are proposed and/or implemented for several systems. This technique is referred to as "SweepSAR" or "Scan-on-Receive" (SCORE). It may be employed with both ESA and reflector antennas. The transmit beam is broadened to fill the desired ground swath using techniques discussed in Chapter 10. While the receive beam could also be broadened, it is preferable to scan the narrow receive beam in the direction of the reflected signal across the full ground swath for a better *S*/*N* ratio. This may require multiple simultaneous receive beams. Such scanning is an inherent ESA capability. A reflector antenna may scan its beam by utilizing multiple feed locations at the focal point of the reflector. Huber [14] provides an excellent description. As we will see in Chapter 18, this is a little more complicated.

Figures 17.8 and 17.9 depict the radar timing required to implement scan on receive. Figure 17.8 shows the initial state at time zero. Figure 17.8a shows a cross section of the radar-ground-target geometry; the sensor (satellite) velocity vector is normal to the page. Figure 17.8b shows the received radar pulse as a function of

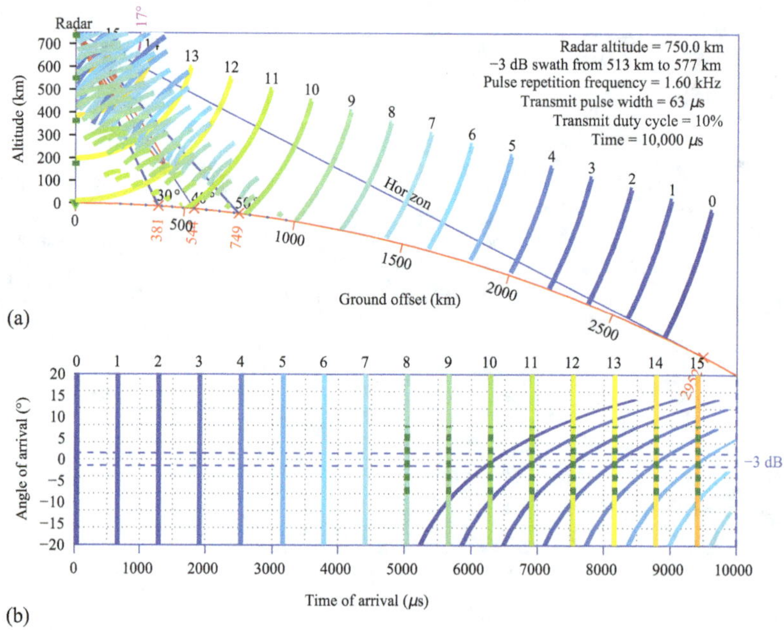

Figure 17.9 *SweepSAR at = 10, 000 μs: (a) radar geometry and (b) received signal*
SweepSAR.m

time of arrival and angle of arrival. The horizontal dashed lines represent the –3 dB elevation gain points of one receive beam.

The antenna pattern is shown in red, overlaid on the line of sight from the radar to the ground point corresponding to 40° incidence angle. The antenna pattern has a 3° elevation beamwidth and uniform illumination sidelobes at −13 dB. The surface of the earth is represented by the curved red line and the ground distance from nadir is shown in red numerals. The bold blue lines show the elevation access region of the radar, which can be selected by steering the beam. The boresight beam is shown pointing at a ground point for 40° incidence angle. Incidence angles of 30°–50° are indicated corresponding to ground offsets of 381 and 749 km and a swath width of approximately 360 km. Corresponding sensor elevation scan range of ±17° is shown in magenta. Ground targets are shown as blue dots ranged within and near the preferred range of incidence angles. The horizon line of sight is indicated and its ground offset is 2,952 km, slightly less than the distance of 3,183 km of the radar line of sight to the horizon.

Figure 17.9 shows the situation at time equal 10, 000 μs. There are 15 outgoing radar pulses (at a 10% transmit duty cycle) coded by color, and the associated radar returns from the blue dots. The magnitude of the returns would vary somewhat, as a function of the transmit beam pattern and the distance to the Earth's surface.

The transmit beam illuminates the full swath. Figure 17.9b shows reflected radar signals appearing at the receiver. The first radar pulse takes over 5,000 μs to return due to the distance to the ground. The nadir return is shown in green. The PRF was selected, so the time of return of the nadir direction coincides with the transmit window, as shown in Figure 17.9b. Figure 18.10 shows the detail of Figure 17.9b.

Because the received pulses overlap at the receiver with up to four pulses received at one time depending on the PRF, multiple low-sidelobe, receive beams are required. These may be created by a reflector antenna with multiple feeds or by an ESA with multiple feed networks or with digital beamforming. This complexity in the feed and in selecting radar parameters is repaid by the benefit of increasing the scan coverage by the same factor as the number of receive beams.

17.2.3 Satellite SAR systems

Next, we consider some radar satellites on orbit using the six examples introduced in Section 1.4.3 and illustrated in Figure 1.8. Figure 17.10 shows the corresponding antennas drawn to the same scale indicating element layout. They are all ESA antennas, at L-band, C-band, and X-band as indicated. Each T/R module drives more than one element. At X-band and C-band, the ensemble of elements connected to a single T/R module is shown as a thin line. The vertical white spaces show the boundaries of the subarrays although there are no gaps between the subarrays. At L-band, each element is represented individually. Table 17.2 describes the subarray architecture.

Table 17.3 summarizes the array factor performance of these satellites. the number of phase centers in the array is generally equal to the number of T/R modules except when there are T/R modules for each polarization. The phase center spacing

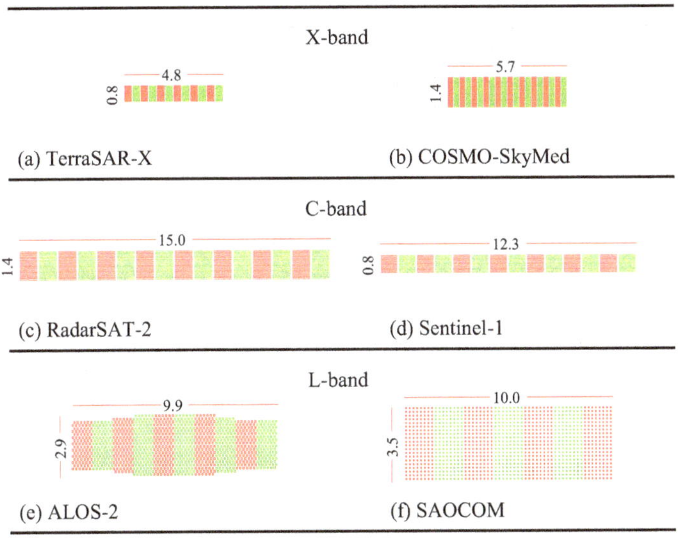

Figure 17.10 X-, C-, and L-band ESAs

Table 17.2 Antenna subarray architecture

System	Elements per subarray	Subarrays per column	Number of columns
TerraSAR-X [15]	16 slots	32	12
COSMO-SkyMed [16]	12 patches	64	20
RadarSAT-2 [17]	20 patches	32	16
Sentinel-1 [18]	20 slots	20	14
ALOS-2 [19]	6 patches	16–20	10
SAOCOM [20]	8 patches	20	7

Each subarray connects to a single T/R module

Table 17.3 Satellite ESAs optimized for SAR

Satellite	Number of modules	Dx (λ)	Dy (λ)	Grating lobe equality (°)	
				Azimuth	Elevation
TerraSAR-X	384	12.81	0.75	2.2	42
COSMO-SkyMed	1,280	9.17	0.70	3.1	46
RadarSAT-2	512	16.56	0.83	1.7	37
Sentinel	280H/280V	15.83	0.74	1.8	43
ALOS-2	180	4.19	0.68	6.9	47
SAOCOM	140	6.07	0.74	4.7	43

Azimuth and elevation angles represent grating lobe equality using (7.4)

is derived from the antenna dimensions and the arrangement of T/R modules. It ranges from 0.61 wavelengths to 17.77 wavelengths. All spacings are greater than one-half wavelength, so scan limits are less than 90° because of grating lobes. The azimuth and elevation angles are computed with (7.4). Angles represent the beam direction relative to boresight and correspond to the angle at which the grating lobe is equal in magnitude to the mainlobe. Accordingly, usable scan limits are considerably less than these angles. Scan limits are quite asymmetric with tighter lattice with elevation scan boundaries much greater than azimuth. These limits are consistent with the intended use for SAR.

17.2.4 RF power and thermal densities

RF power density is a primary input into the thermal design of the ESA. Table 17.4 summarizes some representative RF powers. The RF peak power numbers are similar despite the 10:1 ratio of wavelength. Array sizes are not vastly different as

Table 17.4 RF power densities

Satellite	Band	T/R module power (W)	RF peak power (kW)	Average RF power (W/m^2)
TerraSAR-X	X-band	6	2.26	129
COSMO-SkyMed	X-band	4	5.0	90
RadarSAT-2	C-band	5	2.28	13
Sentinel-1	C-band	16	4.37	50
ALOS-2	L-band	30	5.1	12
SAOCOM	L-band	30	3.9	10

Numbers are approximate

indicated in Figure 17.9 but X-band arrays are distinctly smaller than the C-band or L-band. The T/R module power increases with longer wavelength as the module density is less. RF power generation creates considerable waste heat as it is not particularly efficient. Efficiencies between 10% and 30% are typical. The corresponding waste heat ranges from 9× and 2× the average transmit power. The RF and heat generation in the array is uniformly distributed as it corresponds to the location of the T/R modules.

The T/R module is the primary heat source in an ESA, so its efficiency is quite important. If we consider its efficiency an independent variable, we may redraw the thermal plots from Figure 17.7 as those of Figure 17.11. These figures allow us to compute the equilibrium temperature of an ESA which relies upon radiative cooling. Figure 17.11a shows the results for an ESA which incurs only self-heating due to electronics waste heat, while Figure 17.11b is a more realistic case including the thermal input from earth radiation. Solar illumination also represents a heat input but the spectral distribution is different than the thermal radiation of the array, so it is possible to largely reject the solar input without compromising the desired thermal radiation. So, these approximations provide a lower bound on the probable array temperature but are no substitute for a detailed thermal model. Gilmore [21] provides more insight into the issues.

These predictions represent the equilibrium array face temperature. The component temperature will always be higher because the thermal path from the component to the array face introduces a temperature difference which is very much design dependent. This difference may be quite substantial, especially for the MMIC components which are very concentrated heat sources. The chart assumes that only one side of the array is able to radiate heat to free space and that in normal operation it will be facing the earth, so the earth IR radiance must be accounted for. The X-band arrays are mounted on the spacecraft body, so this assumption is substantially accurate. The C-band and L-band arrays, however, are deployed away from the spacecraft body, so two sides are available for heat removal, and one of the two sides will always be in the shade providing more opportunity for heat removal.

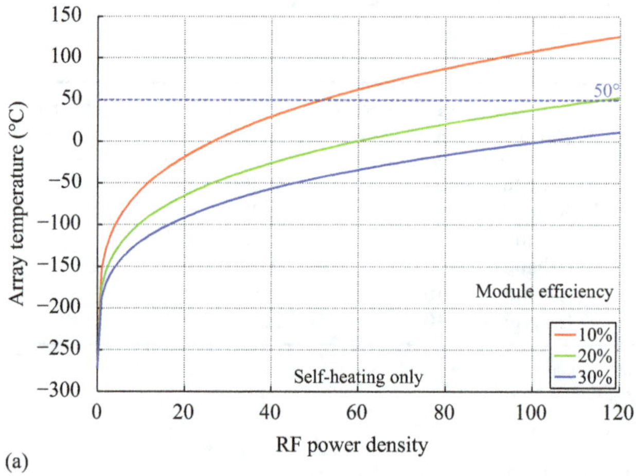

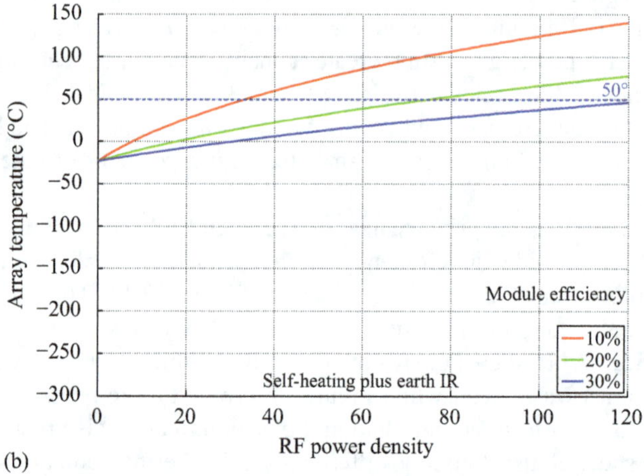

Figure 17.11 *Array equilibrium temperature: (a) self-heating only and (b) self-heating + earth IR*
Thermal.m

If we determine that a 50° array face temperature using single-sided passive cooling techniques is required, we see that, e.g., a 20% efficient module meets that limit for power densities up to 70 W/m^2. This combination rejects heat from a thermal radiator without special provisions and corresponds to the L-band and C-band systems in Table 17.4. For such cases, the ESA face itself may serve as the thermal radiator reducing cost and complexity. Above that number, increasingly detailed analysis is required to account for wavelength-dependent emission, the amount and timing of earth and solar illumination on the ESA, and heat distribution within the antenna.

17.3 X-band systems

X-band provides the highest resolution due to wide bandwidth and short wavelength. Because of its higher frequency, relatively small satellites can generate narrower beams but it does not provide as much ground coverage.

17.3.1 TerraSAR-X

The TerraSAR-X is illustrated in Figure 1.8a. Its performance is summarized in Table 17.5. It uses the SMTR® T/R modules, as described in Chapter 14, originally developed for terrestrial use. The module feeds a low-loss dual-polarized radiating element based on slotted waveguide. Grafmüller *et al.* [15] provide an excellent discussion of the array construction.

The next-generation TerraSAR-X will provide improved performance, notably increased resolution with a higher bandwidth radiating element, as described in Section 1.3.3.2. The antenna size does not require folding for launch.

Two satellites are in orbit flying in tandem providing additional coverage and digital elevation maps derived from interferometric data collection.

Table 17.5 X-band system performance

	TerraSAR-X* 2007–	**COSMO-SkyMed† 2007–**
Altitude (km)	514	619.6
Frequency (GHz)	9.65	9.6
Bandwidth (MHz)	300	300
Peak transmit power (kW)	2	5
Transmit duty factor	20% max	5%–10%
Average transmit power (kW)	0.4	0.5
NESZ (spec; dB)	–19	–22
Resolution (m)	1	1
Incidence angle	15°–60°	20°–59°
Electronic scan (el)	±20°	±20°
Electronic scan (az)	±2.5°	±2°
Array size (m)	4.8 × 0.8	5.7 × 1.4
Number of modules	384	1,280
Element configuration	32 × 192	64 × 240
Number of elements	6,144	15,360
Number of phase centers (el)	32	64
Phase center spacing (el)	0.75λ	0.70λ
Number of phase centers (az)	12	20
Phase center spacing (az)	12.8λ	9.2λ
T/R module power (W)	6	4

*Stangl *et al.* [22], Grafmüller *et al.* [15], Adolph *et al.* [23], Pitz and Miller [24]
†Capece [16], Capuzi *et al.* [25], Torre and Capece [26]

17.3.2 COSMO-SkyMed

The COSMO-SkyMed is illustrated in Figure 1.8b. Its performance is summarized in Table 17.5. It uses a waveguide-fed patch element subarray and provides azimuth time delay beamforming at the panel level with four TDUs providing up to 15λ TTD, discussed in Section 14.5.1.

It achieves improved revisit time by operating four satellites in a single orbital plane. A follow-on system is in development.

17.3.2.1 Antenna beams in (u, v) space

COSMO-SkyMed uses an electronic steered array with limited azimuth coverage. Figure 17.4 shows the projection of possible beams; approximately 400 beam positions are shown. The beam footprint is much smaller than Iridium since it operates at X-band. As previously, the figure axes are u and v.

Boresight is depicted by the red dot with scan angle circles centered on the red dot. The scan angle is up to about 20° off boresight. The incidence angles are shown as circles centered on nadir although the circles are more elliptical. The incidence angle specification is 15°–60°.

17.3.2.2 Antenna beams on globe

Figure 17.4b shows the same data as Figure 17.4a on an Earth-sized globe with the satellite over the equator. The velocity of the satellite is left–right, i.e., zero inclination which is not the actual orbit. Two additional lines in red are drawn representing the Doppler shift resulting from satellite motion of the return signal. The earth's rotation provides an additional Doppler shift, smaller than the satellite-caused shift which is not shown. The range of ± 10 kHz would require a high radar PRF to discern which would in turn compromise the azimuth swath width.

17.3.2.3 Antenna beams on globe (detail)

Figure 17.4c replots the data in latitude and longitude including the Doppler shift lines and demonstrating how small the beamwidth is on the earth.

17.3.2.4 Antenna beams projected to ground

Finally, Figure 17.4d expands the axes in kilometers. The azimuth scan range translates to ± 20 km cross range. A single beam spot is 4.2 km wide and 21.6 km long. The Doppler shift contained in the central beam is less than ± 1 kHz. Additional horizontal arcs represent the range ambiguity resulting from the PRF. As the PRF increases, the returns from adjacent regions come sooner so that at 4-kHz PRF the nonambiguous range is about 400–460 km. The tradeoff between range and azimuth ambiguities limits the maximum nonambiguous region on the earth. In the case shown, the Doppler bandwidth should be sampled at about 4 kHz (two samples per cycle), so the spot size is reasonably well optimized.

17.4 C-band systems

C-band offers a compromise between resolution and area coverage.

17.4.1 RadarSAT-2

RadarSAT-2 is a Canadian SAR satellite launched in 2007 and still operating. Its performance is summarized in Table 17.6. It replaced the RadarSAT-1 which was launched in 1995 and operated through 2013. The latest generation, the RadarSAT Constellation Mission (RCM), was launched in 2019 and is a three-satellite constellation.

17.4.2 Sentinel-1

Sentinel-1 is a project of the European Union and part of its Global Monitoring for Environment and Security. Its performance is summarized in Table 17.6. It is a C-band SAR operating with stripmap and wideswath mode. Two satellites were launched in 2014 and 2016.

Table 17.6 C-band system performance

	RadarSAT-2* **2007–**	**Sentinel-1†** **2014–**
Altitude (km)	798	693
Frequency (GHz)	5.405	5.405
Bandwidth (MHz)	100	100
Peak transmit power (kW)	2.28	4.368
Transmit duty factor	6%	12%
Average transmit power (kW)	0.3	0.524
NESZ (spec; dB)	–20	–22
Resolution (m)	1 × 3	5 × 5
Incidence angle	10°–60°	20°–47°
Electronic scan (el)	±12°	±12°
Electronic scan (az)	±12°	±0.8°
Array size (m)	15 × 1.37	12.3 × 0.84
Number of modules	512	280 × 2
Element configuration	320 × 32	280 × 20
Number of elements	10,240	5,600
Number of phase centers (el)	512	280
Phase center spacing (el)	0.83λ	0.7λ
Number of phase centers (az)	16	14
Phase center spacing (az)	16.56λ	15.83λ
T/R module power (W)	10	15.8

*Raney *et al.* [27], Riendeau and Grenier [17], Brule and Baeggli [28], Morena *et al.* [29], Chabot *et al.* [30]

†Rostan [31–33], Snoeij *et al.* [18], Potin [34]

17.5 L-band systems

L-band provides the largest coverage and is used for earth resources monitoring. The longer wavelength reduces ambiguities in repeat pass interferometry and improves foliage penetration in order to measure earth deformation. It was utilized for the first earth observation SAR satellite, SeaSAT, and the shuttle radars.

A variety of technical approaches, both ESA- and reflector-based, have been evaluated by organizations including JPL, JAXA, DLR, and ISRO.

The Japanese space agency (JAXA) has launched JERS-1 (Japanese Earth Resources Satellite) which used a planar array and two ESA systems, ALOS (Advanced Land Observation Satellite) and ALOS-2. Argentina's space agency CONAE launched an ESA system called SAOCOM (SAtélite Argentino de Observaciốn COn Microondas) in 2018 and a second satellite is planned in 2020.

The European Space Agency (ESA) is evaluating SAOCOM imagery [35] in their Earthnet Data Assessment Pilot (EDAP).

Several reflector-based L-band radar satellites have been proposed: DESDynI, NISAR, and TanDEM-L. JPL and DLR have developed system concepts for reflector-based systems. JPL proposed the DESDynI system which was incorporated in a NASA/ISRO partnership for an L-band and S-band satellite called NISAR. DLR proposed the TanDEM-L and is working with JAXA.

The satellite altitudes are all low earth orbit but ALOS-2 is about 10% lower than ALOS and DESDynI is about 10% higher than ALOS. The orbits are all sun synchronous dawn–dusk with repeating ground tracks. The larger physical size of the DESDynI must be derated by the efficiency and we will see that the useful fraction is similar to the ALOS satellites.

Its performance is summarized in Table 17.7.

ALOS-2 provides better performance than its predecessors with an increased number of higher power modules providing higher resolution and sensitivity. This increased number provides better scan capability and the increased power provides a 5 dB improvement in PG^2.

17.5.1 SeaSAT

The first SAR satellite was a planar array, not an ESA. It had a short operational life because of an on-orbit failure. The program was managed by JPL, and Ball Brothers built the antenna. McCandles [45] provides a historical summary. Figure 17.12 illustrates the satellite elements and the system performance parameters.

The antenna required folding for launch and on-orbit deployment. The deployment sequence of the SeaSAT antenna is shown in Figure 17.13.

17.5.2 Space shuttle radars

Following the SeaSAT, a system with similar capabilities was built and flown on the space shuttle. SIR-A (1981) and SIR-B (1984) were developed from SeaSAT technology, using a similar planar array with improved performance, notably much higher power.

Table 17.7 L-band system performance

	Planar array SeaSAT* 1978	**SIR-C† 1994**	**ESA ALOS-2‡ 2014–**	**SAOCOM§ 2018–**
Altitude (km)	794	215	628	620
Frequency (MHz)	1,274.8	1,277	1,250	1,275
Bandwidth (MHz)	19	10, 20	84	50
Peak transmit power (kW)	0.8	12.6	6.1	3.9
Transmit duty factor	5.5%	6%	7%	15%
Average transmit power (kW)	0.055	0.76	0.43	0.59
NESZ (spec; dB)	–21	–40	–24 to –28	–25
Resolution (m)	25	20	1–100	10–100
Incidence angle	17°–24°	15°–55°	8°–70°	20°–50°
Electronic scan (el)	–	±23°	±30°	tbs
Electronic scan (az)	–	±2°	±3.5°	tbs
Array size (m)	11 × 2.3	12.1 × 2.8	9.9 × 2.9	10 × 3.5
Number of modules	–	252	180	140
Element configuration	–	54 × 28	60 × 20	56 × 20
Number of elements	–	972	1,080	1,120
Number of phase centers (az)	–	9	10	7
Phase center spacing (az)	–	5.68λ	4.19λ	6.07λ
Number of phase centers (el)	–	28	18	20
Phase center spacing (el)	–	0.42λ	0.68λ	0.74λ
T/R module power (W)	–	50	34	28

*Thompson and Laderman [36], Jordan [37]
†Jordan [38, 39]
‡Kankaku *et al.* [19], Okada [40, 41], Yokota *et al.* [42], Kankaku *et al.* [43], SFCG [44]
§Frulla [20]

The last shuttle system was the SIR-C/X-SAR shown in Figure 17.14 and described by Jordan *et al.* [38]. The SIR-C/X-SAR comprised three separate radars: the L-band and C-band ESA instruments were designed by JPL and the X-band planar array radar designed by DLR. It flew twice, in April and October 1994, occupying most of the payload bay. Hirsch *et al.* [50] provide details concerning the two T/R modules, L-band and C-band, developed for the mission.

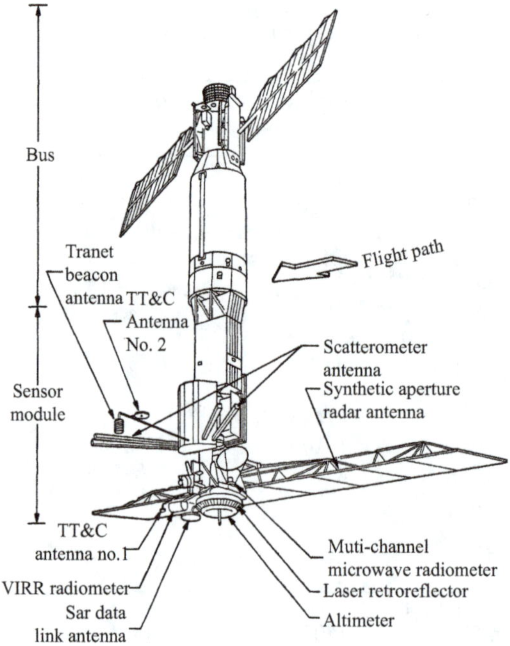

(a)

Attribute	Value
Dimensions	1.5 m D × 21 m L
Mass	2,290 kg
Array type	planar
Array size	10.74 m × 2.16 m
Frequency	1.275 GHz
Bandwidth	19 MHz
Peak power	1.5 kW
Transmit duty	≈ 5%

(b)

Figure 17.12 SeaSAT system: (a) SeaSAT L-band satellite and (b) SeaSAT parameters
Source: (a) Courtesy NASA [46] and (b) Jordan [37]

Finally, the system was enhanced with an adjunct X-band radar receiver to provide interferometric terrain height measurement; this mission flew in February 2000 and generated a digital elevation map of most of the earth. It was a joint project with DLR, who were responsible for data analysis.

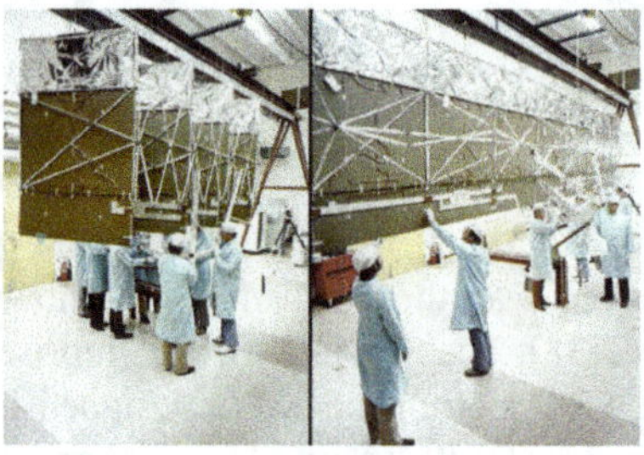

Figure 17.13 SeaSAT antenna deployment
Courtesy JPL [47]

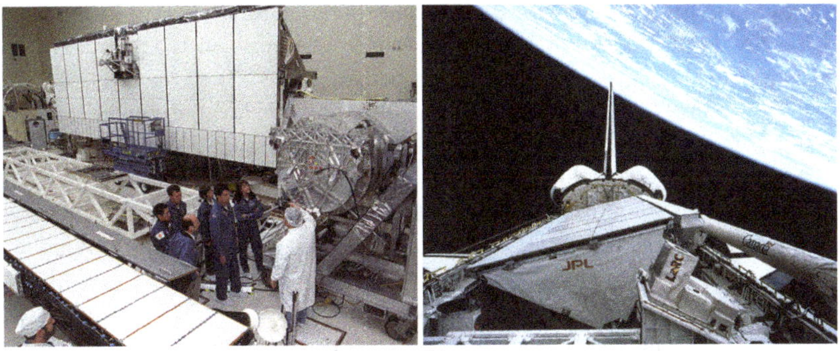

(a) (b)

Figure 17.14 *SIR-C/X-SAR L-, C-, and X-band antennas: (a) in assembly and checkout. Courtesy NASA/JPL-Caltech [48]. (b) In use. Courtesy NASA/JPL-Caltech [49]*

17.5.3 Advanced land observing satellites

These are two generations of ALOS, both ESAs. The antenna size is only slightly larger than SeaSAT/SIR/SRTM. Those systems had about 6 MHz of bandwidth compared to 28 MHz for ALOS [51] and 85 MHz for ALOS-2 [52], respectively, providing 4× and 10× improvement in range resolution. ALOS-2 retains the same size antenna but more than double the transmit power. The PALSAR antenna is strictly rectangular, while the PALSAR-2 antenna is elliptical in outline. Both antennas use (stacked) patch radiating elements. The PALSAR is a square lattice, while the PALSAR-2 is a triangular lattice.

17.5.3.1 ALOS-2 PALSAR array

Figure 17.15 shows the layout and parameters of the ALOS-2 PALSAR antenna. Figure 17.15a shows the position of radiating elements in the ALOS-2 PALSAR array. The array is generally elliptical in outline with an equilateral triangular lattice.

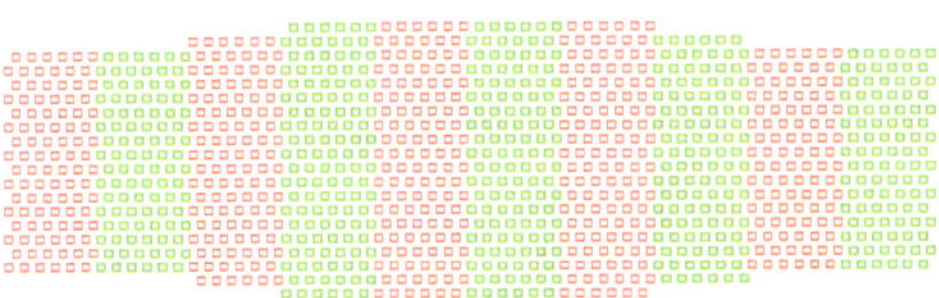

Attribute	Value
Array width	9.90 m
Array height	2.90 m
Array area	25.84 m^2
Wavelength	0.229 m
Number of elements	1,080
Gain ($4\pi A/\lambda^2$)	37.9 dBi
Delta X	16.5 cm
Delta Y	14.5 cm
Number of elements	1,080
Triangular angle	60.4°

(a) (b)

Figure 17.15 *ALOS-2 PALSAR array: (a) ten panels and (b) configuration*

The array is divided into panels as indicated by the colors. These panels could be used for advanced modes with appropriate beamforming and receivers. There are 1,080 radiating elements, so each of the 180 modules feeds 6 elements (1 row of the panel). This reduces the azimuth scan capability but the elevation scan capability is fully supported.

The array can be configured with a split aperture to form two beams on receive. Splitting the aperture on receive supports two receive beams (for increased area rate or increased azimuthal resolution in SAR and also some rudimentary moving target modes).

It is possible to reduce the aperture width from five to three panels to broaden beam in azimuth which provides better along-track resolution but reduces total radiated power. Beam spoiling on transmit could serve a similar purpose.

17.5.4 SAOCOM

SAOCOM is an L-band polarimetric SAR satellite developed as a joint project of the spac agencies of Argentina and Italy. SAOCOM-1A was launched in October 2018 and SAOCOM-1B in August 2020. It has one of the largest ESA arrays in space.

References

[1] Long E. *Iridium Test Article*. NASM2006-934. Washington, DC, USA: Smithsonian National Air and Space Museum; 2006. Available from https://ids.si.edu/ids/dynamic?id=NASM-A19990005000-NASM-2006-934.

[2] Schuss J.J., Upton J., Myers B., *et al*. 'The Iridium main mission antenna concept'. *IEEE Transactions on Antennas and Propagation*. 1999, vol. 47(3), pp. 416–24.

[3] Pattan B. 'The versatile ButlerMatrix'. *Microwave Journal*. 2004, vol. 47(11), pp. 126–35.

[4] van 't Klooster C.G.M. 'On space-based synthetic aperture radar antennas'. *7th European Conference on Synthetic Aperture Radar*; 2008. pp. 1–4.

[5] Ouchi K. 'Recent trend and advance of synthetic aperture radar with selected topics'. *Remote Sensing* 5 February 2013. 2013. Available from www.mdpi.com/journal/remotesensing.

[6] Rudenko S., Neumayer K.-H., Dettmering D., *et al*. 'Improvements in precise orbits of altimetry satellites and their impact on mean sea level monitoring'. *IEEE Transactions on Geoscience and Remote Sensing*. 2017, vol. 55(6), pp. 3382–95.

[7] Tanelli S., Haddad Z.S., Im E., *et al*. 'Radar concepts for the next generation of spaceborne observations of cloud and precipitation processes'. *2018 IEEE Radar Conference (RadarConf18)*. 2018, pp. 1245–9.

[8] Peral E., Im E., Wye L., *et al*. 'Radar technologies for earth remote sensing from CubeSat platforms'. *Proceedings of the IEEE*. 2018, vol. 106(3), pp. 404–18.

[9] National Academies of Sciences, Engineering, and Medicine. *Handbook of Frequency Allocations and Spectrum Protection for Scientific Uses*. Washington, DC, USA: The National Academies Press; 2015. Available from https.//www.nap.edu/catalog/21774/handbook-of-frequency-allocations-and-spectrum-protection-for-scientific-uses.

[10] ITU. *Radio Regulations Articles Edition of 2016*. Geneva, Switzerland: International Telecommunication Union; 2016.

[11] Freeman A., Johnson W.T.K., Huneycutt B., *et al.* 'The "myth" of the minimum SAR antenna area constraint'. *IEEE Transactions on Geoscience and Remote Sensing*. 2000, vol. 38(1), pp. 320–4.

[12] McCandless, Jr. S.W. in Jackson C.R., Apel J.R. (eds.) *Synthetic Aperture Radar Marine User's Manual*; 2005. pp. 1–23. Available from http.//www.sarusersmanual.com/ManualPDF/NOAASARManualCH01pg001-024.pdf.

[13] Younis M. 'Digital beam-forming for high resolution wide swath real and synthetic aperture radar [online]'. *Karlsruhe 2004. (Forschungsberichte aus dem Institut f¨ur H¨ochstfrequenztechnik und Elektronik der Universit¨at Karlsruhe (TH). 42.) Fak. f. Elektrotechnik und Informationstechnik, Diss. v. 22.07.2004*; 2004.

[14] Huber S. 'Advanced spaceborne SAR systems with array-fed reflector antennas'. *2015 IEEE Radar Conference*; May 2015. pp. 0253–8.

[15] Grafmuller B., Herschlein A., Fischer C., *et al.* 'The TerraSAR-X antenna system'. *IEEE International Radar Conference*; 2005. pp. 222–5.

[16] Capece P. 'COSMO SkyMed active phased array SAR instrument'. *2008 IEEE Radar Conference*. 4; 2008. p. 1.

[17] Riendeau S., Grenier C. 'RADARSAT-2 antenna'. *2007 IEEE Aerospace Conference*. 9; 2007. p. 1.

[18] Snoeij P. Sentinel-1 instrument overview. *SEASAR 2012, the 4th International Workshop on Advances in SAR Oceanography*. 2012. Available from https://earth.esa.int/c/documentlibrary/getfile?folderId=233696&name=DLFE-2149.pdf.

[19] Kankaku Y., Suzuki S., Osawa Y., *et al.* 'The overview of the L-band SAR onboard ALOS-2'. *Progress in Electromagnetics Research Symposium (PIERS) Proceedings 2009*; 2009. Available from http://piers.org/piersproceedings/piers2k9Proc.php.

[20] Frulla L. 'SAOCOM mission overview'. *2011 CEOS SAR Calibration and Validation Workshop*; 2011. pp. 1–30.

[21] Gilmore D.G. *Satellite Thermal Control Handbook – Volume Fundamental Technologies. Progress in Astronautics and Aeronautics Series*. The Aerospace Press; 2002.

[22] Stangl M., Werninghaus R., Zahn R. '*The TERRASAR-X active phased array antenna*'. *IEEE International Symposium on Phased Array Systems and Technology, 2003*; 2003. pp. 70–5.

[23] Adolph M., Hackenberg U., Reber R., *et al.* 'High-precision temperature drift compensated T/R-module for satellite based SAR applications'. *2005 European Microwave Conference*. 2005, vol. 2, pp. 1–4.

[24] Pitz W., Miller D. 'The TerraSAR-X satellite'. *IEEE Transactions on Geoscience and Remote Sensing*. 2010, vol. 48(2), pp. 615–22.

[25] Capuzi A., Rivola S., Lazzaro M.D. 'COSMO SKYMED experience – synthetic aperture radar activities in Italy'. *Technologies and Applications of Synthetic Aperture Radar (SAR)*. Madrid, Spain; 2011.

[26] Torre A., Capece P. 'COSMO-SkyMed: The advanced SAR instrument'. *Proceedings of 5th International Conference on Recent Advances in Space Technologies–RAST2011*; 2011. pp. 865–8.

[27] Raney R.K., Luscombe A.P., Langham E.J., *et al.* 'RADARSAT'. *Proceedings of the IEEE*. 1991, vol. 79(6), pp. 839–49.

[28] Brule L., Baeggli H. 'RADARSAT-2 mission update'. *IGARSS 2001 Scanning the Present and Resolving the Future. Proceedings. IEEE 2001 International Geoscience and Remote Sensing Symposium (Cat. No. 01CH37217)*. 6; 2001. pp. 2581–3.

[29] Morena L.C., James K.V., Beck J. 'An introduction to the RADARSAT-2 mission'. *Canadian Journal of Remote Sensing*. 2004, vol. 30(3), *eprint*), pp. 221–34.

[30] Chabot M., Decoust C., Lambert C., *et al.* 'RADARSAT-2 system operations and performance'. *2015 IEEE International Geoscience and Remote Sensing Symposium*. 2015, pp. 1512–5.

[31] Rostan F. 'The C-SAR instrument design performance for the GMES Sentinel-1 mission'. *7th European Conference on Synthetic Aperture Radar*; 2008. pp. 1–4.

[32] Rostan F. 'The Sentinel-1 C-SAR instrument design performance'. *8th European Conference on Synthetic Aperture Radar*; 2010. pp. 1–3.

[33] ESA. *Sentinel-1 ESA's Radar Observatory Mission for GMES Operational Services. Tech. Rep. SP-1322/1*. European Space Agency; 2012. Available from https://sentinel.esa.int/documents/247904/349449/S1SP-13221.pdf.

[34] Potin P. 'Sentinel-1 mission status'. *2015 IEEE International Geoscience and Remote Sensing Symposium*; 2015. pp. 2820–3.

[35] ESA's Earthnet Data Assessment Pilot (EDAP) – Saocom. Available from https://earth.esa.int/web/sppa/activities/edap/sar-missions/saocom.

[36] Thompson T., Laderman A. 'SEASAT-A synthetic aperture. radar system implementation'. *OCEANS'76*; 1976. pp. 247–51.

[37] Jordan R. 'The Seasat-A synthetic aperture radar system'. *IEEE Journal of Oceanic Engineering*. 1980, vol. 5(2), pp. 154–64.

[38] Jordan R.L., Huneycutt B.L., Werner M. 'The SIR-C/X-SAR synthetic aperture radar system'. *Proceedings of the IEEE*. 1991, vol. 79(6), pp. 827–38.

[39] Jordan R.L., Huneycutt B.L., Werner M. 'The SIR-C/X-SAR synthetic aperture radar system'. *IEEE Transactions on Geoscience and Remote Sensing*. 1995, vol. 33(4), pp. 829–39.

[40] Okada Y. 'Hardware performance of L-band SAR system onboard ALOS-2'. *2011 IEEE International Geoscience and Remote Sensing Symposium*; 2011. pp. 894–7.

[41] Okada Y. 'System design of wide swath, high resolution, full polarimetric L-band SAR onboard ALOS-2'. *2013 IEEE International Geoscience and Remote Sensing Symposium–IGARSS*. 2013 IEEE International Geoscience and Remote Sensing Symposium - IGARSS; 2013. pp. 2408–11.

[42] Yokota Y., Okada Y., Nakamura S., *et al.* 'PALSAR-2 initial mission check 2014'. *IEEE Geoscience and Remote Sensing Symposium*. 2014, pp. 3413–4.

[43] Kankaku Y., Sagisaka M., Suzuki S., *et al.* 'PALSAR-2 Launch and early orbit status'. *IEEE Geoscience and Remote Sensing Symposium*. 2014, pp. 3410–2.

[44] Space Frequency Coordination Group. *Technical Information of EESS (Active) Sensors in the Band 1215–1300 MHz*. Tech. Rep. SFCG 37-2R1. SpaceFrequency Coordination Group (SFCG). 2018. Available from https://www.sfcgonline.org/Reports/REP%2037-2R1%20(Technical%20Info%20of%20EESS%20Active%20Sensors).pdf.

[45] McCandles S.W. 'The origin, evolution and legacy of SEASAT. *IGARSS 2003*'. *2003 IEEE International Geoscience and Remote Sensing Symposium. Proceedings*. 2003, vol. 1, pp. 32–4.

[46] NASA. *NASA Satellite to Study Earth's Oceans from Space*. 1978. Available from https://ntrs.nasa.gov/archive/nasa/casi.ntrs.nasa.gov/19780016311.pdf.

[47] JPL. *New Directions – Ocean and Land Observations*. 2015. Available from https. //www.jpl.nasa.gov/jplhistory/newdir/ocean-t.php.

[48] NASA/JPL-Caltech. *Sir-C in Assembly and Checkout*. 2014. Available from htt.//www2.jpl.nasa.gov/srtm/images/photos/srtm22hi.jpg.

[49] NASA/JPL-Caltech. *Sir-C in Space*. 2020. Available from https://www.jpl.nasa.gov/spaceimages/details.php?id=PIA01321.

[50] Hirsch V.A., Miers T.H., Coffin J.R., *et al.* 'Automated T/R module testing for the spaceborne imaging radar C (SIR-C) antenna subsystem'. *33rd ARFTG Conference Digest*. 1989, vol. 15, pp. 46–55.

[51] JAXA. *About Advanced Land Observing Satellite "DAICHI" (ALOS)*. 2020. Available from http://global.jaxa.jp/projects/sat/alos/.

[52] JAXA. *About Advanced Land Observing Satellite-2 "DAICHI-2" (ALOS-2)*. 2020. Available from http://global.jaxa.jp/projects/sat/alos2/.

Chapter 18
ESA and reflector comparison

This chapter discusses some of the design considerations for an L-band earth observation radar based on either an ESA or a reflector.

A variety of technical approaches, discussed in Section 17.5, have been proposed and/or built, so there is a large body of data available to provide insight into the design trades. Section 18.2 summarizes published design approaches, Section 18.3 provides an analysis of a representative reflector design, Section 18.4 provides an analysis of a comparable ESA design, and Section 18.5 discusses the benefits and detriments of the two approaches.

In 1982, Robert Mailloux analyzed ESA-fed reflectors (which he called hybrid antennas) in Chapter 5, "Hybrid antennas," of *The Handbook of Antenna Design* [1]. He noted that:

> Hybrid antennas would be unnecessary if phased arrays could be made very inexpensively. If the system designers' dream of a low-cost array with thousands of little elements, each costing a few dollars and controlled by some central processor had happened or would soon happen, there would be little need to expend much time or effort in the development of hybrid antennas.
>
> The key feature of such systems is that the objective must be large because the scanned array illuminates a spot that moves across the main aperture as a function of scan. This is a direct result of the need to use the whole array at all times ...; the economical choice is to use the array efficiently and the main aperture inefficiently.

Arguably, we have achieved the low-cost module postulated by Dr Mailloux 40 years ago.

18.1 ESA-fed reflector design approach

So far we have discussed ESA design challenges. An ESA-fed reflector experiences those challenges and as well as some unique challenges.

A major challenge is aperture efficiency defined in Section 4.2.1. As noted in Chapter 11, reflectors are not uniformly illuminated and so their aperture efficiency is less than unity. This is still the case with an ESA feed although the ESA provides more flexibility than a feed horn in devising optimal illumination of the reflector. As we saw in Chapter 10, however, shaped beams also have less than unity aperture efficiency. Accordingly, the challenge is to optimize the combination of ESA feed efficiency and reflector illumination efficiency. Typically, the solution is to use an oversize ESA to improve the beam shape and an oversize reflector to reduce the edge illumination, with both choices intended to reduce sidelobes and increase electronic scan capability.

A reflector provides a relatively small electronic field of regard relative to an ESA because the reflector magnifies the effective area of the ESA which is the desired effect but it also reduces the electronic steering volume by a similar ratio.

The ESA feed is smaller and has higher power density than the conventional ESA, increasing thermal challenges. The ESA-fed reflector has roughly the same power consumption and thermal dissipation as an ESA because module efficiencies are not strongly dependent on transmit power. A smaller, more powerful array used by the reflector concentrates the waste heat in a smaller volume and it may require a remote heat sink.

18.2 Reference designs

Several reflector-based L-band radar satellites have been described in the literature, notably DESDynI, NISAR, and TanDEM-L. JPL and DLR have developed system concepts for reflector-based systems. JPL proposed the DESDynI system which was incorporated in a NASA/ISRO partnership for an L-band and S-band satellite called NISAR. DLR proposed the TanDEM-L and is working with JAXA.

These designs transmit a beam with a large elevation coverage and receive a number of beams, higher gain, and smaller elevation coverage to increase swath width, as described in Section 17.2.2 and illustrated in Figure 17.8.

All of these designs describe beam defocusing which enlarges the power contours on the feed at extreme scan angles requiring a larger feed to capture sufficient power. The larger feed also allows a better conjugate field match for transmit. The focal plane power distribution was discussed in Section 11.6 and results from off-axis aberration of parabolic reflector.

Table 18.1 summarizes these three L-band designs and adds a fourth column for a notional ESA with comparable performance.

18.2.1 DESDynI

DESDynI (Deformation, Ecosystem Structure, and Dynamics of Ice) is a NASA/JPL concept for an L-band SAR system optimized for studying hazards and global environmental change. The DESDynI antenna is a 15 m reflector system with switchable beams. It has an offset feed with 24 elements located in the focal plane to support the wide swath.

It provides multiple polarization: single, dual, or fully polarimetric and strip-map or SweepSAR (SCORE) modes with a viewable swath of 340 km and 35 m ground resolution. The DESDynI project completed a preliminary design

Table 18.1 *L-band concept performance*

	Reflector systems			**ESA system**
	DESDynI*	**NISAR†**	**TanDEM-L‡**	**Concept §**
Altitude (km)	761	747	745	750
Frequency (GHz)	1.2575	1.2575	1.2575	1.2575
Bandwidth (MHz)	80	80	80	80
Peak transmit power (kW)	3.2	3	10.9	3.2
NESZ (spec; dB)	–35	< –20	–20 to –25	
Resolution (m)	3–100	3–10	1–10	
Incidence angle(°)	30–50	34–48	38–48	30–50
Electronic scan (el)	±9°	±7°	±8°	±8°
Electronic scan (az)	None	None	±2°	±9° or ±2°
Feed size or ESA size (m)	0.5 × 4.0	0.5 × 1.5	1.0 × 4.6	4.0 × 12.8
Reflector diameter (m)	15	12	15	–
Reflector focal length (m)	15	9	13.5	–
Number of modules	64	12	192	1,280
Element configuration	1 × 24	2 × 12	6 × 32	20 × 64
Number of elements			240	1,280
Number of phase centers (el)	32	12	32	20
Phase center spacing (el)	0.52λ	0.52λ	0.6λ	0.85λ
Number of phase centers (az)	2	2	6	64
Phase center spacing (az)	1.05λ	1.05λ	0.68λ	0.85λ
T/R module power (W)	50	250	57	2.5

*Rosen *et al.* [2, 3], Chamberlain [4, 5]

†Chamberlain and Vacchione [6], Xaypraseuth *et al.* [7], Rosen *et al.* [8, 9] Focardi and Vacchione [10]

‡Younis *et al.* [11], Huber *et al.* [12], Moreira *et al.* [13], Moreira and Krieger [14], and Huber *et al.* [15, 16]

§Section 18.4

review and subsequently has been incorporated into a joint project called NISAR (NASA/ISRO SAR) with the Indian space agency ISRO, .

18.2.2 NISAR

The NISAR project incorporates the designs and capability of DESDynI, albeit with a slightly smaller reflector (12 m instead of 15 m diameter), and adds a second frequency band (S-band) to the original L-band capability. The question of dual band systems, either reflector or ESA, is an interesting one and well covered in the literature but is not discussed here.

18.2.3 TanDEM-L

TanDEM-L is a DLR concept with some input from JAXA. It uses a 15 m reflector and an ESA feed with azimuth beam steering in addition to the elevation beam steering required for scan-on-receive. It plans two satellites flying in formation in order to obtain interferometric data.

18.3 Offset reflector model

Figure 18.1 shows two L-band antennas, one an ESA-fed reflector and the other an ESA, to the same scale. Figure 18.1a is a feed for a reflector and corresponds to the example shown in Figure 11.6. Figure 18.1b represents an equivalent ESA. In Section 18.4, we will derive additional array details to yield similar performance to the reflector. This section uses a design tool called GRASP, developed by the TICRA company, to assess the reflector performance.

Figure 18.2 shows two GRASP renderings of the feed and reflector. Design parameters are shown in Table 18.2. The parabola is constructed in an offset fashion such that the incident rays are not obstructed by the feed. Feed obstruction would represent a hole in the illumination function which would generate unwanted sidelobes and reduce aperture efficiency. Struts and other obstructions, which also tend to raise sidelobes, are omitted; the software readily supports struts and other realistic physical attributes of the antenna. These would alter the results somewhat but have been omitted here for simplicity.

The feed structure is critical to the function of the system and the number of elements in the feed is a critical design decision. The spacing must be close enough to adequately sample the beam as described in Section 11.5 and the number of modules must yield the total desired transmit power. The number of elements will ultimately be considerably more than shown, so the calculated T/R module power is an upper bound.

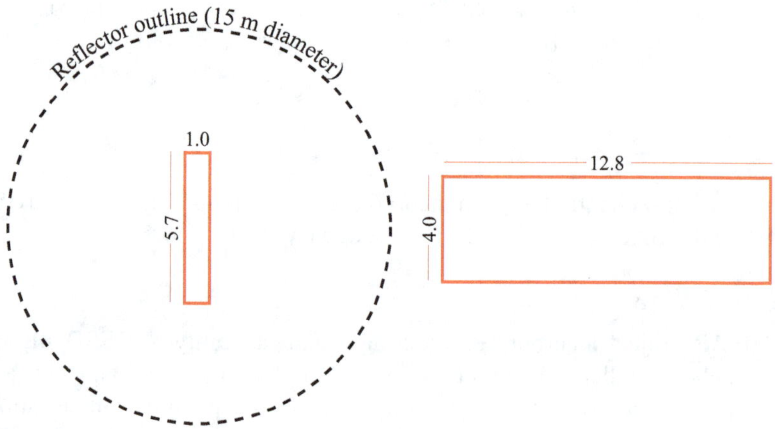

Figure 18.1 L-band arrays: (a) reflector case and (b) ESA case

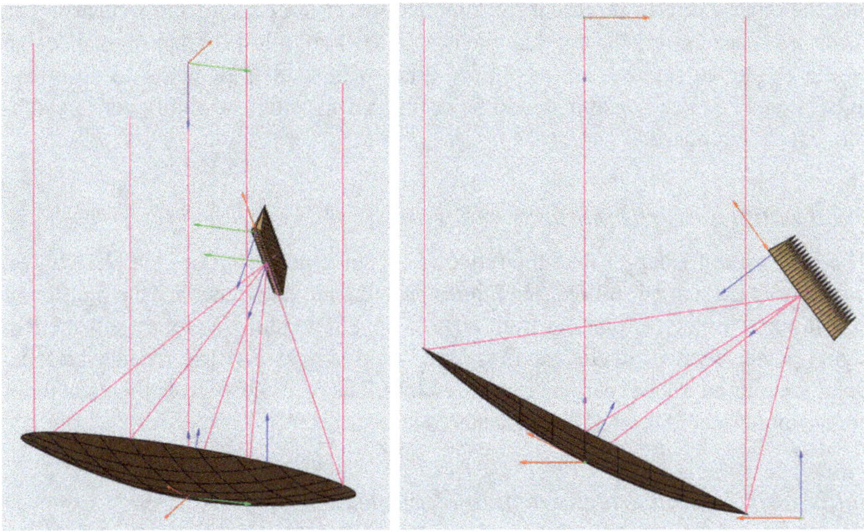

(a) (b)

Figure 18.2 GRASP rendering of offset reflector model: (a) isometric view and (b) Y–Z *view*

The reference designs incorporate feeds capable of both elevation and azimuth scanning. However, for this analysis, the feed has been simplified to a linear array of 24 separate elements located at the focal point to illustrate some basic points and to minimize computer run times. Since the array is one element wide, there is no azimuth scan capability. This analysis uses the GRASP feed horn with a Gaussian beam shape. The taper angle is the full angle of the beam and the taper value is the relative power at the edge of the taper. The taper angle is selected to fill the reflector

Table 18.2 Offset reflector model parameters

	Value
Reflector	
Reflector diameter (m)	15
Focal length (m)	10
Feed offset (m)	10
Feed array	
Number of elements	24
Element taper angle	40°
Element edge taper (dB)	–12
Element spacing (m)	0.2
Peak transmit power (kW)	3
Element power (W)	125

and the taper value to represent the illumination at the edge of the reflector. The feeds are spaced more closely than the size of the horn allows. The ultimate design would be implemented as an overlapped array with smaller elements, as described in Section 11.5. Feed design should be optimized for efficiency/spillover in a subsequent design phase.

Feed pattern overilluminates reflector

The feed pattern is larger than the reflector, so some power misses the reflector on transmit. This is called spillover. In Figure 18.3, the red line illustrates the fraction of signal power reflected, ranging from 80% to 86%. The blue line represents the lost signal power in terms of dB, ranging from –0.65 to 0.95 dB. The feed at the focal point is centered, so it experiences the least loss. The feeds away from the focal point are asymmetric and their spillover tends to increase.

Individual beam patterns principal planes cut

Figure 18.4 shows individual beam patterns in the elevation plane (blue) and the azimuth plane (red) for selected elements (1, 12, 13, and 24). The array feed is oriented in the elevation plane and the beam moves accordingly in elevation. The azimuth beam shape also varies with elevation scanning. The beam shapes represent a transformation of the 40° cone angle of the feed; the beam becomes narrower and the gain increases. The beam is well proportioned and the sidelobes are minimal. The fine structure in the far out sidelobes represents diffraction effects between the edges of the reflector.

The gain changes significantly as the beam is scanned, representing different aperture efficiencies for the different feed positions. The calculated beamwidths are shown in the bottom left corner of the figures.

The effective antenna dimensions shown in the bottom right corner of the figure represent the effective width and height of a rectangular source which would yield the same beamwidth, in effect the dimensions of an equivalent ESA. Near boresight we have an equivalent source dimension of a 12 m^2 source (144 m^2). The two extreme positions (1 and 24) have equivalent dimensions of 3.1 m × 6.8 m (21 m^2)

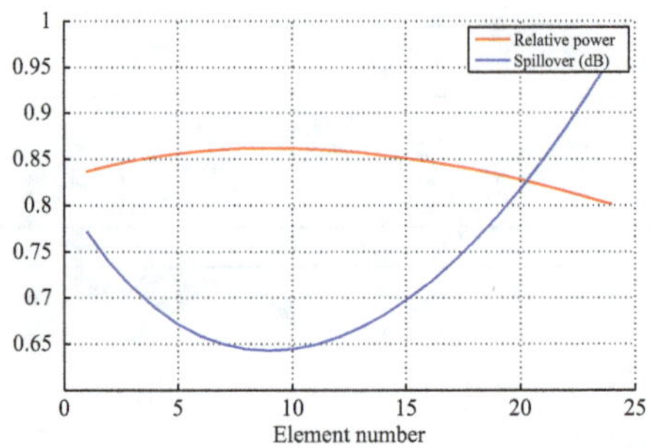

Figure 18.3 Feed pattern overilluminates reflector

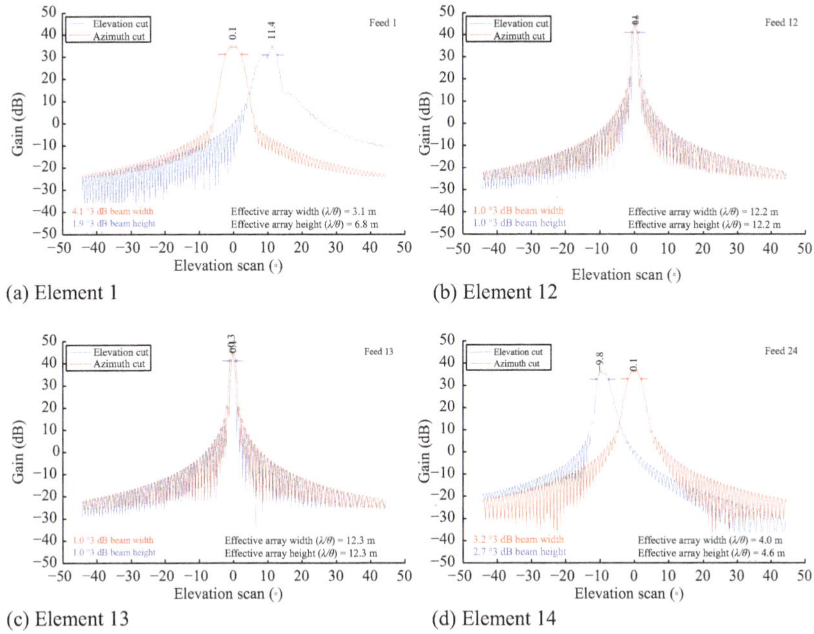

Figure 18.4 Individual element pattern cuts

and 4 m × 4.6 m (18 m^2) which are about 8 dB less than the center. Feed optimization would improve these results considerably as discussed in Section 11.6.

Transmit beam comprises sum of 24 feeds

Figure 18.5 shows the combined effect of all elements operating together and phase. It shows the individual beams as well as their sum which is several dB higher than the individual beams due to the significant beam overlap.

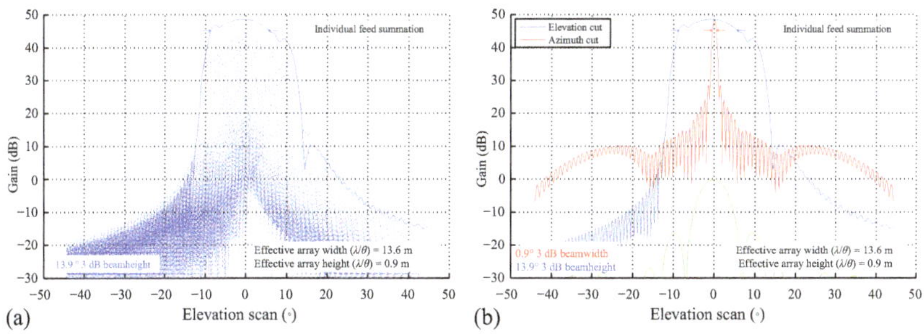

Figure 18.5 Transmit performance: (a) transmit beam sums all 24 elements and (b) principal plane pattern cuts

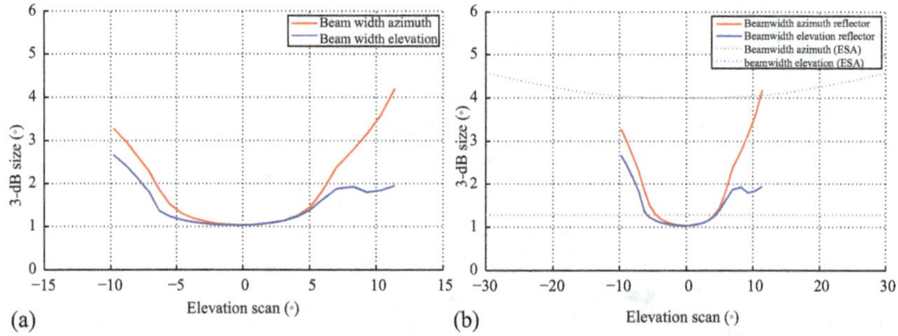

Figure 18.6 *Reflector and ESA beamwidth variation as a function of scan angle: (a) azimuth scan range ±10° and (b) azimuth scan range ±40°*

In SweepSAR mode, the system would transmit on all 24 elements and receive with one or more individual beams.

24 element sum principal plane cuts

Figure 18.5b shows the sum beam in the two principal planes. The elevation pattern is fan beam illuminating ±10° and a narrow azimuth pattern of roughly 1° full width at –3 dB. The azimuth beamwidth increases to almost 3° when scanned in azimuth, as shown in Figure 18.6. Figure 18.6a shows the performance of the reflector system. Figure 18.6b adds ESA scan performance which has much broader scan limits.

Reflector beam gain variation

Figure 18.7 shows the beam shape and peak gain for individual beams. Beam broadening and gain reduction are directly related. Figure 18.7a shows the reflector beam

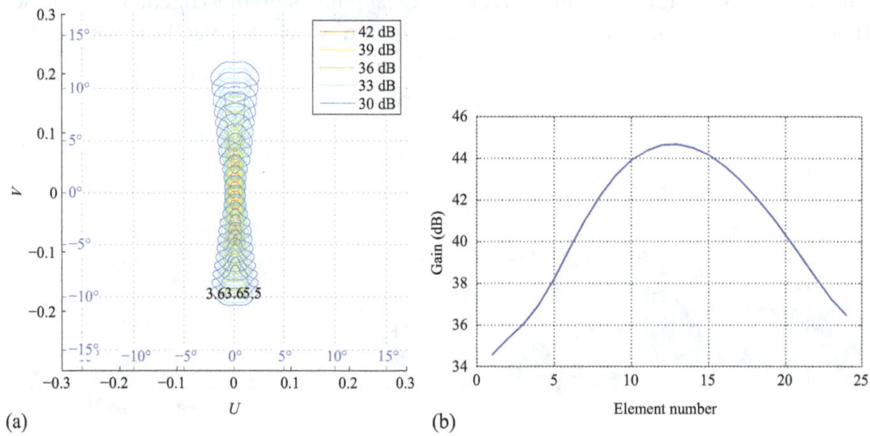

Figure 18.7 *Individual beam peak gain [fix y-axis]: (a) individual beam contours and (b) contours*

contours in azimuth–elevation ($u - v$) coordinates described in *It is sometimes convenient to use the coordinate transformation represented by (C.1) in Appendix C.1* and on the ground in latitude–longitude coordinates. Element gain ranges from 30 to 42 dB as shown in Figure 18.7b.

Equivalent aperture sizes for reflector

Another perspective is the equivalent aperture size. This was approximated based on beamwidth and equivalent aperture in Figures 18.4 and 18.5. GRASP (full version) provides another analytic technique. Electrical currents are induced in the reflector proportional to the tangential magnetic field at the surface. The currents generate the reflected electromagnetic fields. So, the current distribution represents the effective size of the reflector. The effective size is smaller than the physical size, corresponding to a reflector efficiency less than unity.

Figure 18.8 provides several views of the current for the transmit (sum) beam. Figure 18.8a and b shows the magnitude of the currents with the largest currents toward the center of the array. The 3 dB limits for the current are about 8.5 m in the azimuth direction and 3.5 m in the elevation direction. This compares with 13.5 m by 1 m in Figure 18.5.

Currents in reflector

Figure 18.8 shows the currents on an element-by-element basis and the illuminated portion moves across the reflector as different feeds are selected. The element spacing is 0.2 m or 83% of a wavelength (0.24 m). The transition to farfield for a single element is D^2/λ or 0.167 m. Accordingly, the reflector is in the farfield of a single element. Figure 18.8a and b shows the sum of currents for all of the feeds active. The illuminated area of the reflector is much smaller due to the summation of the individual feed currents. The smaller active area results in a wider main-beam as desired, which is the sum of the 24 individual beams.

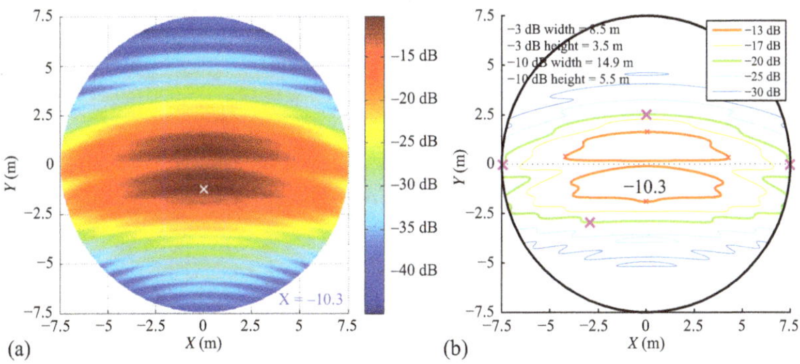

Figure 18.8 Transmit mode currents in reflector: (a) magnitude and (b) contours

The feed is 4.8 m × 0.4 m, so the farfield transitions (D^2/λ) are 96 and 0.67 m, respectively. The current distribution in the Y dimension ranges from 3.5 to 5.5 m for the –3 and –10 dB points, comparable to the Y dimension of the feed, while the current distribution in the X dimension (at the –10 dB point) is almost equal to the reflector diameter, as intended.

18.4 ESA conceptual design

Our design objective is to configure an ESA, whose gain, azimuth, and elevation scan limits and beamwidths and transmit power match the reflector performance. The array size is a fundamental configuration which together with lattice spacing and subarray size determines scan limits. The resulting design will resemble Figure 18.9.

18.4.1 Array lattice

Our scan limits are ±8° in elevation and ±8° in azimuth. We require a lattice with no grating lobes for our scan limits. The lattice spacing is determined by (18.1) which is derived from (7.2). Choosing $\theta = 10°$ for the onset of grating lobes, the lattice spacing should not exceed 0.85 λ. A lattice spacing of 0.20 m = 0.83 λ will be used:

$$d \le \frac{\lambda}{1 + \sin\theta} \tag{18.1}$$

18.4.2 Array height

In the reflector case described in Section 18.2, the elevation beamwidth in receive varies according to beam direction; it is 1° at boresight and 2.7° at the scan limits. For an ESA, beamwidth is relatively constant and primarily determined by the dimensions of the array.

The elevation beamwidth must discriminate against power from the adjacent pulses. Figure 18.10 shows a detail of Figure 17.8b and illustrates the problem. The transmit pulses and the nadir return are shown in green bars with 10% transmit duty. A 35° angle between the antenna boresight and satellite nadir, electronic scan of ±8°, and an altitude of 750 km equate to incidence angles between 30° and

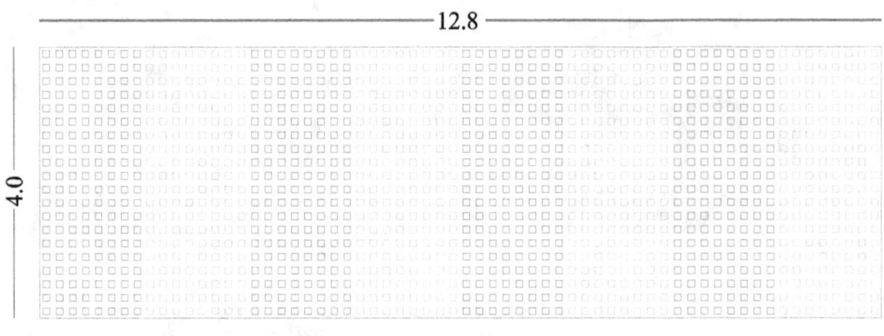

Figure 18.9 Conceptual ESA array

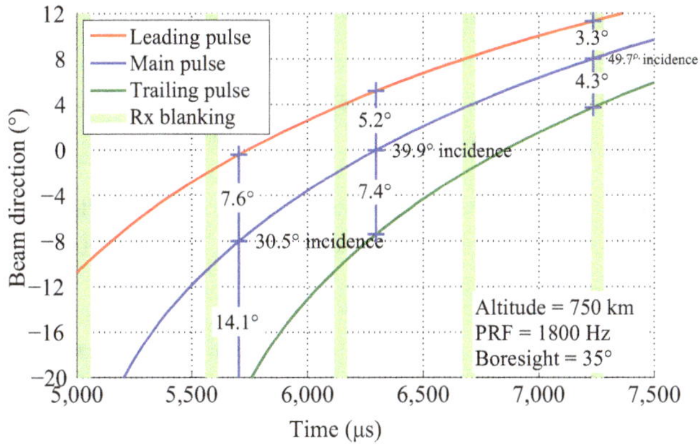

Figure 18.10 Elevation ambiguity
Geometry.m

50°. The illustration depicts the antenna beam direction (in antenna coordinates) of three received pulses transmitted at intervals of PRI. The main pulse is transmitted at time zero and the adjacent pulses are transmitted at ±PRI. SweepSAR operation requires that all three pulses be received simultaneously and separately. The number of simultaneous beams represents the factor by which the swath width is increased. There are three to five simultaneous returns within the electronic scan limits requiring a similar number of receivers.

At boresight, the leading and trailing pulses are separated by about 12° in elevation and at the upper scan limit, the separation is only about 7°. It appears that 3° elevation beamwidth may be inadequate to deal with range ambiguities at high incidence angles. A relatively small reduction in PRF (increasing the PRI) would probably suffice although this degrades the azimuth resolution because of the reduced Doppler sampling. It is also apparent that the Rx blanking hides a few ranges and PRF diversity is required to fill those gaps.

Using (18.2), derived from (2.17) with θ now redefined to represent the full angle ($\theta = 3°$ and $\lambda = 0.24$ m), requires an array height of about 4.0 m representing 20 elements:

$$\begin{aligned} \sin(\theta/2) &= \frac{0.443\lambda}{D} \\ \delta_r D &= \frac{0.443\lambda}{\sin(\theta/2)} \\ &\approx 17\lambda \end{aligned} \tag{18.2}$$

18.4.3 Beam weighting in elevation

As noted, the receive beams require weighting to minimize sidelobes. Figure 18.11 shows unspoiled and spoiled elevation beam patterns. These patterns represent only the array factor. The element factor would play an important role in decreasing the far out sidelobes.

The unweighted elevation beam is shown in Figure 18.11a. It has the usual −13 dB sidelobes. Because of the relatively large lattice, the sidelobes increase at the extreme scan angles, representing sidelobes of invisible grating lobes. It would be good practice to apply an amplitude weighting in elevation to reduce the sidelobes to −30 dB or better. Such weighting would decrease gain by about 0.7 dB and broaden the main beam to 3.8°. The beam broadening might require other mitigation such as a lower PRF at maximum range.

The transmit beam requires spoiling to illuminate the full ±8° swath. Figure 18.11a shows a pattern based on the quadratic phase weighting discussed in Section 10.3. There are ripples in the mainbeam and relatively high sidelobes, higher than for the unspoiled case. The element factor would reduce the sidelobes considerably but would not affect the mainlobe ripple. There is a reduction of gain of 7.4 dB compared to the uniform case. This is a factor of 5.5 which corresponds to the beam broadening ratio 18.9/3 = 6.3.

18.4.4 Array length

For SAR in stripmap mode, the best along-track resolution is one-half of the length of the physical array. In order to achieve this resolution, the PRF must be high enough to sample the full Doppler spectrum of the azimuth beamwidth. This relationship is discussed in Section 17.2.1 (17.3). For our case with an 80-MHz maximum bandwidth, (17.3) supports a best resolution of about 2 m, so our array should be 4 m in length. For our application, however, the resolution requirement is relaxed, so an array of 10–15 m length, comparable to the reflector, is more appropriate.

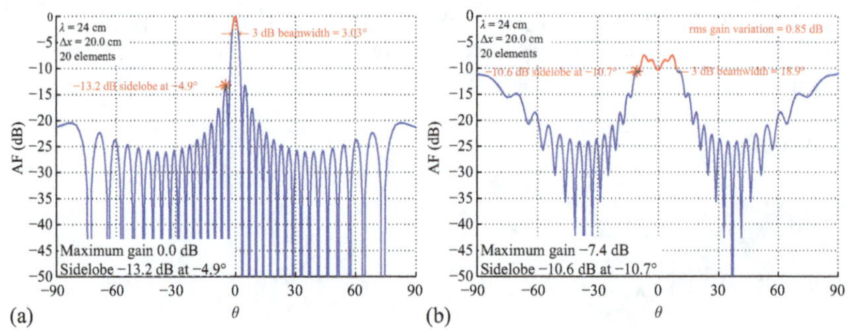

Figure 18.11 Elevation beam patterns: (a) uniform illumination elevation beam and (b) phased spoiled elevation beam
Quadratic2D.m

A second PRF consideration is to ensure that the nadir return from the transmitted pulse arrives during a subsequent transmission period as it is generally large enough to interfere with reception of desired signals. For our case, the round trip time is $2 \cdot h/c = 5$ ms. Accordingly, a PRF of 200 Hz (or any integral multiple) is suitable. For this analysis, we selected 1.8 kHz consistent with 4 m azimuth resolution in accordance with the Doppler criteria shown in Figure 17.5.

18.4.5 Unthinned ESA

With these calculations, we have established the array size and the lattice spacing of the elements providing the number of elements as well as the number of phase centers. There is one T/R module per element, so the transmit power of the module is simply the total power divided by the number of phase centers.

18.4.6 Subarray size

Using azimuth subarrays would reduce the array complexity and parts count, notably T/R modules. However, it also limits azimuth scan by introducing grating lobes. The choice requires a detailed system design to quantify the costs and benefits.

The necessary T/R module power is relatively modest and the reflector-fed systems already provide for T/R modules of up to 10 times the ESA module power. Accordingly, there is a straightforward path to higher ESA power, if the power supply and cooling provisions permit and if the system requires more performance.

While our comparison used a feed of 1×24 elements, the reference designs use feeds from 2×24 to 6×32, so the ESA T/R module count for a reflector feed is not insignificant. The relative array sizes are interesting. The feed array has an area of 5.7 m^2, while the ESA has an area of about 50 m^2, so the reflector:ESA ratio of module count, RF power density, and thermal power density are all 1:9.

From a physical design point of view, the subarrayed system is likely to require somewhat less mass than the full ESA but a detailed design is necessary to quantify the improvement. For space applications, it is necessary to fold the array for launch. The subarray boundaries are natural locations for the hinge and the subarray size may need to be adjusted to accommodate physical constraints.

We may consider a subarray in azimuth, elevation, or both but the subarray must have a beamwidth which encompasses the scan limits within its mainlobe, preferably with some margin against the onset of grating lobes. We established a scan range of $\pm 10°$ at the outset of this analysis, but the reference designs provide relatively little azimuth scan, so an azimuth subarray is an option. All of the L-band ESA satellite listed in Table 17.7 uses azimuth subarrays.

Choosing $\theta = 7°$ which provides margin to a $\pm 2°$ capability and using (2.17) and (17.3), we obtain a subarray width of about 1.64 m as shown in (18.3). With our lattice spacing of 0.2 m, an eight-element subarray is the best choice and requires eight subarrays for our array. This reduces our module quantity and simplifies somewhat the associated cables and feeds:

$$\sin(\theta/2) = \frac{0.443\,\lambda}{D}$$

$$\begin{aligned} D &= \frac{0.443\,\lambda}{\sin(\theta/2)} \\ &\approx 1.75\,\text{m} \end{aligned} \qquad (18.3)$$

Subarrays facilitate the use of time delay for beam steering by reducing the number of time delay units required. The time delay reduces the grating lobe issue and also beam squint with frequency, both useful if azimuth scan is a priority. Section 2.7 provides more information.

Table 18.3 summarizes our preliminary design which represents an ESA with similar performance to the reference reflector design. There are two configuration choices, one with no azimuth subarrays and wide azimuth scan capability and the other with azimuth subarrays and restricted azimuth scan capability.

18.4.7 Power and noise figure

Transmit power and noise figure determine the system NESZ (discussed in section 4.5.1) but are independent of the other parameters of the ESA design although they affect power and cooling as well as system availability. Based on the reference designs, it is clear that the required values are not difficult to achieve.

18.4.8 Beam projection

Figure 18.12 shows the ESA beam projection in the formats used previously. The beam on the ground is indicated by a red ellipse and is about 18 km wide and 74 km long. It is surrounded by other beam positions indicated by black ellipses. The red outline rectangle in Figure 18.12d corresponds to the ground portion which falls within an unambiguous swath and unambiguous azimuth Doppler bandwidth at a PRF of 2,400 Hz. The 3 dB falls within the unambiguous rectangle with some margin, validating the choice of ESA dimensions.

Table 18.3 L-band ESA parameters

ESA parameters	No subarrays	Azimuth subarrays
ESA width × height (m)	12.8 × 4.0	
Lattice spacing (m)	0.2	
Elements	64 × 20	
Subarray size	–	8 × 1
Phase centers	64 × 24	8 × 20
Number of T/R modules	1,280	160
T/R module peak power (W)	2.5	20

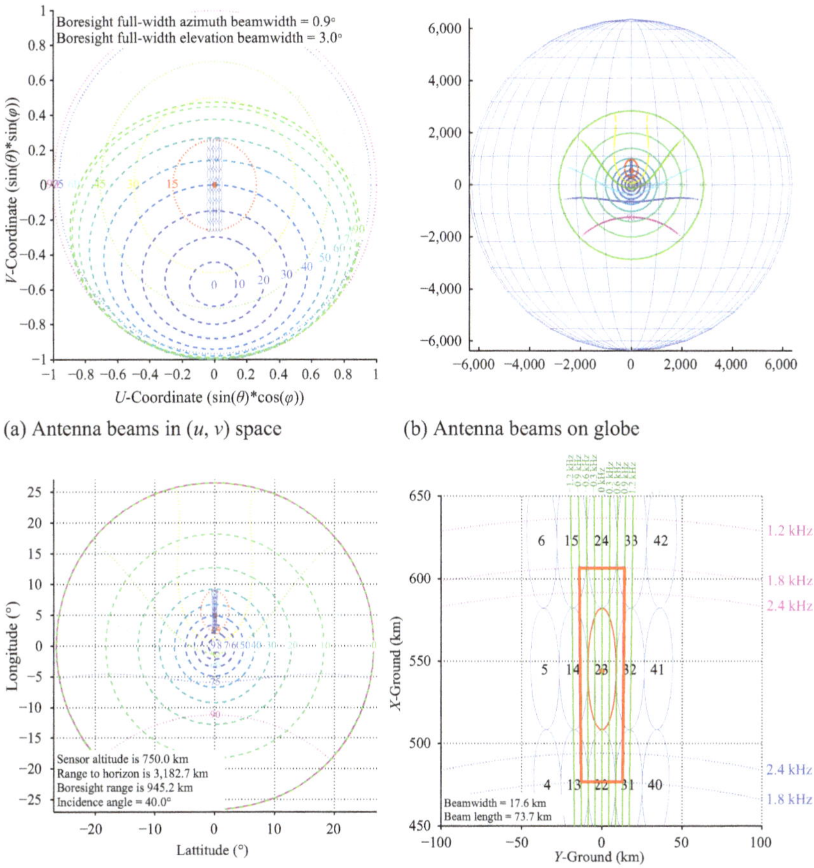

(a) Antenna beams in (u, v) space

(b) Antenna beams on globe

(c) Antenna beams on globe (detail)

(d) Antenna beams projected to ground

Figure 18.12 ESA beam projection
BeamProjection.m

18.4.9 Subarray effects

Figure 18.13 illustrates the array grating lobes and the subarray pattern which suppresses grating lobes on boresight.

18.4.10 Antenna pattern – boresight and steered

Figure 18.14 shows the product of the array and subarray patterns with patterns plotted in *UV* space. On boresight there are no grating lobes but these appear as we scan in elevation and azimuth. The boresight case (Figure 18.14a) and the elevation scan case (Figure 18.14c) show that the grating lobes are suppressed. For the azimuth scan case, Figure 18.14b shows that the grating lobes are a significant limitation on scan.

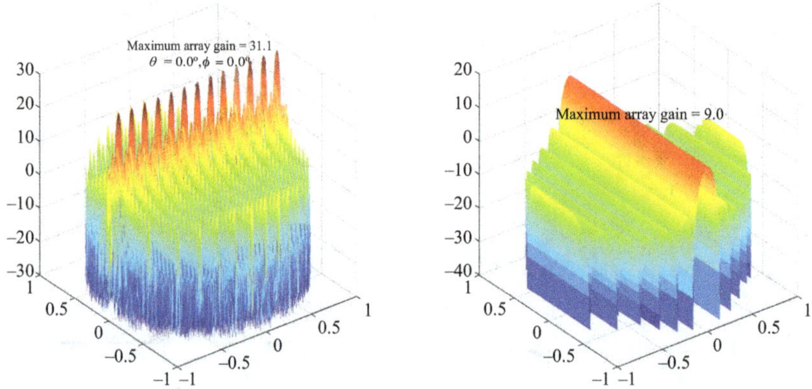

(a) (b)

Figure 18.13 Subarray technique introduces grating lobes: (a) array pattern and (b) subarray pattern
L-BandArray.m

To select cardinal planes, we note that $\phi = 0°$ is the azimuth scan plane and $\phi = 90°$ is the elevation scan plane. A principal plane cut shows the grating lobe onset more clearly. On boresight, the subarray (eight element) suppresses the array factor grating lobes but as the scan increases up to 4°, the array factor grating lobes move while the element factor nulls are stationary so the grating lobes pop up out of the nulls.

18.4.11 Azimuth cut steered in azimuth

Figure 18.15 shows the azimuth plane for three values of azimuth scan.

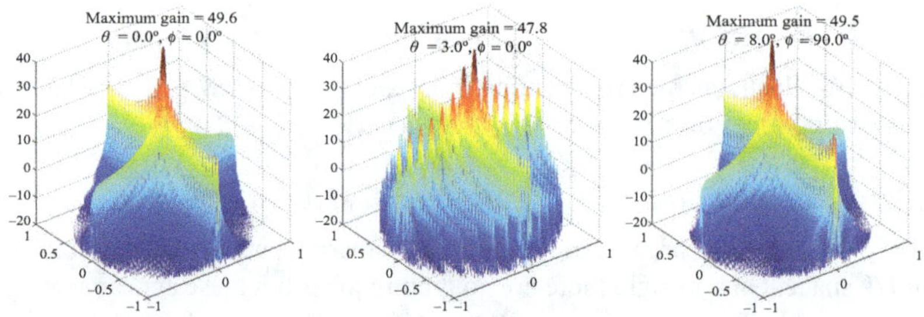

(a) Boresight (b) $(\theta, \phi) = (3°, 0°)$ (c) $(\theta, \phi) = (8°, 90°)$

Figure 18.14 Antenna pattern boresight and steered
L-BandArray.m

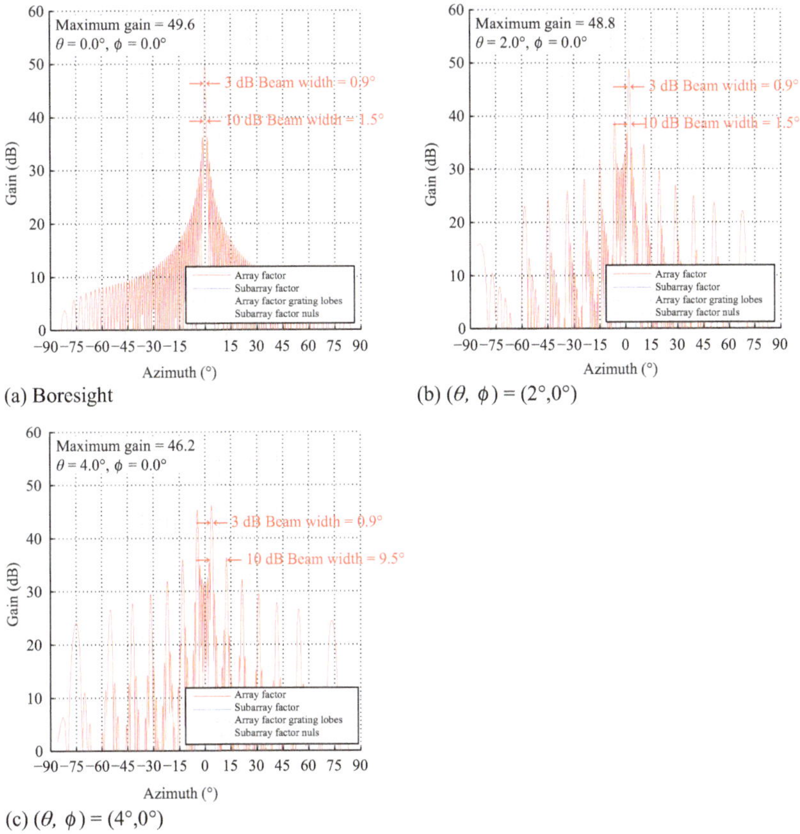

Figure 18.15 Azimuth cut pattern boresight and steered
L-BandArray.m

18.4.12 Elevation cut steered in elevation

Figure 18.16 shows the elevation plane for three values of elevation scan. In the elevation plane, the expected grating lobe starts to emerge at 8° in accordance with our lattice spacing choice.

18.4.13 ESA beamwidth fairly constant with scan

Elevation and azimuth beamwidths change with $\cos^{-1}\theta$. Figures 18.15 and 18.16 show the beamwidths for each scan angle. Figure 18.17 plots them separately as a function of scan angle. The azimuth scan is so small, hence very little effect is seen. The elevation scan follows an inverse $\cos\theta$ relationship.

18.4.14 Gain as a function of scan

Figure 18.18 shows that the gain falloff follows the subarray pattern in azimuth and the element pattern in elevation. The elevation pattern falloff is about 2 dB at 40°.

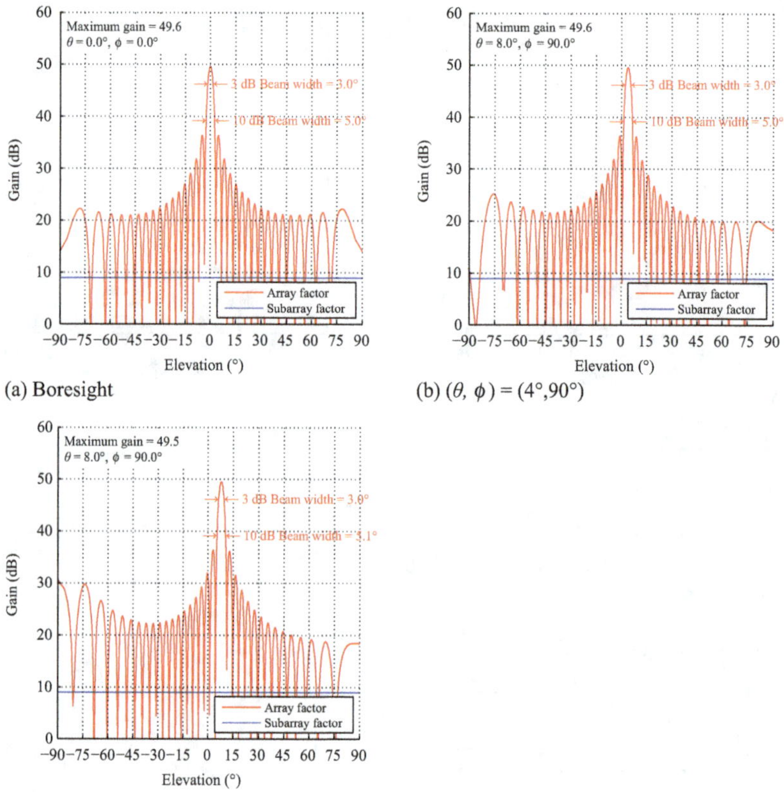

(a) Boresight

(b) $(\theta, \phi) = (4°, 90°)$

(c) $(\theta, \phi) = (8°, 90°)$

Figure 18.16 Elevation cut pattern boresight and steered
L-BandArray.m

18.5 Comparison

This comparison has shown some of the attributes and tradeoffs of the two approaches. Each approach has advantages and disadvantages but one important aspect was not included – the technology base. Costs are driven in large part by development, so existing technologies and experience may determine the choice. Technology continues to improve, based mostly on commercial products, so the assessed cost of both will continue to decrease and most likely both will be produced and used.

Both approaches can satisfy system requirements and can be tailored for better performance. For the two cases examined, the primary difference is the reflector feed and the ESA which differ in size (area) by a factor of 9. The reflector feed considered was a column feed of 24 elements but a better design would incorporate more columns (between 2 and 6) and more rows, e.g., 32, so the module comparison would be between 48 and 192 for the reflector feed and 160 for the ESA.

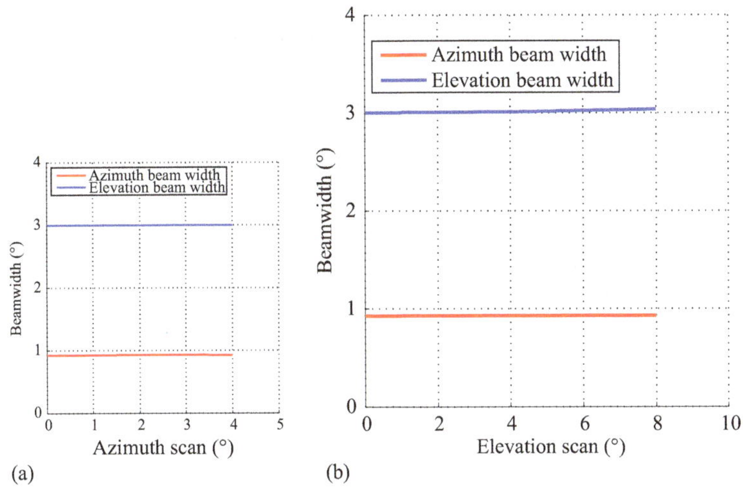

Figure 18.17 *Beamwidth as a function of scan: (a) azimuth scan and (b) elevation scan*
L-BandArray.m

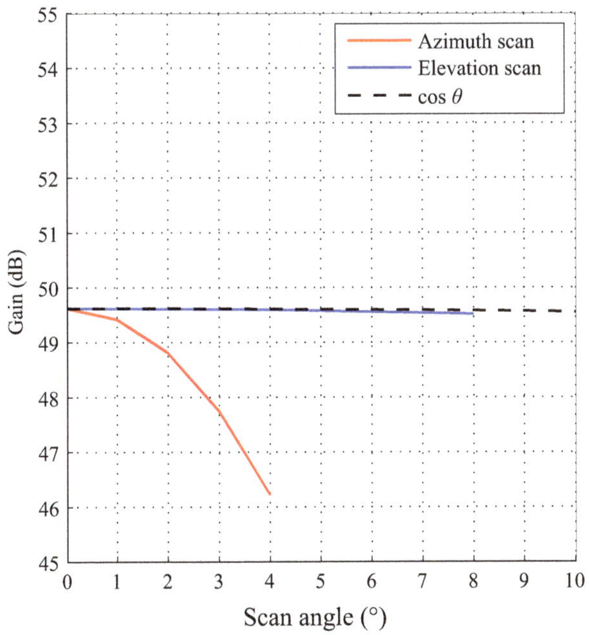

Figure 18.18 *Gain as a function of scan*
L-BandArray.m

Thus, the primary difference becomes the physical size. Reflectors are a mature technology with several commercial suppliers. The ESA requires folding for launch and could be constructed as eight panels, each 1.6 m × 4.0 m in size comprising eight phase centers in azimuth. This is slightly larger than SAOCOM (see section 17.5.4) which is constructed as seven panels, each 1.4 m × 3.5 m in size and uses 160 T/R modules. SAOCOM has seven phase centers in azimuth. ALOS-2 (see section 17.5.3) is constructed with five panels, each approximately 2.0 m × 2.9 m with 180 T/R modules. ALOS-2 has 10 phase centers in azimuth, producing roughly twice the azimuth scan capability of the other systems. So, the proposed parameters have been largely demonstrated with prior programs.

18.5.1 Array

The array length of ~10 m is compatible with azimuth resolution of ~3–10 m.

The array height of 4 m provides a high-gain Rx beam comparable to the reflector. Both designs require elevation spoiling for transmit and the elevation beam has some gain variation. More analysis is required to establish the requirement for the elevation beam and determine if it can be met. Elevation beam agility is required to implement the wide swath SAR modes. Azimuth beam agility enables spotlight modes and other modes such as TOPSAR. ESAs provide greater azimuth steering.

18.5.2 Feeds

The reflector feed design to optimize reflector efficiency requires multiple elements per beam with amplitude and phase control to produce a focused beam while the ESA is relatively straightforward. Array feeds have considerably higher power density than ESAs complicating cooling. The larger number of T/R modules in an ESA provides additional degrees of freedom supporting advanced beam concepts.

18.5.3 Launch constraints

For satellite applications, reflector antennas are more amenable to folding required for launch and provide a lower mass density for larger antenna, providing higher gain in receive. ESA antennas up to 3.5 m × 10 m have been demonstrated on orbit. Both configurations have been proved in use.

18.6 Summary

There does not appear to be a conclusive advantage for ESA or ESA-fed reflectors for the earth observation mission. At X-band, ESAs seem to be the dominant choice, in part because the antennas are smaller and also because of the huge technology base developed for airborne and ground-based radars. At lower frequencies, the reflector benefits are more compelling as the antennas are larger. In many cases, the experience base at the manufacturer will determine the outcome for cost and risk benefits. Relative merits will continue to evolve.

References

[1] Mailloux R.J. 'Hybrid antennas' in Rudge A.W., Milne K., Olver A.D., Knight P. (eds.). *The Handbook of Antenna Design*. London, UK: The Institution of Engineering and Technology; 1982. pp. 415–63.

[2] Rosen P., Thomson M., Jordan R., *et al.* DESDynI SAR/INSAR'. *2010 UNAVCO Science Workshop*. 2010. Available from http://www.unavco.org/community/meetings-events/2010/sciworkshop10/presentations/UNVSW10/31010/1.3/PlenaryRosen/NextGenerationGeodeticImaging.pdf.

[3] Rosen P., Thomson M., Jordan R., *et al.* 'DESDynI SAR/INSAR'. *DESDynI Science Study Working Group Meeting*. 2008. Available from http://desdyni.jpl.nasa.gov/files/DESDynI/RadarDes/&PerfV4a.pdf%20[bad%20link].

[4] Chamberlain N. 'The DESDynI synthetic aperture radar array-fed reflector antenna'. *2010 IEEE International Symposium on Phased Array Systems and Technology*; 2010. pp. 381–6.

[5] Chamberlain N. 'The proposed DESDynI array-fed reflector feed'. *Proceedings of the 2012 IEEE International Symposium on Antennas and Propagation*; 2012. pp. 1–2.

[6] Chamberlain N., Vacchione J. 'Electromagnetic modeling of the proposed DESDynI synthetic aperture radar antenna'. *2014 IEEE Aerospace Conference*; 2014. pp. 1–14.

[7] Xaypraseuth P., Satish R., Chatterjee A. 'NISAR spacecraft concept overview: Design challenges for a proposed flagship dual-frequency SAR mission'. *2015 IEEE Aerospace Conference*; 2015. pp. 1–11.

[8] Rosen P. 'An update on the NASA-ISRO dual-frequency DBF SAR (NISAR) mission'. *2016 IEEE International Geoscience and Remote Sensing Symposium (IGARSS)*; 2016. pp. 2106–8.

[9] Focardi P., Vacchione J.D. 'NISAR flight feed assembly: evolution of the design from initial concept to final configuration'. *2019 13th European Conference on Antennas and Propagation (EuCAP)*; 2019. pp. 1–4.

[10] Younis M. 'Tandem-L instrument design and SAR performance overview'. *2014 IEEE Geoscience and Remote Sensing Symposium*; 2014. pp. 88–91.

[11] Rosen P. 'The NASA-ISRO SAR (NISAR) mission dual-band radar instrument preliminary design'. *2017 IEEE International Geoscience and Remote Sensing Symposium (IGARSS)*; 2017. pp. 3832–5.

[12] Huber S., Younis M., Krieger G., *et al.* 'Advanced spaceborne SAR systems with array-fed reflector antennas'. *2015 IEEE Radar Conference (RadarCon)*. 2015. Available from https://elib.dlr.de/95385/1/Huber2015RadarConpaper.pdf.

[13] Moreira A. 'ALOS-Next/TanDEM-L: a highly innovative SAR mission for global observation of dynamic processes on the earth's surface'. *2015 IEEE International Geoscience and Remote Sensing Symposium (IGARSS)*; 2015. pp. 1253–6.

[14] Moreira A., Krieger G. *Tandem-L – Satellite Mission Proposal for Monitoring Dynamic Processes on the Earth's Surface*. 2016. Available from http://www.dlr.de/dlr/en/Portaldata/1/Resources/docments/2016/Tandem-L-Brochure-2016.06.pdf.

[15] Huber S. 'Tandem-L: design concepts for a next-generation spaceborne sar system'. *Proceedings of EUSAR 2016: 11th European Conference on Synthetic Aperture Radar*; 2016. pp. 1–5.

[16] Huber S., Younis M., Krieger G. 'Tandem-L: SAR system design aspects'. *5th Workshop on Advanced RF Sensors and Remote Sensing Instruments, ARSI'17*. DLR. 2017. Available from https://elib.dlr.de/113660/1/ARSI2017paper.pdf.

Appendix A
Further reading

A.1 Books

- "Phased array radar antennas," Joe Frank and John D. Richards [1], Chapter 13 [Skolnik]
- *Antenna Theory: Analysis and Design*, Constantine Balanis [2]
- *Phased Array Antenna Handbook*, Robert J. Mailloux [3]
- *Introduction to Airborne Radar*, George W. Stimson [4]
- *Phased Array Antennas*, R.C. Hansen [5]
- *Electronically Scanned Arrays MATLAB® Modeling and Simulation*, Arik D. Brown [6]
- *Practical Phased Array Antenna Systems*, Eli Brookner [7]
- "Reflector antennas," Michael Cooley and Daniel Davis [8], Chapter 12 [Skolnik]
- *Electronically Scanned Arrays*, R.J. Mailloux [9]
- *Radar Handbook*, Third Edition, Merrill Skolnik [10]
- *Antenna Handbook: Theory, Applications, and Design*, Y.T. Lo and S.W. Lee [11]
- *Antenna Arrays: A Computational Approach*, Randy L. Haupt [12]

A.2 Web based

- *EW and Radar Handbook* [13]
- Dr. David C. Jenn lecture slides and MATLAB® code [14]
- Jet Propulsion Laboratories, http://southport.jpl.nasa.gov/
- Microwaves 101 [15]
- *Electromagnetic Waves and Antennas*, Sophocles J. Orfanidis [16]
- *Notes on Antenna Engineering 2018,* Natalia K. Nikolova [17]

References

[1] Frank J., Richards J.D. 'Phased array radar antennas' in Skolnik M. (ed.). *Radar Handbook*. New York, NY, USA: McGraw-Hill Education; 2008. pp. 13.1–13.74.

[2] Balanis C.A. *Antenna Theory: Analysis and Design*. 4th edn. New York, NY, USA: Wiley-Interscience; 2016.

[3] Mailloux R.J. *Phased Array Antenna Handbook*. 3rd edn. Boston, MA, USA: Artech House; 2018.

[4] Stimson G.W. *Introduction to Airborne Radar*. 2nd edn. Raleigh, NC, USA: SciTech Publishing; 1998.

[5] Hansen R.C. *Phased Array Antennas*. 2nd edn. New York, NY, USA: John Wiley & Sons, Inc; 2009.

[6] Brown A.D. *Electronically Scanned Arrays MATLAB® Modeling and Simulation*. Boca Ratan, FL, USA: CRC Press; 2012.

[7] Brookner E. 'Antenna array fundamentals: Part 2' in Brookner E. (ed.). *Practical Phased Array Antenna Systems*. Chapter 13. Norwood, MA, USA: Artech House; 1991. pp. 1–28.

[8] Cooley M., Davis D. 'Reflector antennas' in Skolnik M. (ed.). *Radar Handbook*. Chapter 12. New York, NY, USA: McGraw-Hill Education; 2008. pp. 12.1–12.43.

[9] Mailloux R.J. *Electronically Scanned Arrays*. 1st edn. San Rafael, CA, USA: Morgan & Claypool Publishers; 2007.

[10] Skolnik M. *Radar Handbook*. 3rd edn. New York, NY, USA: McGraw-Hill Education; 2008.

[11] Lo Y.T., Lee S.W. 'Antenna handbook: theory, applications, and design' in Lo Y.T., Lee S.W. (eds.). *Antenna Handbook: Theory, Applications, and Design*. New York, NY, USA: Van Nostrand Reinhold Company; 1988.

[12] Haupt R.L. *Antenna Arrays: A Computational Approach*. Hoboken, NJ, USA: John Wiley; 2010.

[13] EW and Radar Handbook. Available from https://ewhdbks.mugu.navy.mil/home.htm.

[14] Jenn D.C. *Faculty at the Naval Postgraduate School*. 2018. Available from http://faculty.nps.edu/jenn/.

[15] Anonymous. *Microwave and RF Information for Engineers — Microwave Calculators, Encyclopedia,* Discussion Forum. 2018. Available from http://www.microwaves101.com/index.cfm.

[16] Orfanidis S.J. *Electromagnetic Waves and Antennas*. 2018. Available from http://www.ece.rutgers.edu/Orfanidi/ewa.

[17] Nikolova N.K. *Notes on Antenna Engineering*. McMaster University, Canada. 2018. Available from http://www.ece.mcmaster.ca/faculty/nikolova/antennadload/currentlectures/LectureNotesAntennasNikolov.pdf.

Appendix B

Comments on MATLAB® code

MATLAB® scripts were used to create most of the figures in this book and for the associated short course. The name of the script is shown beneath its associated figure. The programs are available for download from the publisher's website at https://digital-library.theiet.org/.

The scripts do not include a user interface or extensive error checking. Parameters are hard-coded but may be altered to assess other cases. The underlying engineering formulae represent relatively few lines of the program; much of the rest relates to graphical output and format. The program run times range from seconds to hours. Run times depend, in large part, on the size of arrays and the number of iterations (cases).

The code was developed with a specific hardware and software configuration. Your configuration may differ. Accordingly, some adjustment may be necessary. The 64-bit version of MATLAB® is beneficial for scripts using large arrays. In general, toolboxes are not required but the Chebyshev and Taylor weighting functions contained in the Signal Processing Toolbox® are useful. In its absence, Orfanidis [1] provides equivalent functions which may be substituted. Note that his input parameter definition differ somewhat. Results from his code match those of MATLAB® for values of M exceeding 100.

Several scripts produce a time series of figures which may be combined to create animations. The open source program *ffmpeg.exe* creates avi files from saved frames. This program is available from https://www.ffmpeg.org/ and the associated scripts describe its use.

Several functions support the scripts and should be copied to a directory on your path:

- *ISLR.m*
- *TotalPower.m*
- *arrowhead.m*
- *cisi.m*
- *echelon.m*
- *Sat2EarthTransform.m*
- *saveaspdf.m*
- *showbeamwidth.m*
- *showsidelobes.m*

The remaining programs are best copied into unique directories, as they create intermediate and final files that may be numerous and large. Each script creates and uses a subdirectory for output. Output file names are encoded in the scripts.

The saveaspdf function is called with several parameters: *(handle, savefilename, newboundingbox, savefiles)*. The *handle* parameter is the handle associated with the figure of interest, *savefilename* is the destination file name without extension, *newboundingbox* is an optional parameter to resize the pdf output (default is no change), and *savefiles* is an optional parameter to specify output format (default is pdf). The available output file formats are as follows:

- 1 – pdf only
- 2 – pdf and png
- 3 – pdf and emf (transparent emf)
- 4 – pdf and emf (opaque emf)

The pdf output format is created from the native MATLAB® encapsulated postscript output using the executable program *epstopdf.exe* distributed with TeX and run in a DOS shell.

B.1 Web resources

Numerous authors offer downloadable MATLAB® scripts for radar analysis. Table B.1 lists some resources that are readily available.

Table B.1 MATLAB® resources for radar analysis

Author	Title
Sophocles J. Orfanidis [1]	*Electromagnetic Waves and Antennas*
Natalia K. Nikolova [2]	*Modern Antennas in Wireless Telecommunications*
Constantine Balanis [3]	*Antenna Theory: Analysis and Design*, fourth edition
Svante Björklund and David Rejdemyhr [4]	*DBT Release 2.14. A MATLAB® Toolbox for Radar Array Processing*
Arik D. Brown [5]	*Electronically Scanned Arrays MATLAB® Modeling and Simulation*
Thomas W. Jeffrey [6]	*Phased-Array Radar Design: Application of Radar Fundamentals*
Bassem R. Mahafza [7]	*Radar Systems Analysis and Design Using MATLAB®*
Mehrdad Soumekh [8]	*Synthetic Aperture Radar Signal Processing*

Source: as noted

References

[1] Orfanidis S.J. *Electromagnetic Waves and Antennas.* Rutgers University. August 2016. Available from http://eceweb1.rutgers.edu/Orfanidi/ewa/ewa-1up.pdf.

[2] Nikolova N.K. *Notes on Antenna Engineering*. McMaster University, Canada. 2018. Available from http://www.ece.mcmaster.ca/faculty/nikolova/antennad-load/currentlectures/LectureNoteAntennasNikolova.pdf.

[3] Balanis C.A. *Antenna Theory: Analysis and Design*. 4th edn. New York, NY, USA: Wiley-Interscience; 2016.

[4] Björklund S., Rejdemyhr D. *DBT release 2.156. A MATLAB Toolbox for Modeling, Simulation & Signal Processing in Radar*. 2012. Available from http://denver.bth.se/ing/wideband/sar.nsf/attachment/FOI-D–448/1/pdf//$file/FOI-D–448/1.pdf.

[5] Brown A.D. *Electronically Scanned Arrays MATLAB® Modeling and Simulation*. Boca Ratan, FL, USA: CRC Press; 2012.

[6] Jeffrey T.W. *Phased-Array Radar Design: Application of Radar Fundamentals*. Raleigh, NC, USA: Scitech Publishing; 2009.

[7] Mahafza B.R. *Radar Systems Analysis and Design Using MATLAB®*. Boca Raton, FL, USA: Chapman & Hall/CRC Press; 2000.

[8] Soumekh M. *Synthetic Aperture Radar Signal Processing with MATLAB® Algorithms*. New York, NY, USA: Wiley-Interscience; 1999.

Appendix C
Geometry

C.1 Coordinate systems

Antenna analysis uses polar coordinates (θ, ϕ) as illustrated in Figure C.1.

- Plus z-direction is normal to the array face.
- Theta (θ) is measured relative to the $+z$-axis.
- Phi (ϕ) is measured in the xy-plane relative to the x-axis.
- Array is represented by the lattice of circles in the xy-plane.

The array is shown in green. It lies on the xy-plane ($z = 0$) and radiates in the $+z$-direction (hemispheric). Radiation in the $-z$-direction is usually undesirable and is prevented by a ground plane in the plane of the array. Generally, the ground plane need extend only a few wavelengths away from the array.

For analysis purposes, these coordinates may be used directly or converted to an orthogonal representation termed (u, v) according to (C.1):

$$u = \sin\theta \cos\phi$$
$$v = \sin\theta \sin\phi$$
$$w = \cos\theta \qquad \text{(C.1)}$$

Both coordinate systems are illustrated in Figure C.2. The hemisphere is partitioned into quadrilateral facets (Figure C.2a and C.2b) whose edges align with the coordinate system. The (θ, ϕ) system includes a third coordinate, ρ, the radius, whose value is constant (1). The (u, v) system includes a third coordinate, w, whose value varies between 0 and 1, such that $u^2 + v^2 + w^2 = 1$.

Both coordinate systems are convenient for numerical analysis such as integration over a hemisphere. Figure C.2c and C.2d show a single facet used to compute the solid angle associated with a grid point. The facet solid angle is not constant in either coordinate system. Each system is rotationally symmetric, while the solid angle has a θ dependence shown in (C.2) and graphically in Figure C.3:

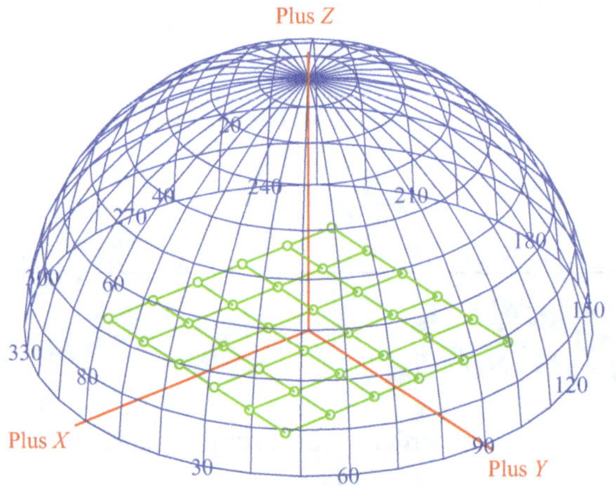

Figure C.1 Spherical coordinate system
CoordinateSystems.m

$$\text{Facet size }(\theta,\phi) = \sin\theta\Delta\theta\Delta\phi$$
$$\text{Facet size }(u,v) = \frac{\Delta u\Delta v}{\cos\theta} \tag{C.2}$$

The (θ,ϕ) coordinate system maps the scan volume exactly and all pointing directions are meaningful. The size of the facet solid angle varies considerably and approaches zero on boresight which provides excessive resolution near boresight and increases the size of the mesh for numeric integration. The (u,v) coordinate system maps approximately 21% of the resolution cells outside the scan volume, representing unavailable scan directions, which also increases the size of the mesh for numeric integration. The size of the resolution cell is relatively constant in the central scan volume.

The $\sin\theta$ factor resulting from concentric rings whose circumference is proportional to $\sin\theta$ is constant so that the number of cells is the same for all rings. The (u,v) case is perhaps less obvious. The resolution cell size is elongated by the slope of the cell which is equal to the inverse of $\cos\theta = w$. Figure C.3 illustrates both cases:

$$\text{Area} = \frac{1}{2}\left|D_1 \times D_2\right| \tag{C.3}$$

To compare the two approaches, we compare the number of facets for the case where the facet size is equal for $\theta = 45°$. Equation (C.4) shows this calculation:

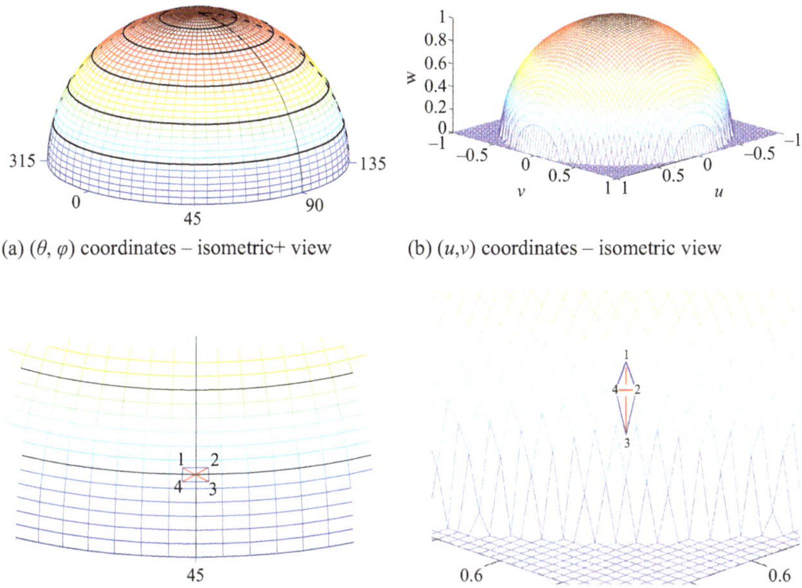

(a) (θ, φ) coordinates – isometric+ view

(b) (u,v) coordinates – isometric view

(c) (θ, φ) coordinates – facet detail

(d) (u,v) coordinates – facet detail

Figure C.2 Two useful systems defining pointing direction
CoordinateSystems.m

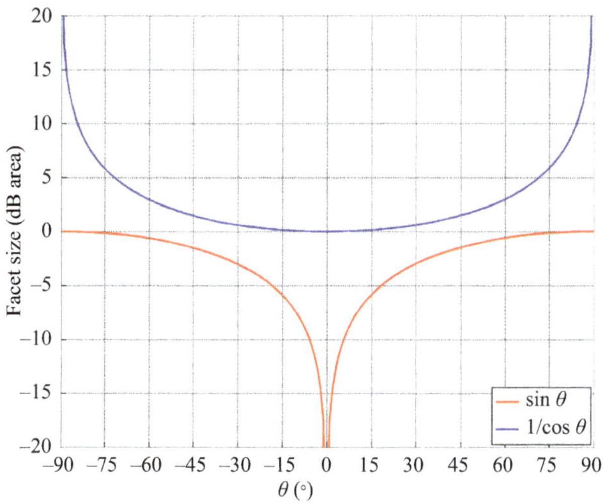

Figure C.3 Resolution cell size
CoordinateSystems.m

$$\sin\theta\ \Delta\theta\ \Delta\phi = \frac{\Delta u \Delta v}{\cos\theta}$$

$$\Delta\theta\ \Delta\phi = \frac{\Delta u \Delta v}{\sin\theta\cos\theta} \tag{C.4}$$

$$\Delta\theta\ \Delta\phi = 2\Delta u \Delta v$$

The number of facets for the (θ, ϕ) system is

$$\begin{aligned}\text{Number of facets }(\theta, \phi) &= \frac{\pi/2}{\Delta\theta} \times \frac{2\pi}{\Delta\phi} \\ &= \frac{\pi^2}{\Delta\theta\Delta\phi}\end{aligned} \tag{C.5}$$

and the number of facets for the (u, v) system is

$$\begin{aligned}\text{Number of facets }(u, v) &= \frac{2}{\Delta u} \times \frac{2}{\Delta v} \\ &= \frac{4}{\Delta u \Delta v}\end{aligned} \tag{C.6}$$

Substituting (C.5) and (C.6) into (C.4) yields (C.7):

$$\frac{\pi^2}{\text{Number of facets }(\theta, \phi)} = 2\frac{4}{\text{Number of facets }(u, v)} \tag{C.7}$$

So we find that

$$\text{Number of facets }(\theta, \phi) = \left(\frac{\pi^2}{8}\right)\text{Number of facets }(u, v) \tag{C.8}$$

The (θ, ϕ) system requires about 23% more facets than the (u, v) for the situation described. While hardly extravagant, it will increase the computation time commensurately with little benefit.

Some approximations are inherent partitioning the hemisphere into facets:

- The area of the facet is smaller than the area of the spherical surface, just as a chord is shorter than its corresponding arc, so the solid angle tends to be underestimated. This error goes to 0 as the size of the facet goes to 0. Additionally, the facet is not a planar surface as the four corners are not coplanar. Accordingly, the area of the facet is understated. This error also goes to 0 as the size of the facet goes to 0.
- As is apparent in Figure C.2, for the (u, v) analysis, the grid points falling in the (u, v)-plane ($w = 0$) fall outside the unit circle, leading to a small excess in the computed area. This error can be avoided by adjusting the mesh, moving points adjacent to but outside the unit circle radially inward to lie on the unit circle, as shown in Figure C.4.

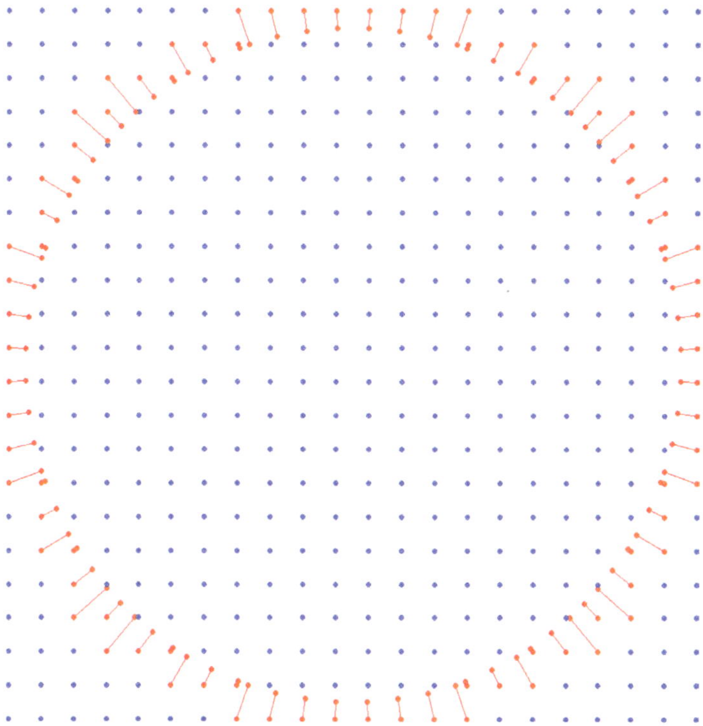

Figure C.4 Meshpoints adjustment
CoordinateSystems.m

To assess the magnitude of the error associated with the second approximation, we may compare the sum of the facet solid angles to the solid angle of a hemisphere, 2π. Table C.1 shows the results for (u, v) between 21×21 and 633×633 facets. The column titled "Without adjustment" does not correct grid points outside the unit circle; the column titled "With adjustment" moves grid points to the unit circle. This correction improves accuracy.

For our purposes, these approximations are sufficiently accurate. However, the grid should be selected with sufficient granularity to accommodate the fine detail of the antenna pattern being analyzed.

C.2 Satellite geometry

C.2.1 Geometric relationships

Satellite radar is governed by the usual equations, with some additional constraints due to curvature of the Earth's surface and to orbital mechanics. Angles and lengths are easily computed with trigonometric identities. Platform speed is a direct function of platform altitude.

Table C.1 Hemisphere solid angle computation

Facets	$\Delta u = \Delta v$	**Without adjustment**	**With adjustment**
21 × 21	0.1	$1.02349 \times 2\pi$	$0.99236 \times 2\pi$
63 × 63	0.032	$1.00679 \times 2\pi$	$0.99874 \times 2\pi$
201 × 201	0.01	$1.00212 \times 2\pi$	$0.99992 \times 2\pi$
633 × 633	0.003	$1.00086 \times 2\pi$	$1.00005 \times 2\pi$

SolidAngle.m

Some of the figures represent the projection of the beam pattern onto the surface of the Earth. Ruff [1] provides the details of the intersection of a cone and a sphere. Figure C.5 shows the viewing geometry from a satellite at altitude h whose velocity is normal to the page and whose beam is oriented normal to the velocity vector, i.e., in the plane of the page.

Equation (C.9) applies the law of sines to Figure C.5. There is an interesting ambiguity associated with the incidence angle. At any satellite altitude, there are two triangles with the same incidence angle. A line segment from the observer has two points of intersection, the near (visible) side and the far (over the horizon) side. These coincide for incidence angle of 90°, i.e., grazing. Figure C.6 shows solutions for incidence angles from 0° to 90° for the line-of-sight as well as the through-earth solutions. The angle β is the same but α differs by $\pi - 2\theta_i$:

$$\frac{\sin\alpha}{r} = \frac{\sin(\pi - \theta_i)}{r_e + h} = \frac{\sin\beta}{r_e} \tag{C.9}$$

Equation (C.10) applies the law of cosines to Figure C.5. This equation is ambiguous as it has two solutions corresponding to the included angle being acute and obtuse and these two solutions are inherent in the solution to the quadratic equation:

$$\begin{aligned} rl(r_e + h)^2 &= r^2 + r_e^2 - 2\, r\, r_e \cos(\pi - \theta_i) \\ &= r^2 + r_e^2 + 2\, r\, r_e \cos(\theta_i) \end{aligned} \tag{C.10}$$

Equation (C.10) may be rewritten in standard form as (C.11):

$$\begin{aligned} &r^2 + 2(r_e \cos\theta_i)\, r + r_e^2 - (r_e + h)^2 = 0 \\ &r^2 + 2(r_e \cos\theta_i)\, r - h^2 - 2hr_e = 0 \end{aligned} \tag{C.11}$$

and we may compute the roots of (C.11) using the quadratic formula in (C.12). The solution has two roots with valid physical interpretations.

We select the near swath point r_1 by choosing the positive coefficient of the square root:

$$\begin{aligned} r &= -r_e \cos\theta_1 \pm \sqrt{r_e^2 \cos^2(\theta_1) + h^2 + 2hr_e} \\ r_1 &= -r_e \cos\theta_1 + \sqrt{r_e^2 \cos^2(\theta_1) + h^2 + 2hr_e} \end{aligned} \tag{C.12}$$

We may use (C.9) to find the associated value of β_1 as shown in (C.13):

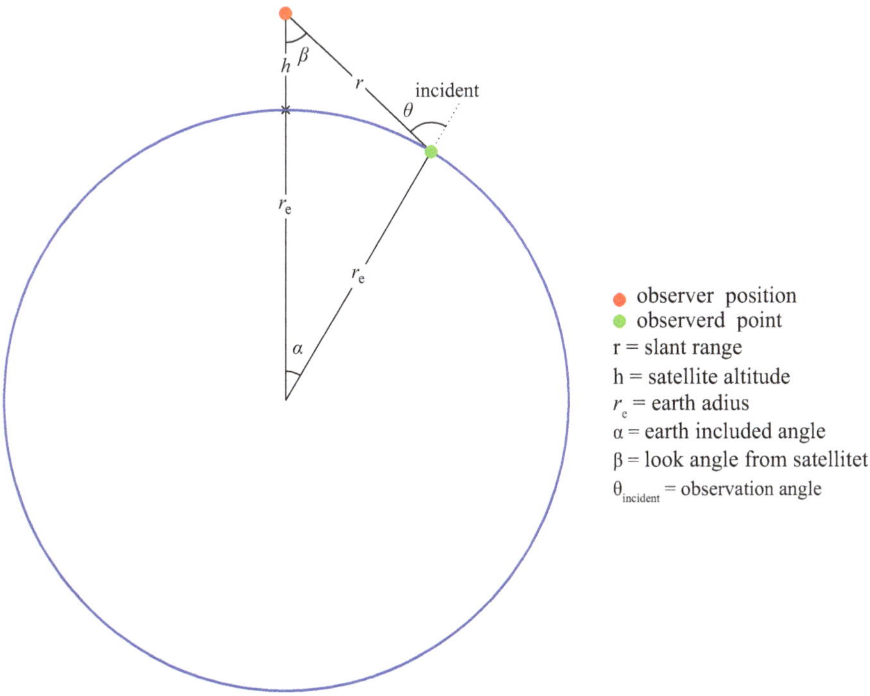

Figure C.5 Satellite viewing geometry
Geometry.m

$$\sin(\beta_1) = r_e \frac{\sin\theta_1}{r_e + h} \tag{C.13}$$

We find the value of α_1 using the relation that the sum of the interior angles of a triangle is π radians as shown in (C.14). This equation would differ if we solved for the far intersection point as we would use θ_1 directly instead of using its complement $\pi - \theta_1$:

$$\begin{aligned}\alpha_1 &= \pi - \beta_1 - (\pi - \theta_1)\\ &= \theta_1 - \beta_1\end{aligned} \tag{C.14}$$

C.2.2 Ground swath

The ground swath is the target region illuminated by the transmitter within a single pulse repetition interval to preserve a unique relationship between the time of returned radar power and the position of the associated scatterer. Good design practice restricts the ground illumination to the range swath because returns from a scatterer outside the swath are indistinguishable from those of a scatterer inside the swath. We compute the range increment that fits within the radar transmit duty cycle on the basis that receive is not possible during transmit:

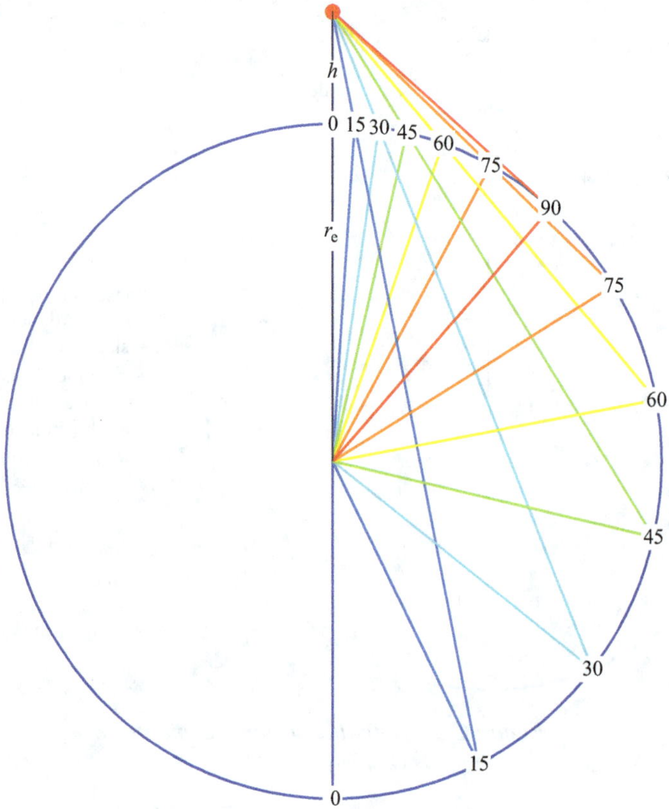

Figure C.6 Viewing geometries for some values of θ_i
Geometry.m

$$\Delta r = \frac{c(1 - DF)}{2\,PRF} \tag{C.15}$$

We use the range increment to compute the slant range, r_2, to the far swath point:

$$r_2 = r_1 + \Delta r \tag{C.16}$$

Using the slant range and the law of cosines (C.17), we may compute the associated angle, α_2 (C.18):

$$r_2^{\,2} = (r_e + h)^2 + r_e^{\,2} - 2\,r_e\,(r_e + h)\cos(\alpha_2) \tag{C.17}$$

$$\cos(\alpha_2) = \frac{(r_e + h)^2 + r_e^{\,2} - r_2^{\,2}}{2\,r_e\,(r_e + h)} \tag{C.18}$$

With α_2, we may use the law of sines to compute θ_2 (C.19) and β_2 (C.20):

$$\sin(\theta_2) = \sin(\alpha_2)\left(\frac{r_e + h}{r_2}\right) \tag{C.19}$$

$$\sin(\beta_2) = \sin(\alpha_2)\left(\frac{r_e}{r_2}\right) \tag{C.20}$$

Finally, the width of the ground swath is simply the difference in α multiplied by the earth's radius:

$$\Delta R = r_e\,(\alpha_2 - \alpha_1) \tag{C.21}$$

An approximate solution is found by multiplying the slant range Δr by the sine of the incidence angle θ_i:

$$\Delta R = \Delta r\,\sin\theta_i \tag{C.22}$$

$$\text{Resolution} = \frac{\text{Antenna length}}{2} \tag{C.23}$$

$$\text{PRF} = \frac{2\ \text{Velocity}}{\text{Antenna length}} \tag{C.24}$$

$$\text{PRF} = \frac{\text{Velocity}}{\text{Resolution}} \tag{C.25}$$

Reference

[1] Ruff I. 'The intersection of a cone and a sphere: a contribution to the geometry of satellite viewing'. *Journal of Applied Meteorology*. 1971, vol. 10(3), pp. 607–9.

Index

www.ingramcontent.com/pod-product-compliance
Lightning Source LLC
LaVergne TN
LVHW011842200926
848484LV00002B/389